PHASE TRANSITIONS IN POLYMERS

The Role of Metastable States

PHASE TRANSITIONS IN POLYMERS

The Role of Metastable States

STEPHEN Z. D. CHENG

ELSEVIER

Amsterdam • Boston • Heidelberg • London • New York • Oxford
Paris • San Diego • San Francisco • Singapore • Sydney • Tokyo

Elsevier
Radarweg 29, PO Box 211, 1000 AE Amsterdam, The Netherlands
The Boulevard, Langford Lane, Kidlington, Oxford OX5 1GB, UK

First edition 2008

British Library Cataloguing in Publication Data
A catalogue record for this book is available from the British Library

Library of Congress Cataloging-in-Publication Data
A catalog record for this book is available from the Library of Congress

ISBN: 978-0-444-51911-5

For information on all Elsevier publications
visit our website at books.elsevier.com

Contents

Foreword

Frequently, something new and revolutionary may become so successful that it ends up eventually being taken for granted. This is what has happened to the field of crystalline polymers, which had its "big bang" just over 50 years ago with the discovery of polymer single crystals, lamellae, and chain folding. Explosive growth followed in the ensuing few decades based on the detailed studies of crystallization, annealing, and melting, the determination and control of molecular conformations, crystal structures, and morphologies, the heated debates over crystal-growth models and the validity and extent of chain folding, the elucidation and exploitation of mechanical, optical, and other physical properties—not to mention the centrally important synthesis of all kinds of new crystalline polymers.

Yet, in our own days, the "big bang" has turned (in the words of T. S. Eliot) into a mere "whimper." Not that crystalline polymers are no longer important—quite the contrary! After all, 70% of all commercial polymers are crystalline, including an enormous number and variety of materials in our daily lives and in all kinds of high-technology applications. Crystalline polymers have become so pervasive that we just take them for granted. Polymer researchers have moved on to newer areas of more current appeal (e.g., block copolymers, hybrids, nano-everything, bio-, . . .).

But, have the major questions in crystalline polymers been solved? Has it been time to move on? No, not by a long shot. What happened is that some of the unsolved problems have led to intense controversies or dead ends, and researchers in crystalline polymers have focused their attention more and more on ever narrower problems and ever more specialized materials. Most of us have been cultivating our own "trees of knowledge," working on our favorite polymers, properties, or techniques. Sometimes our efforts extend beyond these individual "trees" and build a little "grove" as we address, for example, families of related polymers or properties of greater breadth. But, as the saying goes, we generally keep missing the forest for the trees.

This is precisely why this book is so refreshing and useful. It focuses on the "big picture": it shines a giant floodlight over the entire forest, not only illuminating it brightly, but revealing a number of hidden connections and unknown paths among the trees. While most books look at our field "statically" (e.g., discuss their various sub-topics individually and separately), this book takes a "dynamic" approach that examines crystalline (but also other) polymers in terms of their phase transformations. Professor Cheng has devoted more than two decades of his own research on thinking and developing this area. While making a number of major discoveries about individual polymers and classes of polymers, he has always looked for connections, phenomena, and common threads that underlie their solid-state behavior.

And in so doing, he has discovered those connections in the phase transitions of polymeric materials and, more specifically, in the role of metastable states.

The concept of metastability is intimately linked with the existence of energy barriers in thermodynamics. A metastable state describes a phase in which an energy barrier must be surmounted before that phase can be transformed to one of lower free energy. While first- and second-order transitions are typically reversible by changing the temperature, transitions from a metastable to a more stable phase are not. In polymers, their macromolecular nature and the accompanying entanglements, persistence lengths, conformational characteristics, and kinetic factors assure that almost any accessible solid phase will be in a metastable state. This results in a wealth of morphological features, polymorphism, phase-transition characteristics, and interconversion aspects that cumulatively have no rival among other material classes.

It is this highly complex and diverse menagerie that this book comes to address and to which it brings a systematic order, interconnection, and interpretation. Following three introductory chapters that refresh the necessary background on thermodynamics and kinetics of phase transitions in general (including the basic concepts of metastability), the fourth chapter goes into the phase transitions and metastable states of polymers. The various competing models proposed to explain nucleation and growth, the intricate details of melting under different conditions, and the extremely broad and rich literature on metastable states in phase-separated blends and copolymers are examined here.

The fifth chapter focuses on kinetically obtained phases across the polymer realm and brings together the remarkable richness of polymorphism, growth competition, transient states, and pressure- and field-induced morphologies. Particularly welcome is the exposition on the frequently surprising—yet dominating—effects of surface and interfacial factors on some of the more intriguing structural and morphological features that have been identified in so many polymers over the last 50 years. It is here perhaps more than anywhere that this book lifts us high above the treetops and enables us to visualize the full forest, its pathways, its connections, and its common underlying features. The sixth chapter then effectively assembles the various metastabilities that are seen on different length scales, explores their interdependence, and brings together research topics not frequently examined jointly, such as liquid–liquid phase separation coupled with crystallization and even gelation.

The unifying, integrative approach and the fully up-to-date distillation of the many hundreds of separate studies published in this broad field are going to be the most immediate benefits to the reader. As far as I know there is no other volume that covers these topics to anywhere near this extent, and certainly none with this richness of recent references discussed integratively and often critically. But perhaps the bigger long-range benefit might stem from revisiting topics and areas that are generally not examined under the light of metastable states, and thus obtaining new insights and understanding across topical interfaces.

The writing itself does not make for easy reading. When I went through the first, unedited, version of this book, I frequently had to go back and re-read certain sentences a second and sometimes even a third time. This is indeed a dense text with lots of involved thoughts and meaning beneath the surface—but the reader's investment in time will be well rewarded with integrative knowledge and new understanding. The book should be particularly valuable not only to researchers and practitioners in polymer physics and technology, but also to graduate students.

In the United States, at least, instruction on crystalline polymers has generally been reduced to a minimum under pressure from other areas and through the need for greater interdisciplinarity in curricula. Regrettably, it is not at all unusual to see structure, crystallinity, crystallization, and morphology dispensed with in one or two lectures in an introductory polymer course. Given the pervasiveness of crystalline polymers in the world of materials and in our daily lives, this appears quite inadequate. Even students who go on to graduate studies in polymers might not get a deeper exposure since such courses are frequently elective. It is therefore hoped that this book, with its fresh and enlightening approach across metastable states, might perhaps help fuel a resurgence in the teaching of the science of crystalline polymers at universities.

Andrew J. Lovinger

Preface and Acknowledgments

The roots of this book go back approximately eighteen years to a time when I attempted to organize a graduate course titled "Phase Transitions in Polymers." During my preparation, two questions continually occurred to me: Can I identify a specific concept to connect different types of polymer phase transitions in a comprehensive way? What is the organic tie which links all knowledge about phase transitions in polymers, providing a platform to understand this topic? This information is particularly important when we deal with complicated phase behaviors involving the interdependence of different types of phase transitions. With the encouragement of two of my mentors, Professor Bernhard Wunderlich and Professor Andrew Keller, I used metastable states and metastability as the central concept connecting the phase transitions in polymers.

Throughout my academic career, I have acquired a great deal of knowledge from my mentors. Notably, I studied for my Ph.D. degree and spent my postdoctoral time with Professor Wunderlich, who taught me the thermodynamics of polymers and, particularly, the thermodynamic concept of metastability. Professor Keller gave me insight into kinetically controlled metastable states. Between 1993 and 1998, Professor Keller and I spent many days and nights discussing and defining these concepts, searching for explanations of experimental observations, and trying to generalize these concepts into universal principles. This endeavor resulted in a review article, published in 1998 in *Polymer*, which summarized our thoughts about these concepts at that time. Today, almost ten years later, many of the principles we described in our review article are still useful. In fact, in the last year of Professor Keller's life, we talked about the possibility of co-authoring a book about the role of metastable states in polymer phase transitions.

Since Professor Keller's sudden death in 1999, I have continuously offered the course "Phase Transitions in Polymers" at The University of Akron, as well as at various research and industrial institutions in China, France, Taiwan, and the United States. Through my teaching, I specifically attempt to connect all phase transition behavior in polymers, using the concept of metastable states and metastability as the central theme. Approximately three years ago, I decided to start writing this book. Although, now I must move forward without Professor Keller's insight, I feel that the attempt to use these concepts to discuss polymer phase transitions is a worthy goal. Throughout this book, readers will find many original contributions made by Professor Keller and his academic associates, which first appeared in the 1998 review article.

The organization of this book takes a different path from that of normal polymer physics books. The first chapter is an introduction, which provides the necessary background needed for more advanced topics discussed later in the book. In this chapter, several important concepts are used to describe the microscopic aspects of phases, including structural symmetry and order, as well as macroscopic aspects of phases using classical thermodynamics. These two worlds are connected by statistical mechanics. I also introduce the thermodynamic definition of phase transitions. The second chapter is a brief overview of phase transitions in materials. In particular, I have structured this chapter to illustrate both the thermodynamic and kinetic aspects of liquid–gas and solid–liquid transitions and transitions involving mesophases in single and multiple component systems. The purpose of the second chapter is to broaden the perspective of phase transitions for the readers who are familiar only with phase transitions in polymers, thus establishing that phase transitions in polymers are a subset of the overall field of phase transitions.

The third chapter introduces the concept of metastable states and metastability. According to classical thermodynamics, superheated and supercooled states are metastable states, but of greater interest are metastable states that appear because of competing kinetics. Utilizing these two origins for metastable states, the following three chapters (four, five and six) deal with different metastable states in polymer phase transitions. Chapter 4 focuses on fundamental and classical cases of metastable states in polymer phase transitions on a single length scale. In particular, metastable states and metastabilities caused by supercooling and superheating single component systems are described. The effect of phase size on metastability is specifically introduced and discussed, as well, with significant attention focused on polymer crystallization, melting, and mesophase transitions. The metastable state concept in multiple component systems is also discussed in the case of liquid–liquid phase separation in blends and block copolymers.

In chapter 5, the concept that metastable states can arise from competing kinetics is introduced. These kinetically trapped metastable states are often short lived. The kinetic concept also brings the role of transformation rates into the discussion, with the rates being solely determined by the transition barriers. This concept explains many experimental observations of phases that are not thermodynamically the most stable, but kinetically form the fastest. Finally, in chapter 6, my intention is to illustrate the interdependence of metastable states on different length scales in polymers. The combination of different types of metastable states leads to complicated phenomena in both single and multiple component systems. A metastable state is always competing with at least one other phase. The intervention of various transition processes, such as vitrification, liquid–liquid phase separation, and/or gelation with crystallization, provides challenges to understand experimental observations. Considering the interdependence of metastable states on different length scales allows us to explain these observations. Chapter 7 is my personal outlook on this research field.

Undertaking the task of writing a book requires background knowledge and experience, along with the contributions of friends, former and present students, and colleagues. The bulk of the material in this book was initially collected and discussed with Professor Keller for our 1998 review article. Many of my friends, including Dr. Andrew J. Lovinger of the National Science Foundation in Arlington, Virginia; Professor Bernard Lotz of Institut Charles Sadron, CNRS in Strasburg, France;

Dr. Freddy Khoury of the National Institute of Standards and Technology in Gaithersburg, Maryland; Professor Goran Ungar of the University of Sheffield in Sheffield, England; Professor Akihiko Toda of Hiroshima University in Hiroshima, Japan; and Professor Sanjay Rastogi of Loughborough University in Leicestershire, England, spent their precious time reading through the manuscript of this book and providing extremely valuable comments, which have led to many significant revisions and corrections. Professor Charles C. Han of the Institute of Chemistry at the Chinese Academy of Science in Beijing, China and Professor Deng-Ke Yang of the Liquid Crystal Institute at Kent State University, Kent, Ohio, also read various parts of the manuscript and provided their valuable comments and corrections. In addition, I would like to express my deepest appreciation to Dr. Andrew J. Lovinger for contributing the foreword for this book.

My former and present students likewise read the manuscript and made their contributions to correct and revise this book. All figures in this book were carefully replotted or reprinted by Mr. Jing Wang of The University of Akron, who has made a remarkable contribution. Financial support for most of the research in my own group presented in this book was from the National Science Foundation, the Air Force Office of Scientific Research, the National Aeronautics and Space Administration, the Ohio State Board of Regents, The University of Akron, and various industries.

My parents have taught me the importance of humanity, hard work, and logical thinking. My wife, Susan, and my daughter, Wendy, have given me unlimited love, support, and understanding and have encouraged me throughout the past twenty years of my academic life. My friends, students, and colleagues have helped me to achieve success in my academic career. I cannot recount how much I owe my parents, family, friends, students, and colleagues, so I dedicate this book to them with my most sincere gratitude.

Stephen Z. D. Cheng
Akron
January 2008

CHAPTER ONE

Introduction

Contents

1. Phases in Single-Component Systems

This chapter provides a brief review of our basic understanding of phases from both macroscopic and microscopic points of view. From the macroscopic perspective, phases can be identified utilizing a series of thermodynamic properties based on classical thermodynamics; from the microscopic perspective, a phase needs to be defined on the basis of its structural symmetry and the types of order found in the phase. These two different perspectives can be reconciled by using statistical mechanics to bridge the length scale differences. When one phase changes to another isobarically or isothermally, this change is a phase transition. Phase transitions are not only scientifically interesting, but also practically important. In this chapter, the Ehrenfest definition of phase transitions in single-component systems is introduced. These transitions can be categorized as discontinuous transitions or critical phenomena. The equilibrium phase and stability can be determined on the basis of classical thermodynamics. By knowing these pieces of information, one can construct a phase transition free energy landscape for a system.

1.1. Macroscopic description of phases

The states of matter and the transformations between them are subjects of extensive research in condensed matter physics and solid state chemistry. They also provide a fundamental basis for today's materials science and engineering fields, as well as play an important role in profoundly improving the world's economy and our quality of life. It is well known that solids, liquids, and gases are the three basic states of matter. Phenomenological definitions of these states are introduced in elementary school, where it is taught that a substance which can fill a container and conform to its shape is a liquid. More specifically, liquids are a class of substances that possess a fixed volume but not a fixed shape. On the other hand, a substance that possesses both a fixed volume and a fixed shape is a solid, while a gas has neither a fixed volume nor a fixed shape. A more complete understanding of these states, based on both macroscopic and microscopic descriptions, comes much later in our education.

Quantitative macroscopic descriptions of the states of matter were first presented approximately two hundred years ago. A series of macroscopic thermodynamic properties, such as temperature, pressure, enthalpy, entropy, and free energy, were used to describe an equilibrium phase, which defines a state of matter. The basis for this kind of description is derived from a few empirical laws of classical thermodynamics which were established in the nineteenth century during the Industrial Revolution. Among them, the most important laws are that heat does not spontaneously move from a low temperature body to a high temperature body and that at equilibrium, the temperatures of the two bodies must be equal. The conservation of energy dictates that although energy can be transformed from one type to another, it cannot be created or destroyed. This principle constitutes the first law of thermodynamics. Next, work can be completely transferred to heat, but heat cannot be transferred to work with 100% efficiency. The energy which cannot be transferred to work is related to entropy. The second law of thermodynamics states more precisely that the heat lost is the product of absolute temperature and entropy. The third law of thermodynamics quantitatively describes the relationship between entropy, a macroscopic property, and statistical thermodynamic probability (the Boltzmann equation).

Thermodynamic descriptions of a phase are based on changes in the thermodynamic properties (or functions) of different systems, such as isolated systems (without energy and mass transfer to the surroundings), closed systems (without mass transfer, but with energy transfer to the surroundings), or open systems (with both energy and mass transfer to the surroundings). The thermodynamic properties, enthalpy (H), entropy (S), and Gibbs free energy (G), of a closed, isobaric, single-component, single-phase system can be calculated with calorimetric isobaric heat capacity data (C_P) at different temperatures. The corresponding equations are given below:

$$H = H_0 + \int_0^T C_P \mathrm{d}T \tag{1.1}$$

$$S = \int_0^T \left(\frac{C_P}{T}\right) \mathrm{d}T \tag{1.2}$$

$$G = H - TS \tag{1.3}$$

The thermodynamic properties of a closed, single-component, single-phase system at constant volume are internal energy (U), entropy (S), and Helmholtz free energy (F). Each can be calculated from isometric heat capacity data (C_V) at different temperatures. The corresponding equations are given below:

$$U = U_0 + \int_0^T C_V \mathrm{d}T \tag{1.4}$$

$$S = \int_0^T \left(\frac{C_V}{T}\right) \mathrm{d}T \tag{1.5}$$

$$F = U - TS \tag{1.6}$$

The relationship between isobaric heat capacity and isometric heat capacity follows the equation given below:

$$C_P - C_V = TV\alpha^2\beta = \left[\left(\frac{\partial U}{\partial V}\right)_{T,n} + P\right]\left(\frac{\partial V}{\partial T}\right)_{P,n} \tag{1.7}$$

In this equation, α is the coefficient of thermal expansion, which is defined as $(\partial V/\partial T)_P/V$, and β is the compressibility, which is defined as $-(\partial V/\partial P)_T/V$. It is evident how important it is to experimentally and/or theoretically know the heat capacity (C_P or C_V) of a system, since all of the thermodynamic properties of a phase can be obtained from Eqs. (1.1) to (1.6) (see, e.g., Wunderlich, 2005).

1.2. Microscopic description of phases

The macroscopic thermodynamic properties of a phase originate from the microscopic molecular interactions and motion in the equilibrium state. Generally speaking, molecular interactions promote structural order, while the random nature of thermal motion promotes disorder. In different types of phases, the microscopic structures and motions are different.

To categorize, quantitatively, the microscopic structural order of different phases, a series of mathematical operations is needed. For this purpose, a set of symmetry operations was introduced to represent a set of arbitrary translational, rotational, and reflection operations (the Euclidean group) to define the microscopic structure of a phase. Since liquid- and gas-phase structures are statistically invariant under all of these operations, their symmetry group is the entire Euclidean group that embodies the largest possible number of symmetry operations, and so, fluids have the highest possible symmetry. This characteristic implies that liquids and gases cannot be distinguished by symmetry. Other equilibrium phases are invariant only under some subgroup of the Euclidean group and thus have a lower number of symmetry operations than do the fluid phases. The ordered structures in those phases reduce the applicable symmetry operations. For example, in a perfect crystalline solid, the symmetry operations become periodic in three-dimensional space, and the particle (atoms or molecules) density is invariant only with respect to translation through lattice vectors.

To practically describe a phase, we need a structural function which represents the average relative positions of particles in a three-dimensional space. The description of a phase begins by defining a number density function. Following the textbook published in 1995 by Chaikin and Lubensky, the number density function specifies the number of particles per unit volume at a point in a three-dimensional space $\boldsymbol{r}(x,y,z)$, denoted as $n(\boldsymbol{r})$. The ensemble average of the density function is the average density $\langle n(\boldsymbol{r})\rangle$ at the position $\boldsymbol{r}$. In a homogeneous fluid, this average is equal to a ratio between the overall number of particles and the system volume, indicating that $\langle n(\boldsymbol{r})\rangle$ is independent of both the magnitude and the direction of $\boldsymbol{r}$ in this system.

This structural function can be experimentally obtained by scattering experiments because scattering vectors represent a Fourier transform of the particle density correlation function. A correlation function specifically describes the relationship between the density functions $n(\boldsymbol{r}_1)$ and $n(\boldsymbol{r}_2)$ at two different positions ($\boldsymbol{r}_1$ and $\boldsymbol{r}_2$) in three-dimensional space. The correlation function is the ensemble average of the products of the density functions at these two points in the space. Several types of density–density correlation functions are commonly used. One of them, called an Ursell function, incorporates the difference between $\langle n(\boldsymbol{r}_1)\cdot n(\boldsymbol{r}_2)\rangle$ and $\langle n(\boldsymbol{r}_1)\rangle\langle n(\boldsymbol{r}_2)\rangle$. Note that the correlation function can be reconstructed from the structure function only if it is invariant with respect to position in the three-dimensional space, as in the case of homogeneous fluids. Since periodic crystalline solids do not meet this criterion, experimental diffraction methods, such as x-ray and neutron diffraction, are required to obtain the correlation function. Furthermore, construction of a density function from a correlation function is impossible (Chaikin and Lubensky, 1995).

The microscopic structures of phases are classified on the basis of their positional, bond orientational, and molecular orientational order. Positional order describes the periodic arrays of atoms or molecules. Bond orientational order describes the alignment of directional vectors along certain periodicities, which are used to determine the lattice axes. Finally, molecular orientational order describes the alignment of atoms or molecules with respect to particular directions.

Each of these three types of order may be short range, quasi-long range, or long range. The terms short range, quasi-long range, and long range order do not refer to the domain size of a structure, but instead to precisely how spatial and orientational correlations decay with distance. For short range order, the decay follows an exponential function; while, the correlation decay for quasi-long range order obeys a power law with respect to distance. For long range order, the correlation decay follows a step function. Figure 1.1 illustrates these three different types of correlation decay in one-dimensional real space and their corresponding diffraction patterns in reciprocal space (Demus *et al.*, 1998; Leadbetter, 1987). These three types of decay in positional, bond orientational, and molecular orientational order are summarized in Fig. 1.2.

We can now classify three-dimensional phases on the basis of their structural order and correlation decay. For a crystalline solid, the average structure is invariant only with respect to certain discrete lattice translations and point group operations comprising the space group, and they possess long range positional, bond orientational, and molecular orientational order. For an isotropic fluid with an average structure that is invariant under arbitrary rotation and translation, these three types of order decay exponentially, and so they are short range. Between a three-dimensional long range ordered crystalline solid and a short range ordered isotropic liquid, there are other classes of phases which exhibit long range decay in some of the three types of

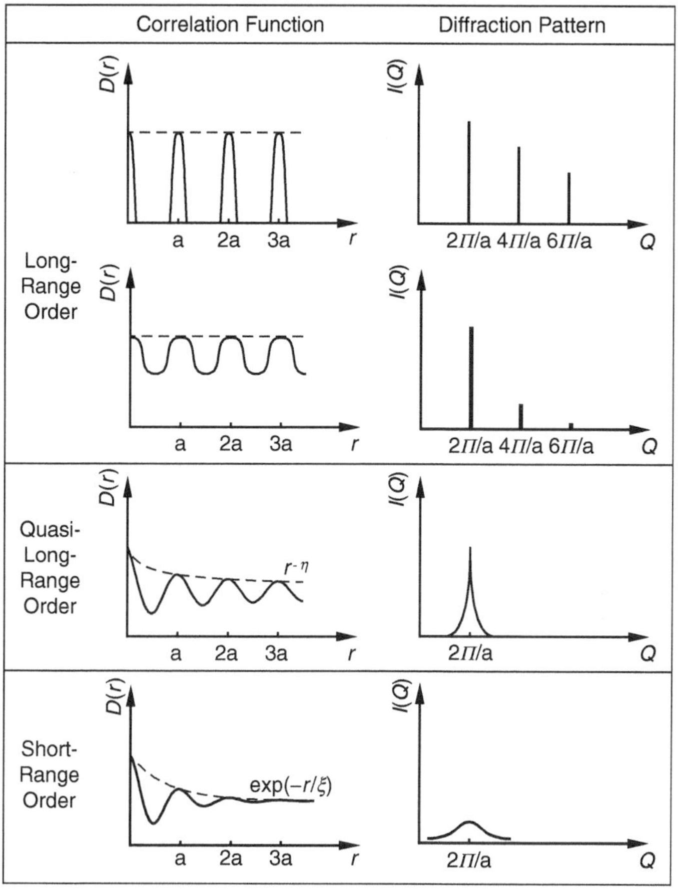

Figure 1.1 Illustrations of three different types of order decay: long range, quasi-long range, and short range order in real space and their corresponding one-dimensional diffraction patterns in reciprocal space. $D(r)$ is a displacement at a distance r, where $r^{-\eta}$ is a temperature-dependent quantity related to the elastic properties of the phase, and η is of the order of 0.1–0.4. I is a reciprocal lattice, ξ is the correlation length and it is usually a few nanometers [replotted from Leadbetter (1987), with permission].

order and short or quasi-long range in the rest. These phase types are called mesophases. The word "meso" originates from the Greek *mesos* and means "in-between," thus mesophases represent those phases which are "in-between" the crystalline solid and the isotropic liquid. Combinations of short range, quasi-long range, and long range decay for the three types of order lead to different phases which can be structurally identified, such as various liquid crystalline and plastic crystal phases.

Let us use liquid crystalline phases as an example of a mesophase. This group of materials was first reported about 120 years ago by Reinitzer when he was studying the physical properties of cholesteryl esters. Although Reinitzer did not give an explanation, he described "... a peculiar, very splendid colour phenomenon ..." before hardening and associated this observation with "two melting temperatures"

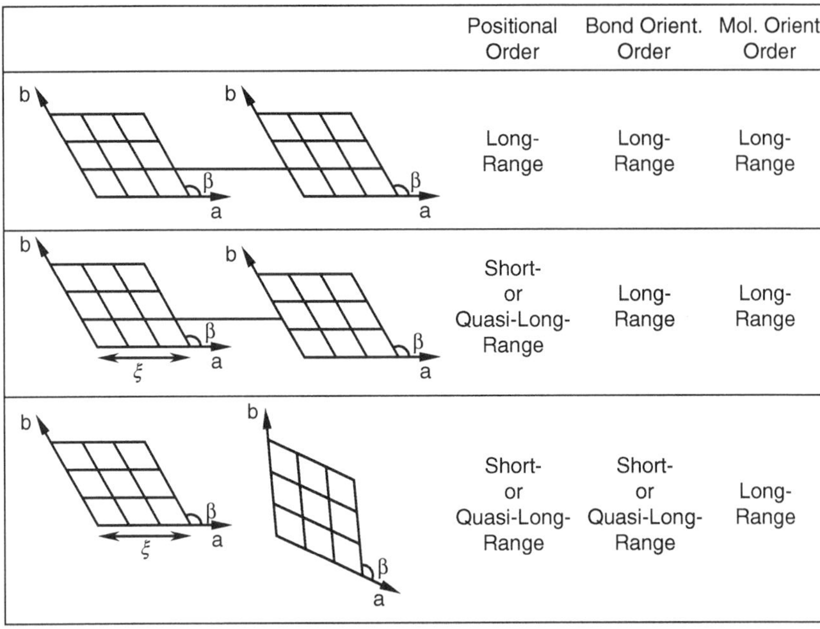

Figure 1.2 Illustrations of positional, bond orientational, and molecular orientational order decay. In this illustration, the molecular orientation is assumed to be perpendicular to the two-dimensional lattice [replotted from Demus *et al.* (1998), with permission].

(Renitzer, 1888). After further investigation of Reinitzer's samples, Lehmann coined the terms "crystalline liquid" and "liquid crystal" to describe this new phase (Lehmann and Verhandl, 1900). The experimental observation was that the crystalline cholesteryl ester sample melted first at a high temperature and flowed like a fluid, yet the fluid retained its birefringence. The birefringence disappeared only when the sample was heated to an even higher temperature. Therefore, this phase exists between two "melting temperatures": one being the crystal melting temperature and the other being the clearing temperature of the liquid. Since at that time the general understanding was that only crystals generated birefringence, the phase between these two temperatures was called a liquid crystal phase—a "crystal" that can flow like a "liquid."

Later, people found that a mesogen, which is usually made with aromatic or other rigid units, and a flexible tail, which is usually made from aliphatic units, are required to form this class of phases. There are many different liquid crystalline subphases that have been identified and characterized (see, e.g., Pershan, 1988). Figure 1.3 illustrates how increasing order in calamitic (from the *Greek* "kalamos" for rod-like) liquid crystalline systems changes the phase classification. In this figure, the nematic (from the *Greek* "nema" for thread-like) phase possesses the lowest order with only long range molecular orientational order and short range positional and bond orientational order. The next class of calamitic liquid crystalline phases is the smectic (from the *Greek* "smektos" for layer-like) phase, which possesses layered structures. Depending upon the mesogen orientation within the layers, they can be

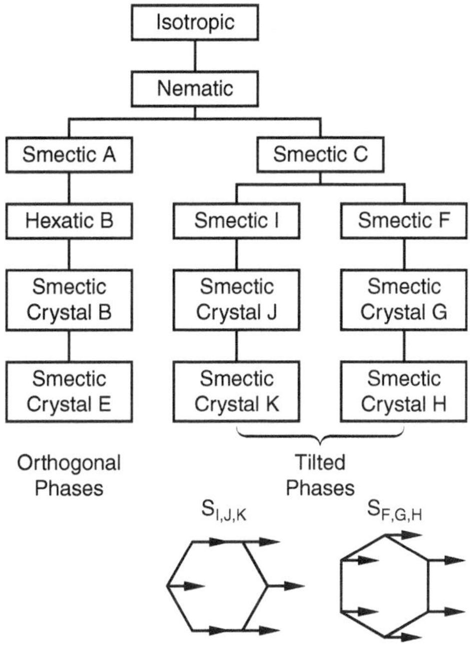

Figure 1.3 Illustration of calamitic liquid crystalline phases defined by their increasing order [replotted from Pershan (1988), with permission].

smectic A (parallel to the layer normal) or smectic C (tilted from the layer normal). Further increasing order leads to a number of highly ordered liquid crystalline phases, as listed in Fig. 1.3. In one series of these phases, the mesogen remains parallel to the layer normal, yet there is an increase in the lateral packing order within and between the layers. In the other two series, the mesogen is tilted away from the layer normal: one is toward an apex; the other toward a side. The lateral packing order within the layers increases moving down the list in Fig. 1.3 where the phase appears. A series of new liquid crystalline phases, parallel to the three series described above but not included in Fig. 1.3, has recently been identified where the mesogens within the layers are tilted between a side and an apex.

A new class of liquid crystals with discotic (disc-like) mesogens was discovered by Chandrasekhar *et al.* (1977). A set of phase classifications for this type of mesophase is listed in Fig. 1.4. These phases are now part of a broader category of liquid crystalline phases known as columnar phases. Detailed descriptions of these phases can be found in many reference books (see, e.g., Demus *et al.*, 1998; Pershan, 1988).

1.3. Connection between microscopic descriptions and macroscopic properties

From the perspective of dynamics, the motions of microscopic particles are vastly different in different phase structures. This difference introduces questions concerning how to quantitatively describe these motions and how the motions of each interacting

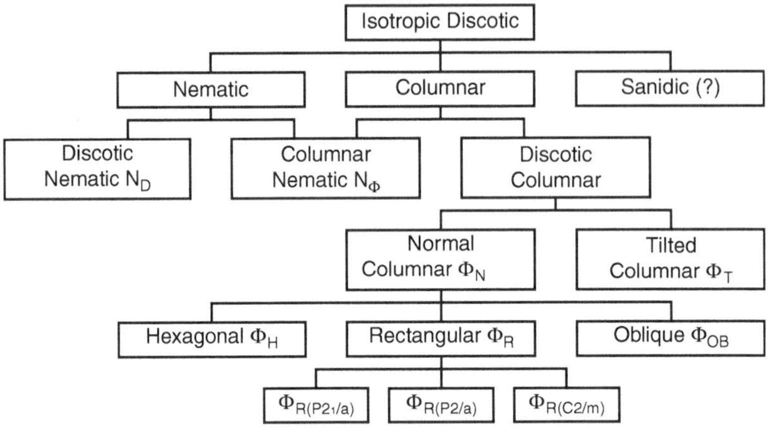

Figure 1.4 Illustration of the evolution of order through liquid crystal phases that are based on discotic mesogens. Although space groups are used in defining the structures, the long range order in these phases is confined to two dimensions. The question mark on the sanidic phase reflects the current debate about its existence [replotted from Demus *et al.* (1998), with permission].

particle contribute to the macroscopic phase properties. One needs to view the macroscopic thermodynamic properties of a system as an average of the microscopic mechanical motions of the particles. Statistical mechanics serves as a bridge which connects the macroscopic properties with microscopic motions. If one tries to describe the motions of a few interacting particles, mechanical descriptions (classical or quantum) may lead to numerical solutions, with the help of significant computational power. However, when one considers a system containing a large number of particles (a "many body" problem), numerical solutions are practically impossible. For example, a drop of water with a diameter of a few millimeters contains over 1×10^{20} water molecules. Vast computational resources would be required to understand this relatively simple system. The main method used to solve this problem has been to simplify it by averaging the interactions with neighboring particles, making them an effective medium. This method is the basis of the "mean-field" approach. In the first half of the twentieth century, a variety of mean-field theories were developed (see, e.g., Landau and Lifshitz, 1969). Statistical mechanics starts with thermal motion at absolute zero (0 K) as a reference. Then, in order to quantitatively describe thermal motion, an energy term is introduced. Using the product of the Boltzmann constant ($k = 1.38 \times 10^{-16}$ erg/K) and absolute temperature, kT, allows a description of the thermal energy (motion) which leads to disordering. Although the basics of statistical mechanics can be found in textbooks (see, e.g., Hill, 1960), a very brief description to illustrate how to connect microscopic thermal motions to macroscopic properties via equilibrium statistical mechanics is given below.

Assuming that we have a system that consists of N particles, where N is an extremely large number, with each particle at an energy level ε_i where $i = 1, 2, 3, \ldots$, and the number of particles at a particular energy level ε_1 is n_1, at ε_2 is $n_2 \ldots$, and at ε_i is $n_i, \ldots$ we can determine that $N = \sum_i n_i$ and $E = \sum_i n_i \varepsilon_i$. For a given distribution of particles occupying different energy levels, $\boldsymbol{n}$, the number of ways to arrange the particles occupying an energy level is given by Eq. (1.8).

$$W_t(\boldsymbol{n}) = N!/(n_1!n_2! \ldots) = N!/(\Pi_i n_i) \tag{1.8}$$

The probability, at a constant temperature, of a particle occupying energy level ε_i can be determined using the equation given below:

$$P_i = \frac{[\sum_i W_t(\boldsymbol{n})n_i(\boldsymbol{n})/(\sum_i W_t(\boldsymbol{n}))]}{N} \tag{1.9}$$

When $N \rightarrow \infty$, P_i obeys the Boltzmann distribution [Eq. (1.10)] (Hill, 1960).

$$P_i = \frac{\exp\{-\varepsilon_i/(kT)\}}{[\sum_i \exp\{-\varepsilon_i/(kT)\}]} \tag{1.10}$$

In this equation, the ratio between ε_i and kT is a comparison of the energy level, ε_i, with respect to the thermal energy, kT. In Eq. (1.10), the denominator, which is a summation over all the energy levels, is the partition function (Q). To be more precise, we need to consider that a few states could possess the same energy level. Therefore, a degeneracy factor, g_i, is introduced into the partition function.

$$Q = \sum_i g_i \exp\left\{\frac{-\varepsilon_i}{(kT)}\right\} \tag{1.11}$$

By knowing the partition function, we can establish the relationships between microscopic thermal motions and macroscopic thermodynamic properties. The description for the Helmholtz free energy is given by Eq. (1.12).

$$F = -kT \ln Q(T) \tag{1.12}$$

This function gives access to other thermodynamic properties. Internal energy (U) can be obtained by taking the derivative of the free energy with respect to temperature.

$$U = kT^2 \left(\frac{\partial \ln Q}{\partial T}\right)_{V,n} \tag{1.13}$$

Entropy (S) is given by Eq. (1.14).

$$S = kT^2 \left(\frac{\partial \ln Q}{\partial T}\right)_{V,n} + k \ln Q \tag{1.14}$$

Isometric heat capacity (C_V) is the second derivative of the free energy.

$$C_V = \left(\frac{k}{T^2}\right)\left[\frac{\partial^2 \ln Q}{\partial(1/T)^2}\right]_{V,n} \quad (1.15)$$

Other properties can be determined using this same approach.

2. Phase Transitions in Single-Component Systems

2.1. Definitions of phase transitions

The classification of phase transitions based on equilibrium thermodynamics was first proposed by Ehrenfest (1933). This macroscopic approach is based on the continuity or discontinuity of the free energy function after taking successive derivatives. It is known that the first derivatives of the free energy are pressure (or volume), entropy (or temperature), and polarizability. The second derivatives of free energy are compressibility, expansivity, heat capacity, and dielectric susceptibility. The Ehrenfest classification defines a first-order transition as one in which the free energy is continuous (the Gibbs free energy in isobaric conditions; the Helmholtz free energy in isometric conditions), but the first derivatives of the free energy are discontinuous. This implies that at the transition temperature, the thermodynamic properties represented by the first derivatives of the free energy at constant pressure or volume exhibit abrupt changes. This is illustrated for the isobaric case in Fig. 1.5. At the transition temperature, the functions that are the second derivatives of the free energy exhibit an infinitely narrow peak of infinite height, while the integration value of this peak height with respect to temperature is finite (a Dirac δ function).

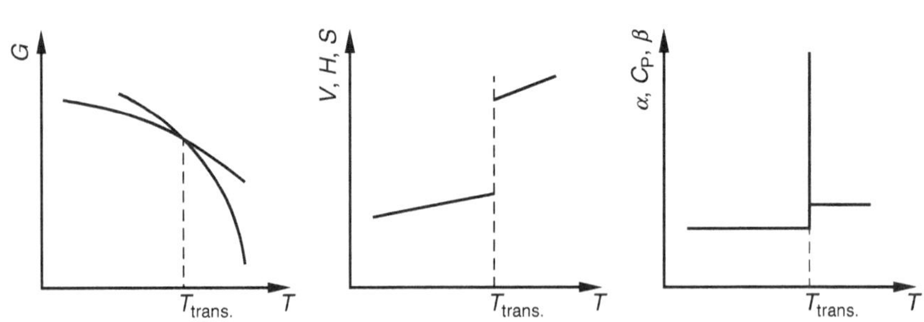

Figure 1.5 Thermodynamic function changes at a first-order transition under isobaric conditions. Note that the enthalpy (H), entropy (S), and volume (V) are discontinuous first derivatives of the Gibbs free energy (G) and that they exhibit a discontinuous change at the transition. The isobaric heat capacity (C_P), coefficient of thermal expansion (α), and compressibility (β) exhibit a Dirac δ function-like behavior.

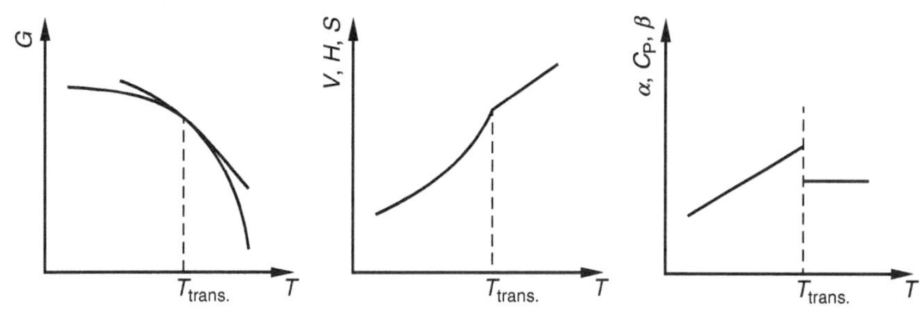

Figure 1.6 Thermodynamic function changes at a second-order transition under isobaric conditions. Note that the enthalpy (H), entropy (S), and volume (V) are continuous first derivatives of the Gibbs free energy (G); while, isobaric heat capacity (C_P), coefficient of thermal expansion (α), and compressibility (β) are discontinuous second derivatives of the Gibbs free energy (G).

A second-order transition, according to the Ehrenfest classification, is defined as one in which the free energy and its first derivatives are continuous, but the second derivatives of the free energy at constant pressure or volume exhibit discontinuous behavior. Figure 1.6 shows an isobaric example of these changes. Only the third derivative of the free energy exhibits a Dirac δ function-like behavior.

In a more general form, a Kth-order transition can be defined as one in which all of the $(K-1)$ derivatives of the free energy are continuous, and the Kth derivatives of the free energy are discontinuous. The higher-order classifications in this system are not absolutely necessary for experimental scientists since practically only first-order transitions (crystallization, crystal melting, most liquid crystal transitions, and others) and second-order transitions (a transformation between liquid and gas phases at the critical point, superfluid and superconducting transitions without an external magnetic field, and several ferromagnetic phase transitions, such as the Curie point) are observed. A few specific two-dimensional systems may exhibit higher than second-order transitions, such as fluid-phase mixtures with more than two components (such as tricritical, tetracritical, or pentacritical points) and some theoretical predictions (such as the Bose–Einstein condensation in an ideal Bose gas). For simplicity, first-order transitions can be recognized as discontinuous transitions; while, second-order and higher-order transitions are classified as continuous phase transitions or critical phenomena.

With respect to phase transitions, a high temperature phase usually possesses a relatively low degree of order and a relatively high degree of symmetry, while a low temperature phase has the opposite trend. A logical question arises as to whether or not statistical mechanics can be used to describe these phase transitions. This issue caused much debate about 70 years ago. The major problem was that statistical mechanics deals with smooth functions, such as Eqs. (1.12)–(1.15). The averaging process to find the partition function [Eq. (11)] requires integration, which make the function even smoother. Phase transitions, however, have discontinuous changes in thermodynamic properties. This issue raised the question: Can statistical mechanics be adapted to describe these discontinuous changes? If we take the "thermodynamic limits," whereby both the number of particles, N, and volume, V, tend toward infinity while the ratio between these two, N/V, is limited, the discontinuous changes can be theoretically predicted. In the past century, the use of mean-field theories to describe phase transitions has developed significantly. People found that the mean-field theory can describe systems with relatively small fluctuations. Near a critical point where large fluctuations are exhibited, another field theory, such as group renormalization, is necessary to illustrate the behavior of the system. A detailed description of this topic can be found in condensed matter physics textbooks (see, e.g., Chaikin and Lubensky, 1995).

In the mean-field theory, an order parameter (Φ) is utilized to describe the changes in order at the phase transitions. For a variety of phase transitions, the order parameter, which may be a number or a vector, can be very different, depending on the system studied. However, if we limit ourselves to classical and simple condensed systems, the order parameter in single-component systems is density (or volume), and in multiple-component systems, it is concentration. At high temperatures, the order parameter is usually reduced to a constant. At a transition temperature, order sets in. Below this temperature, the order parameter rises above the constant. Therefore, at the transition temperature, a discontinuous change in the order parameter can be observed, as shown in Fig. 1.7a. When the order parameter rises continuously from the constant at the transition temperature during cooling, it is a second-order continuous transition as shown in Fig. 1.7b.

2.2. Phase equilibrium and stability

The thermodynamic concepts of equilibrium and stability in phases and at phase transitions are important. In order to understand the difference between these two concepts, one can use classical mechanics as an analogy, as shown in Fig. 1.8. For a state in mechanical equilibrium, the summation of the forces, both translational and

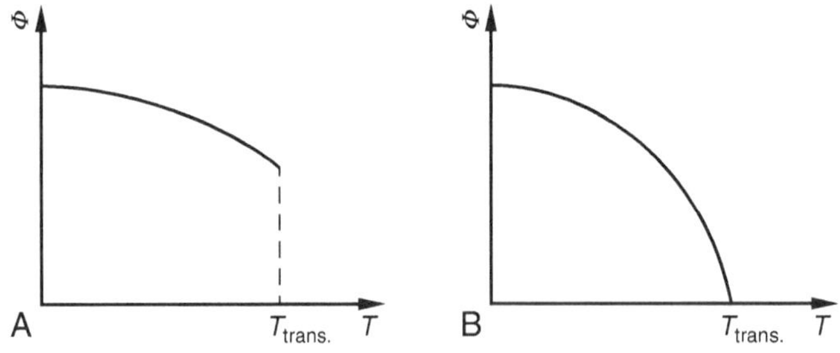

Figure 1.7 Order parameter changes with temperature for first-order (a) and second-order (b) transitions.

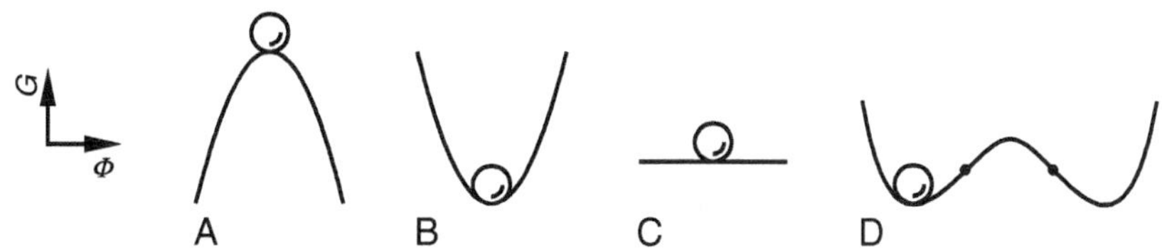

Figure 1.8 Concepts of equilibrium and stability in classical mechanics: (a) an unstable equilibrium; (b) a stable equilibrium; (c) a random equilibrium; and (d) a double-well, which includes one unstable equilibrium point surrounded by two stable equilibrium points. Two filled dots represent the limit of stability.

rotational, on the system is zero. However, the stability of a system reflects how this system responds to an outside perturbation. When a system dampens and dissipates a perturbation, the system is stable. On the other hand, if a system magnifies a perturbation, the system is unstable (Cheng and Keller, 1998). As shown in Fig. 1.8, the first case demonstrates an unstable equilibrium (Fig. 1.8a); the second case shows a stable equilibrium (Fig. 1.8b); and the third case displays a random equilibrium in classical mechanics (Fig. 1.8c). Figure 1.8d includes a double-well case which has one unstable equilibrium point surrounded by two stable equilibrium points (Yu *et al.*, 2005).

These concepts also apply to statistical thermodynamics. Similar to Fig. 1.8, if we consider a plot of free energy (G) versus an order parameter (Φ), a phase in thermodynamic equilibrium requires that the first-order derivative of the free energy, with respect to the order parameter be equal to zero, $dG/d\Phi = 0$. Additionally, for a stable phase, it is also necessary that the second derivative of the free energy with respect to the order parameter be positive, $d^2G/d\Phi^2 > 0$. Violation of these two criteria starts at $d^2G/d\Phi^2 = 0$. These points correspond to the inflection points in Fig. 1.8d represented by the two filled dots. At these two points, the phase reaches the limit of thermodynamic stability. In order to have a stability equal to the limit of stability, it is required that both $d^3G/d\Phi^3 = 0$ and $d^4G/d\Phi^4 > 0$. These are the necessary and sufficient criteria for stability. Generally speaking, a stable phase requires that the lowest-order, nonvanishing, even derivative be larger than zero and all lower-order derivatives be zero (Debenedetti, 1996). It has been found that these higher derivatives of Gibbs free energy are exquisitely sensitive to small variations in the particle interactions constructing the structures from which we build our analysis of the thermodynamic properties.

To understand the pathway used in phase transitions, it is necessary to construct a Gibbs free energy landscape. Generally speaking, this landscape contains different combinations of free energy peaks (barriers—see Fig. 1.9a), free energy valleys (stable minima—see Fig. 1.9b), flat free energy paths (no free energy changes—see Fig. 1.9c), and neighboring free energy valleys (double-wells—see Fig. 1.9d). If we plot the Gibbs free energy with respect to both volume and pressure, as shown in Fig. 1.9, these possible events thus become the elements in constructing the free energy landscape used to describe phase transitions.

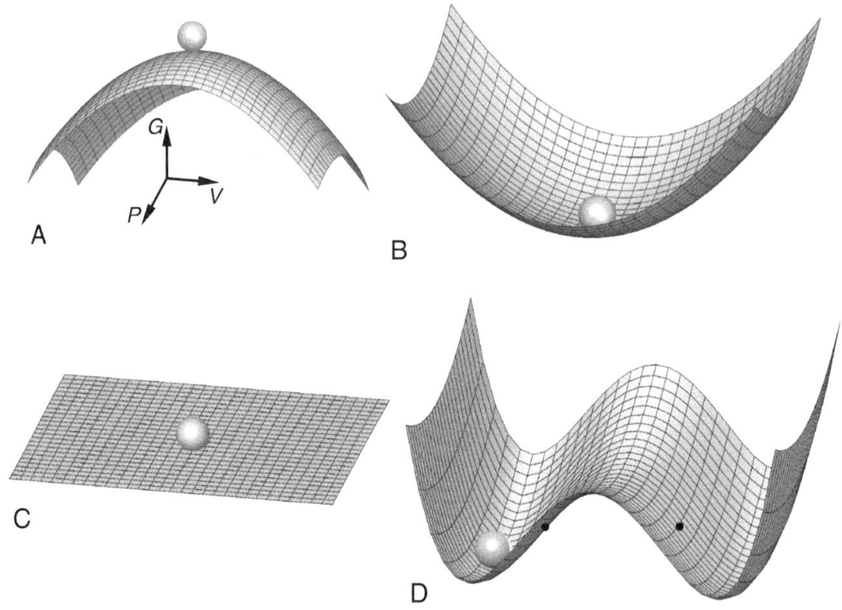

Figure 1.9 Possible elements in constructing a Gibbs free energy landscape in illustrating phase pathways: (a) a free energy peak (a barrier); (b) a free energy valley (a stable minimum); (c) a flat free energy path (no free energy change); and (d) two neighboring free energy valleys (double-wells, see Fig. 1.8d).

REFERENCES AND FURTHER READING

Chaikin, P. M., and Lubensky, T. C. (1995). "Principles of Condensed Matter Physics." Cambridge University Press, New York.

Chandrasekhar, S., Sadashiva, B. K., and Suresh, K. A. (1977). Liquid crystals of disc-like molecules. *Pramana* **9**, 471–480.

Cheng, S. Z. D., and Keller, A. (1998). The role of metastable states in polymer phase transitions: Concepts, principles and experimental observations. *Annu. Rev. Mater. Sci.* **28**, 533–562.

Debenedetti, P. G. (1996). "Metastable Liquids: Concepts and Principles." Princeton University Press, Princeton.

Demus, D., Goodby, J., Gray, G. W., Spiess, H.-W., and Vill, V., eds. (1998). "Handbook of Liquid Crystals." Wiley-VCH, Weinheim.

Ehrenfest, P. (1933). Phase changes in the ordinary and extended sense classified according to the corresponding singularities of the thermodynamic potential. *Proceedings of the Section of Sciences, Koninklijke Akademie van Wetenschappen te Amsterdam* **36**, 153–157.

Hill, T. L. (1960). "Introduction to Statistical Thermodynamics." Addison-Wesley, Reading.

Landau, L. D., and Lifshitz, E. M. (1969). "Statistical Physics." Addison-Wesley, Reading.

Leadbetter, A. J. (1987). Structural classification of liquid crystals. *In* "Thermotropic Liquid Crystals" (G. W. Gray, ed.), Chapter 1. John Wiley & Sons, Chichester.

Lehmann, O., and Verhandl, D. (1900). *Deutschen Phys. Ges. Sitzung.* **16**, 1; from Kelker, H. (1973). History of liquid crystals. *Mol. Crystals Liquid Crystals* **21**, 1–48.

Pershan, P. S. (1988). "Structure of Liquid Crystal Phases." World Scientific, Singapore.

Renitzer, F. (1888). Beiträge zur Kenntniss des Cholesterins. *Monatshefte für Chemie* **9**, 421–441.
Wunderlich, B. (2005). "Thermal Analysis of Polymeric Materials." Springer, Berlin.
Yu, L., Hao, B., and Chen, X. (2005). "Phase Transitions and Critical Phenomena." Science Publication, Beijing.

CHAPTER TWO

Thermodynamics and Kinetics of Phase Transitions

Contents

In this chapter, brief general descriptions are given for the thermodynamics and kinetics of phase transitions in simple, small molecules. This chapter is particularly designed to introduce the background research in this area. The chapter starts by reviewing, from a thermodynamic point of view, phase transitions in single component systems, including liquid–gas, crystalline solid–liquid, crystalline solid–solid, and transitions involving mesophases. The kinetic aspects of crystallization, melting, and mesophase transitions will be discussed specifically. The thermodynamic descriptions are then extended to multiple component systems, with particular attention to liquid–liquid phase separation, crystalline solid–liquid, and mesophase–liquid transitions of binary mixtures. A description of the kinetics of liquid–liquid phase separation in binary mixtures in the initial, intermediate, and late coarsening stages is also provided.

1. Thermodynamics of Phase Transitions in Single Component Systems

1.1. An example of liquid–gas transitions: van der Waals gas

The first question which needs to be answered in discussing phase transitions is whether or not the phase transition is intrinsically possible. The answer to this question is based in thermodynamics. Let us first examine liquid–gas phase transitions. The condensation of a van der Waals gas will be used to illustrate the liquid–gas transition behavior.

Three macroscopic thermodynamic properties—volume, pressure, and temperature—determine the physical state of a phase. We know that these three properties are not independent of each other. When two of these three properties are fixed, the third one is determined. The relationship among these three properties at thermodynamic equilibrium is called an equation of state. For an ideal gas, the most well-known equation of state is:

$$PV = nN_{\mathrm{A}}kT = nRT = \frac{Nm\langle v\rangle^2}{3} \tag{2.1}$$

where P is the pressure, V is the volume, T is the temperature, N_{A} is Avogadro's number, n is the number of moles of particles (atoms or molecules), and R is the gas constant, which is equal to $N_{\mathrm{A}}k$ (k is the Boltzmann constant). The gas constant is used for mole-based calculations of particles; while, the Boltzmann constant is used for single particle calculations. The right side of Eq. (2.1) is a microscopic statistical description of an ideal gas, where N is the number of particles, m is the mass of the particles, and $\langle v\rangle$ is the average particle velocity. According to the description in Eq. (2.1), the gas state can never undergo a phase transition, so it cannot condense into a liquid. The reason is very simple: In an ideal gas, the particles do not interact, so they cannot aggregate. An ideal gas is, therefore, a "permanent" gas.

In order to describe the condensation process of a gas to a liquid, van der Waals, who was studying for his Doctor of Philosophy degree in 1873, suggested introducing two basic non-idealities into the ideal gas system: that every particle occupies a volume (b) and that there are inter-particle interactions in the system. The van der Waals equation of state can be written as:

$$\left[P + \left(\frac{N}{V}\right)^2 a\right](V - Nb) = nN_{\mathrm{A}}kT \tag{2.2}$$

where the inter-particle interactions generate an "internal pressure," expressed as $(N/V)^2a$, where a is a proportionality constant. It is obvious that when N/V is small, this equation is close to that of the ideal gas, and no liquid–gas phase transition will occur. However, as soon as N/V is no longer small, a phase transition can occur. This reflects the fact that above a specific temperature, only the gas phase exists, no matter how much pressure is applied to the system. This temperature is called the

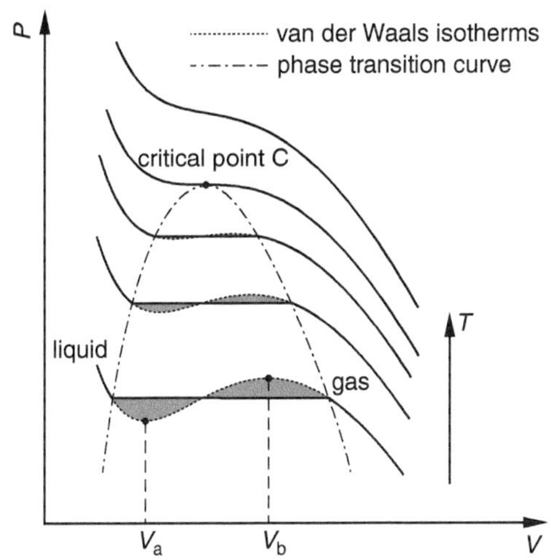

Figure 2.1 Set of van der Waals isotherms at different temperatures in a pressure–volume plot in the region near the critical point (C). The integration of the van der Waals isotherm (dashed lines) pressures ($P - P_c$) with respect to volume between V(liquid) and V(gas) must be equal to zero, since the areas above and below the horizontal lines are equal with opposite signs.

critical temperature. The pressure that is just sufficient enough to liquefy the phase at the critical temperature is called the critical pressure. The volume occupied at the critical temperature and critical pressure is called the critical volume (see e.g., Moore, 1972). If we plot the relationship between pressure and volume, as shown in Fig. 2.1, at temperatures much higher than the critical temperature, the pressure–volume behavior of the gas approximates that of the ideal gas. However, at the critical temperature, a singularity appears (the critical point C in Fig. 2.1) at which both the first and second derivatives of pressure with respect to volume at constant temperature are zero:

$$\left(\frac{\partial P}{\partial V}\right)_{\mathrm{T}} = 0 \tag{2.3}$$

and

$$\left(\frac{\partial^2 P}{\partial V^2}\right)_{\mathrm{T}} = 0 \tag{2.4}$$

Combining Eqs. (2.3) and (2.4) with the van der Waals equation [Eq. (2.2)], we can calculate the pressure, volume, and temperature parameters at the critical point:

$$T_{\mathrm{C}} = \frac{8a}{27bk} \tag{2.5a}$$

$$P_C = \frac{a}{27b^2} \tag{2.5b}$$

$$V_C = 3Nb \tag{2.5c}$$

where T_C, P_C, and V_C are the critical temperature, pressure, and volume, respectively. If we define reduced temperature as $T^* = T/T_C$, reduced volume as $V^* = V/V_C$, and reduced pressure as $P^* = P/P_C$, all of which are unitless, we can obtain a form of the van der Waals equation which supposedly all gases should obey:

$$\left(P^* + \frac{3}{V^{*2}}\right)(3V^* - 1) = 8T^* \tag{2.6}$$

because Eq. (2.6) does not contain any parameters which are specific to the chemical particulars of any gas.

From the van der Waals isotherms (the dashed lines) below the critical temperature in Fig. 2.1, it is evident that there is a volume region between the minimum and the maximum at each temperature where $(\partial P/\partial V)_T > 0$. The compressibility at a constant temperature is defined as $\beta = -(\partial V/\partial P)_T/V$, and in this volume region, it becomes negative. Therefore, in this region, the van der Waals equation predicts an impossible event that the volume increases with increasing pressure. Note that $(\partial P/\partial V)_T < 0$ or $\beta > 0$ is a necessary condition for the phase stability. This region should thus correspond to the existence of an unstable state, and one needs a linear horizontal line to connect both the liquid and gas phase according to the equal area principle (the Maxwell principle). Specifically, this principle states that the integration of pressure with respect to volume, along the line, must equal zero (see Fig. 2.1). This principle indicates that along this line at constant temperature, both the liquid and gas phases coexist.

The van der Waals equation was the first attempt to utilize the "mean-field" approach to analyze the liquid–gas phase transition. As shown in Fig. 2.1, the boundary to the left of the critical point with the smaller volume is the liquid, while the boundary to the right with a larger volume is the gas phase. When the temperature is lower than the critical temperature, there are regions where the liquid and gas can coexist. As the temperature approaches the critical temperature, the physical differences between the liquid and the gas phases gradually disappear. At the critical point $(\partial P/\partial V)_T = 0$, the compressibility, β, goes to infinity. Here the compressibility diverges, indicating that an infinitely small pressure change can lead to large volume fluctuations.

1.2. General descriptions of liquid–gas and crystalline solid–liquid transitions

If we know the phase behavior at each pressure (or volume) and temperature, a phase diagram can be constructed for a single-component system. This phase diagram provides all the information we need to determine whether a phase transition can take place at a given temperature and pressure (or volume). The condensation of a van der Waals gas approximately describes the liquid–gas phase transition under conditions that are close to an ideal gas. The question is: How do we extend our knowledge to an understanding of a crystalline solid–liquid transition in a single-component system?

A complete understanding of the phase transition behavior of the gas, liquid, and crystalline solid states for a single-component system can be achieved only by using a three-dimensional pressure–volume–temperature phase diagram. A two-dimensional phase diagram is a cross section of the three-dimensional phase diagram where one of these three properties is constant. Figure 2.2A represents a typical phase diagram of three states in the pressure–temperature plane. The lines in this phase diagram describe a discontinuous change of thermodynamic properties, and thus, are phase boundary lines. When adding the volume axis perpendicular to the pressure–temperature plane in Fig. 2.2A, these lines become curved planes that construct the three-dimensional phase diagram. The isolated points in Fig. 2.2A represent both simple and multiple critical points. A phase transition which passes through a critical point does not involve volume and enthalpy changes. In the phase diagram shown in this figure, the critical point is for a liquid–gas phase transition. We can repeatedly move the phase from a gas to a liquid by going around the critical point in the two-dimensional pressure–temperature plane, as shown in Fig. 2.2A. The reason that this pathway is possible is because the liquid and gas phases possess identical symmetries. We can also plot the two-dimensional pressure–volume phase diagram of the three states, as shown in Fig. 2.2B. Compared with Fig. 2.1, this figure has an additional solid phase to consider.

Let us analyze the phase diagram in Fig. 2.2A. The QC line represents the liquid–gas transition (vaporization and condensation) at different pressures and temperatures. Point C is the critical point in this phase diagram. The QB line illustrates the

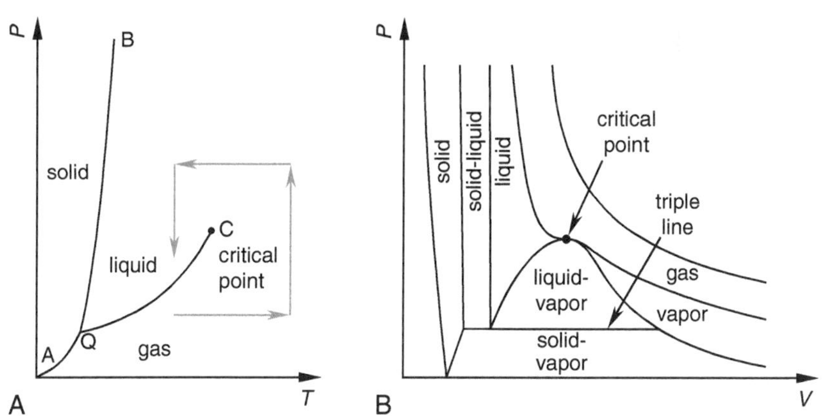

Figure 2.2 Schematic illustration of three phases in the pressure–temperature phase plane. The pathway illustrated in this figure shows a continuous transition from liquid to gas around the critical point (A). Note that the triple point Q in (A) becomes a triple line in the pressure–volume phase plane (B).

crystalline solid–liquid transition, while the QA line is a crystalline solid–gas transition. All of these transitions involve latent heat, which is a clear indication that they are first-order transitions in which particle interactions suddenly change during the phase transition. The crystalline solid–liquid and crystalline solid–gas transitions involve symmetry breaking. In this phase diagram, the three lines merge at one point, Q, which is the triple point where the three phases can coexist. Note that in Fig. 2.2B, the triple point in the pressure–volume plane becomes a horizontal triple line.

Figure 2.3 is a combination of both Fig. 2.2A and B, resulting in a three-dimensional phase diagram. There are three areas in which two phases are observed to coexist, including the crystalline solid and liquid, the crystalline solid and gas, as well as the liquid and gas phases. To analyze Fig. 2.3, we will take five temperatures where $T_4 > T_C > T_3 > T_2 > T_1$, and temperature T_C passes through the critical point. At temperatures higher than T_C, such as T_4, there is a gas phase. By decreasing the temperature to T_3, the system can pass through a liquid–gas transition. A further decrease in the temperature to T_2 causes the system to undergo both a crystalline solid–liquid transition and a liquid–gas transition. Finally, at temperature T_1, there is only a crystalline solid–gas transition.

Here a question is raised: Does a critical point exist for crystalline solid–liquid transitions? A critical point would terminate the phase boundary line dividing the crystalline solid and liquid phases. Based on our current understanding, a critical point for the crystalline solid–liquid transition does not exist because every crystalline solid–liquid transition involves symmetry breaking.

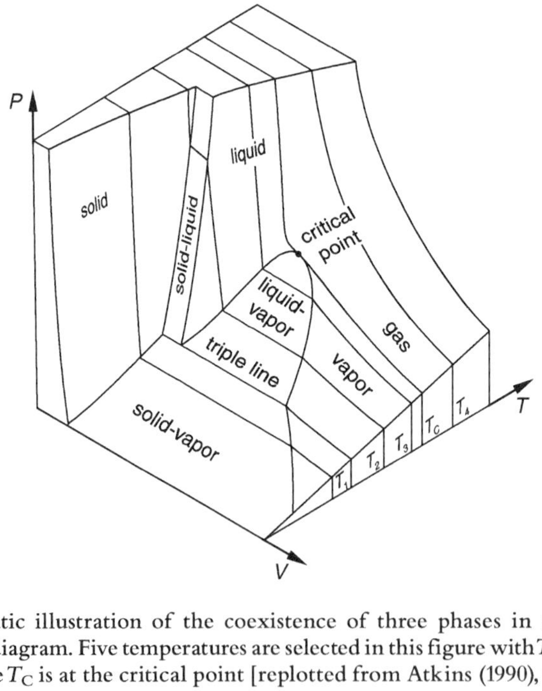

Figure 2.3 Schematic illustration of the coexistence of three phases in a pressure–volume–temperature phase diagram. Five temperatures are selected in this figure with $T_4 > T_C > T_3 > T_2 > T_1$. The temperature T_C is at the critical point [replotted from Atkins (1990), with permission].

Let us examine the crystalline solid–liquid phase transition. Crystal melting is a first-order transition. To provide a simple atomic description of crystal melting, it is postulated that the melting process can be hypothetically broken down into three consecutive steps. First, the crystal lattice expands to the average separation of the particles in the isotropic liquid. Second, rotational disorder is introduced into each lattice site, while the expanded long-range positional order in this "crystalline" structure remains. Third, introducing large amplitude particle motion into the "crystalline" structure finally destroys the remaining long-range positional order.

Based on this hypothetical description, volume and entropy are the two thermodynamic properties most useful in describing equilibrium crystal melting. The volume increase during melting for the majority of atomic, small molecule and polymer crystals is generally a few percent but can reach upwards of 20%. There is a small group of materials in which the volume change upon melting is negative, such as water at atmospheric pressure, which due to the effect of hydrogen bonding, has an 8% decrease in volume during ice melting at 0 °C. A few atomic crystals, such as bismuth and germanium, also undergo a 3% decrease in volume upon melting.

A better way to describe general trends during crystal melting is with the change in entropy. Following Wunderlich's description (1980) of crystals upon melting, simple crystals with spherical mono-atomic motifs melt following Richard's rule (1897). The change of entropy during melting for this class of crystals always ranges between 7 and 14 J/(K mol). This is because there is only a change in positional entropy ($\Delta S_{\text{position}}$) during the melting process. When the shape of the motif is non-spherical, the orientational entropy change ($\Delta S_{\text{orientation}}$) needs to be considered as well, which is described by Walden's rule (1908). This additional contribution to the overall change of entropy has been measured to be between 20 and 50 J/(K mol). Therefore, a general description of entropy change in the equilibrium melting of a simple crystal can be expressed by the equation:

$$\Delta S_{\text{melting}} = \Delta S_{\text{position}} + \Delta S_{\text{orientation}} \tag{2.7}$$

When the crystal is constructed from molecules containing flexible bonds such as polymers, Eq. (2.7) needs to be further modified by adding the conformational entropy change ($\Delta S_{\text{conformation}}$), as described by Wunderlich (1980) and Wunderlich and Grebowicz (1984). Each rotational unit contributes entropy ranging between 7 and 15 J/(K mol). The overall change of entropy for the equilibrium melting of crystals of flexible molecules and polymers can be described using the equation below:

$$\Delta S_{\text{melting}} = \Delta S_{\text{position}} + \Delta S_{\text{orientation}} + \Delta S_{\text{conformation}} \tag{2.8}$$

Note that the last term in Eq. (2.8) is dependent upon the number of rotational units in the molecule or repeat units of the polymer chain. Therefore, we can describe isobaric crystal melting in thermodynamic terms using the change in entropy at the equilibrium melting temperature. The product of these two thermodynamic properties, the change in entropy and the equilibrium melting temperature, represents the energy that is equivalent to the molecular interactions required to construct the crystal, namely the change of enthalpy (the equilibrium heat of fusion).

1.3. Crystalline solid–solid transitions

Crystalline solid–solid transitions are often observed in single component systems. Different from the process of crystallization–melting–recrystallization, crystalline solid–solid transitions take place solely in the solid state. By changing temperature (or pressure), a crystalline solid can be transformed into another crystalline solid without entering an isotropic liquid phase. These transitions result in material polymorphs. In most cases, crystalline solid–solid transitions are first-order transitions which undergo discontinuous changes in volume, enthalpy, and entropy due to crystal packing changes. The magnitude of these changes is usually small compared to the changes occurring from crystalline solid–liquid transitions. Although this transition requires symmetry breaking in structure to qualify as a first-order transition, the positional changes of the molecules to transform from one structure to the next must occur cooperatively, and the displacements cannot be too large.

Two well-known examples of crystalline solid–solid transitions in simple crystalline solids are water and carbon. The transitions between different ice forms at elevated pressure and the transition between graphite and diamond at elevated temperature and pressure in carbon are excellent illustrations of crystalline solid–solid transitions. The two-dimensional pressure–temperature phase diagrams of these systems are shown in Fig. 2.4A and B. In Fig. 2.4A, the normal ice phase (ice I) melting temperature decreases with increasing pressure due to the volume

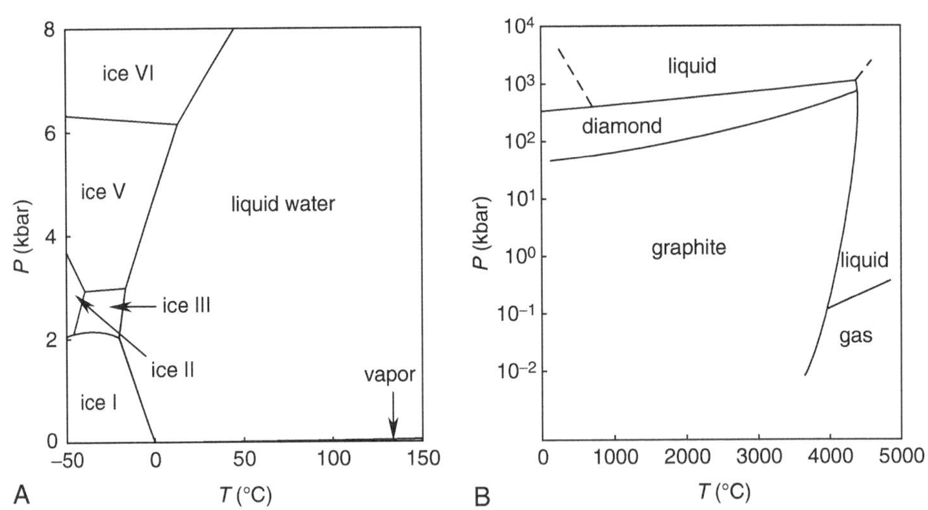

Figure 2.4 Phase diagrams for water (A) and carbon (B) in pressure–temperature phase planes. In water, there are several different ice phases. In carbon, the graphite and diamond phases are well known. They are constructed by the identical elemental carbon with different bonding structures and packing schemes [replotted from Atkins (1990), with permission].

reduction of ice upon melting. With increasing pressure, the ice I phase undergoes crystalline solid–solid transitions to enter other ice forms. These ice phases appear because the applied pressure modifies the water–water physical bonds/interactions. These ice forms possess higher melting temperatures (see e.g., Atkins, 1990; Moore, 1972). On the other hand, the phase behavior of carbon, presented in Fig. 2.4B, was difficult to establish because at extremely high temperatures and pressures other phases may intervene, making the experiments very complicated. Thus, we will focus only on the transition between graphite and diamond. Although diamond becomes stable at over 730 °C and 10,000 atmospheres of pressure, the conversion of pure graphite to diamond only occurs at practically useful rates at 200,000 atmospheres and 3730 °C. From the structural perspective, this transition involves a conversion from the two-dimensional sheet-like bonding structure of graphite to the three-dimensional network bonding structure of diamond (see e.g., Atkins, 1990).

1.4. Transitions involving mesophases

Between crystalline solids with three-dimensional long range order and liquids with three-dimensional short range order, a number of mesophases exist. Since single component mesophases lack long range order in at least one of the three types of order (positional, bond orientational, and molecular orientational), they are defined in terms of their structural order and symmetry (see Section 1.2 of Chapter 1). A specific feature of these transitions is that they all occur at or near thermodynamic equilibrium.

Liquid crystalline phases comprise a majority of all mesophases. The theoretical understanding of mesophase transition behavior started over a half-century ago with the development of scaling concepts in low temperature physics and the calculation of non-Gaussian exponents in statistical mechanics. The liquid crystal director (such as the orientational director) is used as the order parameter to describe liquid crystalline phase transitions. In these systems, there are liquid crystalline–isotropic liquid, liquid crystalline–liquid crystalline, and liquid crystalline–crystalline solid phase transitions. Many of the phase transitions are first-order transitions. Some of the liquid crystalline–liquid crystalline phase transitions are second-order transitions. Detailed theoretical descriptions and experimental overviews of liquid crystalline transitions can be found in the *Handbook of Liquid Crystals* (Barois, 1998; Demus, 1998).

An extensive collection of experimental data on the phase transition properties involving liquid crystals, such as transition temperatures and changes of volume, enthalpy and entropy at the transition temperatures, can be found. This information raises the question whether this data can provide general rules to understand the phase transition behavior of liquid crystals. Qualitatively speaking, since the structural order and symmetry of liquid crystals is in between that of a crystalline solid and an isotropic liquid, we would expect that the thermodynamic properties of the phase transitions should also depend on how close the structural characteristics of the liquid crystals are to those of a crystalline solid or the isotropic liquid. For low-ordered liquid crystalline phases such as nematic, smectic A and smectic C, we know that the density difference between these liquid crystalline phases and isotropic liquids is considerably smaller than the difference between the liquid crystalline phases and the crystalline solid. The changes in enthalpy and entropy associated with these transitions are only a few percent of the overall change in enthalpy and entropy between the crystalline solid and the isotropic liquid. On the other hand, the highly ordered smectic crystals are structurally similar to the crystalline solids, and, therefore, their transition thermodynamic properties and those of the crystalline solid should be similar. Furthermore, although the trend of volume changes at mesophase transitions is qualitatively parallel to the changes of enthalpy, they do not quantitatively match (see e.g., Demus *et al.*, 1983).

Generally speaking, the transition between a nematic phase and an isotropic liquid exhibits a small change in entropy, ranging from a fraction to several J/(K mol); while, the change in entropy for the transition from a nematic to a low-ordered smectic phase usually ranges from several to tens of J/(K mol). This is due to the fact that formation of the low-ordered smectic phases requires introducing an additional ordered layer structure. Note that these transition entropies are still only fractions of the value predicted by Walden's rule, which describes the orientational entropy change during crystal melting.

Finding quantitative rules to describe these liquid crystalline transitions has proven to be difficult. As early as the 1970s, the transition properties of over 200 liquid crystals and other mesophases have been collected. Many attempts have been made to find universal rules describing the phase transition properties of liquid crystals. If we limit ourselves to discussing only first-order transitions in liquid crystals, we may use either the change in volume or the change in entropy to analyze the different phase transitions in a variety of liquid crystals.

A large amount of data concerning the changes in volume (and thus density) at liquid crystalline phase transitions has been published by Bahadur (1976), Beguin *et al.* (1984), Pisipati *et al.* (1989), Tsykalo (1991), and others. However, quantitative correlations between changes in volume and specific types of liquid crystalline transitions are difficult to establish. The changes in entropy at these liquid crystalline first-order transitions can be determined from experimental measurements when the changes of enthalpy at the transitions and the transition temperatures are known. The thermodynamic properties of these liquid crystalline transitions can be found in collections published by Barrall and Johnson (1974), Schantz and Johnson (1978), Huang *et al.* (1985), Garland (1992), and others. The changes in entropy can also be determined using the entropic contributions of positional, orientational, and conformational order, as described by the crystal melting Eq. (2.8) in Section 1.2 (Wunderlich, 1980; Wunderlich and Grebowicz, 1984). The correlations between the change in entropy and specific liquid crystalline transitions, such as the nematic–smectic transition, are also less quantitative than the correlations found for crystal melting.

A significant difficulty in carrying out a quantitative analysis of liquid crystalline transitions is that liquid crystals consist of both a rigid mesogen and flexible tails. Partitioning the thermodynamic contributions of these two dissimilar parts needs to be taken into consideration. When the flexible tail is constructed from methylene units, this part also contributes to the change of entropy during liquid crystalline phase transitions. In contrast, flexible parts constructed from ethylene oxide and dimethylsiloxane units do not contribute to the change of entropy at these phase transitions (Blumstein and Blumstein, 1988; Blumstein and Thomas, 1982; Yandrasits *et al.*, 1992; Yoon *et al.*, 1996a,b). Therefore, the partitioning of the changes in volume and entropy of the mesogen and the flexible tails depends on the composition of the tails. Efforts to figure out the appropriate partitioning have been made based on density measurements (Guillon and Skoulios, 1976a,b), but systematic investigations have not been completed. Another issue is the extent to which structural order in the liquid crystalline phase results from specific intermolecular interactions. In some cases, the mesogens, like cyanobiphenyls, interact through dipole–dipole attraction. These partitioning difficulties may be reflected in semi-quantitative analyses of transition properties involving liquid crystals.

On the other hand, analyses of individual liquid crystal compounds have been more successful. In general, liquid crystalline materials proceed from the isotropic liquid to the crystalline solid through a sequence of liquid crystal phases of increasing order. For a single component, we know the changes in entropy of these first-order transitions from experiment. We can also estimate the overall change in entropy of this compound from its ultimate equilibrium crystal phase to its isotropic liquid. The entropy increase at the transitions during isobaric heating is attributed to the loss of positional, orientational, and conformational order. By comparing the estimated overall change of entropy with the summation of each of the experimentally observed transitional entropy changes, one can determine whether the summation of the entropy changes at the transitions matches the estimated overall change of entropy, or whether there are "hidden" transitions. These hidden transitions are usually continuous phase transitions.

One example that illustrates this principle is *N,N′*-bis(4-N-octyloxy-benzal)-1,4-phenylenediamine, as shown in Fig. 2.5. The chemical structure consists of both a mesogen and two flexible tails. The isobaric heat capacity data in Fig. 2.5 shows eight phases. In a step-wise fashion, the phase structure loses order with increasing temperature as it passes through each phase transition. Note that even below the K_3 crystal phase transition temperature, the molecule still needs to go through a glass transition at 77 °C to reach its solid state heat capacity. Adding up all the changes in entropy of these transitions does not fully account for the estimated entropy change from the crystalline solid to the isotropic liquid. Carbon-13 solid-state nuclear magnetic resonance experimental results indicate that even in the K_3 crystal phase, which is the lowest temperature phase before entering the glassy state, the flexible octyl tails in the molecule are not completely in an all *trans*-conformation (Cheng *et al.*, 1992).

Another interesting area is the effect of pressure on liquid crystalline phase structures and stabilities. At different pressures, the temperature range in which a particular liquid crystalline phase is stable may increase, decrease, or in some extreme cases, the liquid crystalline phase may be completely suppressed or induced where one did not previously exist. These phase stability changes with pressure can be clearly illustrated in a pressure–temperature phase diagram. Among these behaviors, one of the most fascinating observations is reentrant phase transitions. Namely, a higher temperature phase with a lower order can reappear in a temperature region

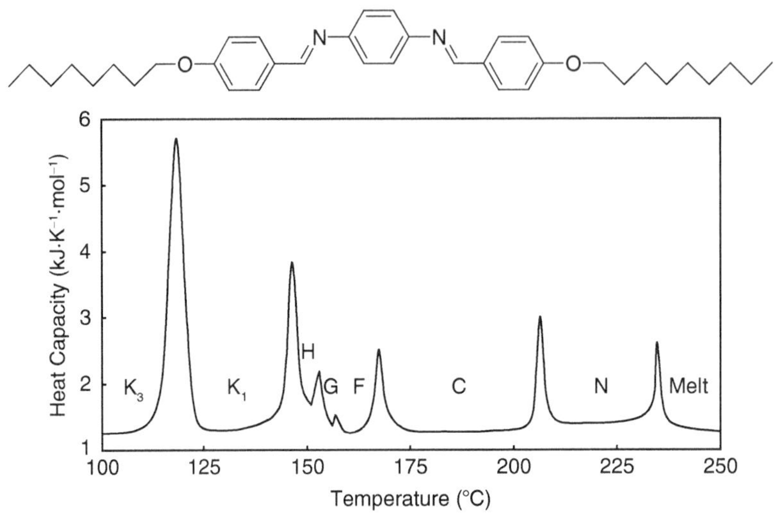

Figure 2.5 Heat capacity data for *N,N′*-bis(4-N-octyloxybenzal)-1,4-phenylene-diamine, as measured in calorimetric experiments. There are eight phases. From the low temperature side, the K_3 crystal, K_1 crystal, highly ordered smectic H, G, and F phases, smectic C, nematic, and isotropic liquid phases were identified [replotted from Wunderlich (2005), with permission].

below a stable liquid crystal phase with a higher order. This phase behavior can be explained by frustration, complex steric factors, or competing fluctuations (Cladis, 1998). For example, in the cases of cyano-Schiff bases and cyanobiphenyls with *n*-alkyl or *n*-alkoxy tails, a nematic phase disappears with increasing pressure and reappears at even higher pressures (Cladis *et al.*, 1977). This reentrant behavior also appears in a mixture of cyanohexyl- and cyanooctyl-oxybiphenyl at atmospheric pressure for different concentrations. Figure 2.6 schematically illustrates the phase diagram of this mixture (Guillon *et al.*, 1980). It is evident that the smectic A phase can only be formed in the relatively low concentration region of cyanohexyl-oxybiphenyl, and more importantly, the derivative of concentration with respect to temperature of the phase stability boundary changes sign. This leads to a maximum at which the derivative of concentration with respect to temperature is equal to zero, creating a reentrant nematic phase.

In addition to liquid crystalline transitions, there are also plastic crystal mesophase transitions. In plastic crystals, the spherical or slightly aspherical motif is rotationally disordered while maintaining three-dimensional long-range positional order. Plastic crystals are thus optically isotropic, not exhibiting birefringence. An early review summarized a number of examples of plastic crystals (see e.g., Smith, 1975). One example is fullerenes, such as C_{60}. The transition entropy of this fullerene from the crystal to the plastic crystal phase is attributed to orientational disorder while the positional order is maintained. This transition takes place with a 27.3 J/(K mol) change in entropy at −14 °C. This value of entropy change fits Walden's rule (Wunderlich and Jin, 1993). The crystal-to-plastic-crystal transition also corresponds to a sudden change of the molecular motion detected by spin-lattice relaxation time

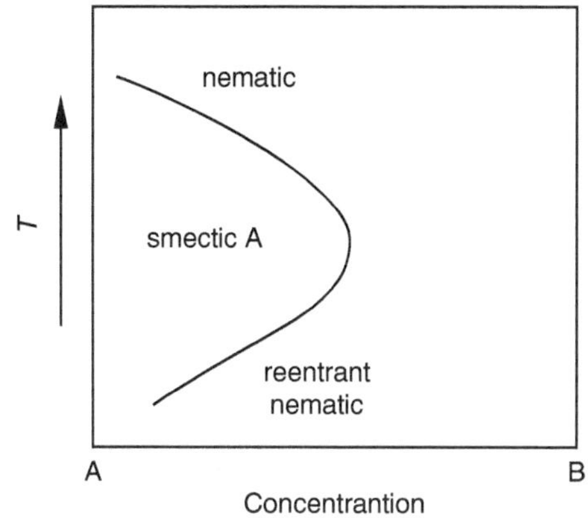

Figure 2.6 Schematic illustration of the temperature–concentration phase diagram for cyanohexyl-oxybiphenyl (A) and cyanooctyl-oxybiphenyl (B) mixtures at atmospheric pressure. The slope of the concentration change with respect to temperature is positive for the upper half of the phase stability boundary line, and it is negative in the bottom half of the line [replotted from Guillon *et al.* (1980), with permission].

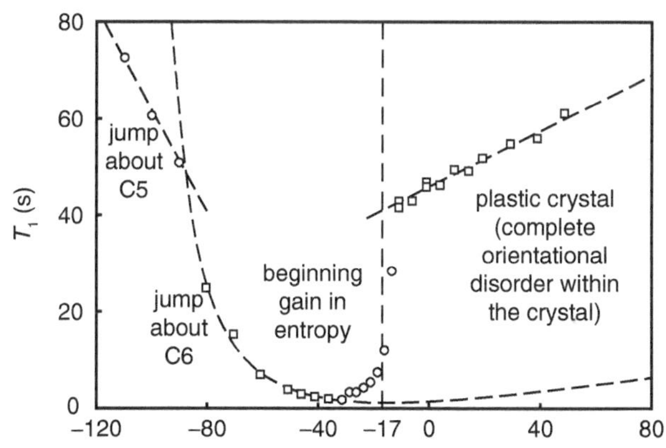

Figure 2.7 The spin-lattice relaxation time (T_1) with temperature for fullerene measured by carbon-13 solid-state nuclear magnetic resonance experiments. Three modes of rotational motion can be identified. The disordering temperature to enter the plastic crystal phase occurs at −17 °C, which corresponds well with the calorimetric measurements which placed the transition at −14 °C [replotted from Tycko *et al.* (1991), with permission].

in carbon-13 nuclear magnetic resonance experiments. Figure 2.7 shows this relaxation time (T_1), which corresponds to the change in molecular rotation with temperature. Three types of rotational modes were detected: rotation around the fivefold symmetry axis (C_5), rotation around the sixfold symmetry axis (C_6), and full rotation (Tycko *et al.*, 1991). The disordering transition temperature for the plastic crystal phase, as detected by this magnetic resonance technique, was −17 °C, which is virtually identical to the calorimetric measurement (Wunderlich and Jin, 1993).

Now we will specifically discuss discotic liquid crystalline phases of small molecules (comparing Figure 1.3 with Figure 1.4). These phases are considered part of the "columnar" class of mesophases. The thermodynamic transition properties of columnar phases are complex with a number of them having different degrees of order and symmetry. The structural characteristic of columnar phases is that no true long range order exists along the column direction (de Gennes and Prost, 1995). The driving force to form columns in many cases is π-π interactions between the aromatic cores. Figure 2.8 illustrates the diffraction pattern resulting from x-ray analysis, which can identify the structural features from oriented mono-domains of a low-ordered liquid crystalline phase, such as a smectic A phase and a hexagonal packed columnar phase. For the calamitic smectic A phase, its long range (or quasi-long range) order lies along the smectic layer normal (presented as sharp diffraction spots along the meridian in Fig. 2.8A), while short range order is along the lateral molecular packing direction (presented as diffuse halos on the equator in Fig. 2.8A). In hexagonally packed columnar phases, the lateral packing of columns has long range order (along the equator in Fig. 2.8B), while along the columns there is only short range order (along the meridian in Fig. 2.8B). This structural feature is maintained in the cases of rectangular, or obliquely packed, columnar phases.

A number of examples of columnar mesophases and their technological applications have been reviewed (see e.g., Cammidge and Bushby, 1998; Hoeben *et al.*, 2005). In polymers, a group of mesophases having identical characteristic structural features as the columnar phases has been identified. This knowledge has become increasingly important in understanding the fundamental classification of the mesophases, as well as realizing their practical use.

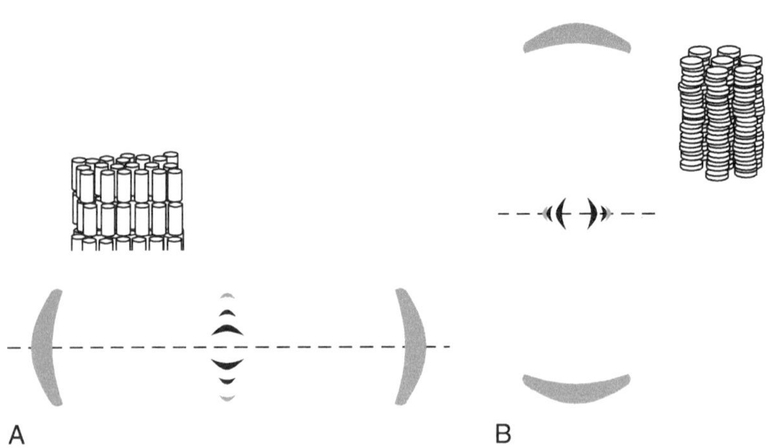

Figure 2.8 Schematic illustrations of x-ray diffraction patterns for mono-domains of a low-ordered calamitic smectic A phase with a long-range order along the layer normal (meridian) direction (A), and a hexagonally packed columnar phase with long range order along the lateral packing (equator) direction (B).

2. Kinetic Aspects of Phase Transitions in Single Component Systems

2.1. Crystallization

Whether a phase transition is possible or impossible is determined by thermodynamics, but how the transition progresses to the final state is determined by the kinetics of the phase transformation. Kinetics, therefore, is important in discussing phase transitions. Kinetics usually assumes a microscopic model of how the particles activate during the transition, then uses this model to understand macroscopic phase transition rates, which are usually measured by diffraction, calorimetric, spectroscopic, or microscopic experimental techniques. Among the transitions described above, we will first focus on crystalline solid–liquid transitions.

Crystallization, the transition to a crystalline solid, from the isotropic liquid does not take place at equilibrium. This principle was recognized as early as the eighteenth century (see e.g., Volmer, 1939). First, let us build a simple particle model for crystallization to help visualize the microscopic processes that occur during crystallization. The basic ideas of this model were established more than one century ago by Gibbs (1878) and continued to be developed by Tammann (1903), Tammann and Mehl (1925), Volmer (1939), Frenkel (1946), and many others. In this model, each particle needs to go through several steps before crystallization, such as adsorption, diffusion, aggregation, and nucleation. The slowest step is the rate-limiting process. This concept is essential to understanding polymer crystallization.

A nucleation-limited process, which is very common, will be used to illustrate crystallization. This construction is based on the assumption that particle density fluctuations caused by the thermal energy in a supercooled liquid can overcome the nucleation barrier caused by the formation of new surfaces of a small crystal (nucleus). Crystallization occurs through both homogeneous (three-dimensional) nucleation and heterogeneous nucleation on an existing growth front (two-dimensional nucleation). When this step takes the most time (the rate-limiting step), crystallization is a nucleation-limited process. For a given size at a constant volume and energy, the probability that a nucleus exists is a function of the change in entropy based on Boltzmann's law ($\propto \exp\{\Delta S/k\}$) where ΔS is the difference in entropies between the liquid and the crystalline solid. At a constant pressure and temperature, the probability that a nucleus of a given size exists is proportional to $\exp\{-\Delta G/(kT)\}$, where ΔG is the difference in free energies between the liquid and the crystalline solid. Turnbull and Fisher derived an equation for the rate of homogeneous nucleation (i) during the beginning of crystal formation. This equation is dominated by two opposing factors: the nucleation barrier free energy (ΔG) and the activation free energy for diffusion (ΔG_η) (Turnbull and Fisher, 1949):

$$i = \left(\frac{NkT}{h}\right)\exp\left[\frac{-(\Delta G + \Delta G_\eta)}{(kT_x)}\right] \tag{2.9}$$

where h is the Planck constant, N is the number of uncrystallized particles that can participate in the nucleation process, and T_x is the temperature at which crystallization takes place. Most discussions about crystallization focus on the nucleation barrier, ΔG. The general understanding is that in the supercooled liquid, the driving force for crystallization is the difference between the free energies of the final (crystal) and initial (liquid) states, and any crystal formation must start at an initial stage with a large surface area-to-volume ratio. The overall free energy of forming a small crystal can thus be represented as:

$$\Delta G = G_{\mathrm{crystal}} - G_{\mathrm{melt}} = G_{\mathrm{bulk}} + \sum\nolimits_{\mathrm{i}} \gamma_{\mathrm{i}} A_{\mathrm{i}} - G_{\mathrm{melt}} = \Delta G_{\mathrm{c}} + \sum\nolimits_{\mathrm{i}} \gamma_{\mathrm{i}} A_{\mathrm{i}} \tag{2.10}$$

where G_{bulk} is the free energy of the crystal disregarding surface effects and ΔG_{c} represents the change in bulk free energy, which is always negative. A_i represents the ith-type overall surface area, and γ_i is the ith-type specific surface free energy of a crystal. When the crystal is small (in the nucleation stage), the surface free energy is a non-negligible, positive term in Eq. (2.10) and leads to a destabilizing of the nucleus. Therefore, there is competition between the surface free energy term and the bulk free energy term. The positive surface free energy creates a nucleation barrier for the transformation from the isotropic liquid to the crystalline solid, as illustrated in Fig. 2.9 for homogeneous nucleation.

Based on different nucleus shapes, we can write detailed analytical surface free energy terms to make Eq. (2.10) quantitatively describe different types of nucleation barriers. Taking the first derivative of Eq. (2.10) with respect to the nucleus size allows one to obtain the critical nucleation size, and therefore, an analytical expression for the maximum nucleation barrier (ΔG^*). If the term ΔG_η can be predicted based on particle relaxation behavior, which is associated with molecular dynamics that can be characterized by such properties as viscosity, we can also write the detailed analytical

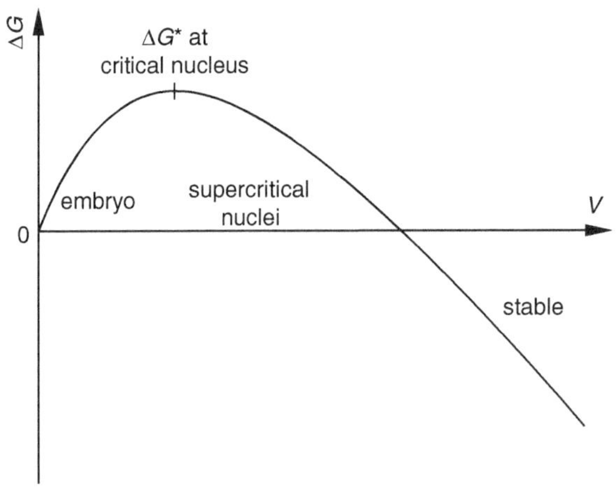

Figure 2.9 Schematic illustration of the free energy barrier for homogeneous nucleation.

form of the activation free energy (ΔG_{η}) in Eq. (2.9). Equation (2.9) represents not only homogeneous (primary) nucleation, but also heterogeneous (secondary and surface) nucleation processes. If the surface nucleation process is the rate-limiting step for crystal growth, its nucleation rate must have a form similar to Eq. (2.9).

The heterogeneous nucleation rate also exhibits an exponential dependence with respect to the inverse of supercooling. Based on nucleation theory, the barrier term ΔG depends on the supercooling. With increased supercooling, the degree of metastability of the liquid increases, and the nucleation barrier decreases. However, the activation term ΔG_{η} has the opposite trend. Therefore, the nucleation rates for both the homogeneous and heterogeneous nucleation processes form bell-shaped curves with respect to supercooling between the equilibrium melting temperature ($\Delta T = 0$) and the glass transition temperature ($\Delta T = T_m^0 - T_g$). At both of these temperatures, the nucleation rates are zero (at $\Delta T = 0$, the nucleation barrier for both nucleation processes goes to infinity, while at $\Delta T = T_m^0 - T_g$, large-scale molecular mobility diminishes). It should also be noted that at the same supercooling, the surface nucleation barrier is lower than that of primary nucleation because less surface area is introduced. Furthermore, the maxima of these two bell-shaped curves may not be located at the same supercooling.

The physical picture of surface nucleation involves a crystal surface that is close to a crystallographically perfect plane with few growth ledges. Most of the particles attached to the crystal surface end up detaching from the surface. In order to have crystal growth, the formation of growth ledges, or "hillocks," is the most time-consuming step. Formation of a growth ledge, or "hillock," requires that a few particles aggregate on the surface to form a nucleus. This nucleation process is one kind of lateral (layer) growth that adequately describes how particles crystallize onto existing crystal surfaces. A clear morphological characteristic of nucleation-limited growth is that it generates single crystals with well-defined, macroscopic faceted crystal habits. Since the faster growth planes are used up first, the observed growth fronts are the crystallographic planes with the slowest growth.

This simple model of crystallization via surface nucleation includes several steps, such as adsorption, one- and two-dimensional transport across the surface, and detailed structural transformations on the growth surface to form the ledges. It is also possible to have diffusion-limited crystal growth where the population of the arriving particles is limited. As a result, molecular transport may be the slowest step, and therefore, the controlling step in the crystal growth process. Sometimes, the ledges initiate screw and edge dislocations which limit crystal growth (defect-limited).

If a crystal growth surface contains a number of ledges creating a rough surface, nucleation may not be the rate-limiting process. Rough surface growth is called continuous (normal) growth (see e.g., Rosenberger, 1982). This model implies that every attachment (or impingement) site is a potential growth site, and one- and two-dimensional transport, as well as the actual surface morphology, can be ignored. If one assumes that in this model $\exp\{-\Delta G_a/(kT_x)\}$ represents the activation energy of an impinging particle (the attachment process) and $\exp\{-\Delta G_d/(kT_x)\}$ is the departure activation energy of a particle (the detachment process), the crystal growth rate at the temperature T_x can be expressed as:

$$R = r_a^* \exp\left[\frac{-\Delta G_a}{(kT_x)}\right] - r_d^* \exp\left[\frac{-\Delta G_d}{(kT_x)}\right] \tag{2.11}$$

Also note that:

$$\Delta G_d = \Delta G_a + \Delta h_f \tag{2.12}$$

where Δh_f is the latent heat of crystallization. At the equilibrium melting temperature (T_m^0), the growth rate $R = 0$, $r_a = r_a{}^*$ and $r_d = r_d{}^*$, we thus have:

$$\frac{r_a^*}{r_d^*} = \exp\left[\frac{-\Delta h_f}{(kT_m^0)}\right] \tag{2.13}$$

Combining Eqs. (2.12) and (2.13), Eq. (2.11) can now take the form:

$$R = r_a^* \exp\left[\frac{-\Delta G_a}{(kT_x)}\right] \left\{1 - \exp\left[\frac{-\Delta h_f \Delta T}{(kT_x T_m^0)}\right]\right\} \tag{2.14}$$

When $\Delta h_f \Delta T/(kT_x T_m^0) \ll 1$,

$$R = r_a^* \exp\left[\frac{-\Delta G_a}{(kT_x)}\right] \left[\frac{\Delta h_f \Delta T}{(kT_x T_m^0)}\right] \tag{2.15}$$

For a narrow T_x range at low supercoolings (ΔT),

$$R \propto \Delta T \tag{2.16}$$

While in the high supercooling (ΔT) range,

$$R \propto \Delta T \exp\left[\frac{-\Delta G_a}{(kT_x)}\right] \tag{2.17}$$

Hence, a linear relationship between the growth rate and supercooling exists in the low supercooling region; while, at high supercoolings, the growth rates exhibit an exponential dependence on the ratio between the arriving activation energy and the thermal energy.

The single crystal habits of this type of rough surface growth are dominated by temperature or concentration gradients during crystal growth. The shape of a single crystal thus replicates the temperature or concentration profile. This generates, in many cases, curved crystal shapes. It should be remembered that the absence of macroscopic facets in curved single crystal habits does not necessarily imply that continuous crystal growth has prevailed. Continuous growth, however, cannot create faceted crystal habits.

The critical criterion differentiating these two different growth mechanisms is the roughness at the crystal growth front. The question is then raised: How much roughness is required to change the crystal growth mechanism from one to another? In simple, small molecules, the roughening temperature is defined by the α factor. In the growth of simple solid crystals from the isotropic liquid, the α factor is $\Delta h_f/(kT_R) \approx 2$ (where T_R is the roughening temperature) (Hunt and Jackson, 1966; Jackson *et al.*, 1967). However, no criterion has been established in polymer crystal growth.

Finally, both of these crystallization mechanisms indicate that crystal formation from the isotropic liquid always occurs at a temperature below the equilibrium melting temperature; thus, it should be emphasized again that crystallization takes place under non-equilibrium conditions.

2.2. Crystal melting kinetics

Melting kinetics is usually considered above but near the equilibrium melting temperature. Melting rates are generally more difficult to experimentally monitor compared with crystal growth rates. This is because the crystal growth rates are measured under supercooling, while the melting rates must be measured under superheated conditions. As a result, fewer investigations into crystal melting kinetics have been conducted and reported. Limited experimental observations of simple crystalline solids show that near the equilibrium melting temperature, melting occurs only at the crystal–liquid phase boundary and is the reverse process of crystallization. It has been illustrated by microscopic models that equilibrium melting can take place only at the corners and ledges of crystals that act as "nuclei" and cannot occur at the center of a perfect crystal, even at temperatures slightly higher than the equilibrium melting temperature (Volmer and Schmidt, 1937).

In order to measure the melting rates, the crystals need to be sufficiently perfect and large in size to allow enough residence time for the crystals in the liquid. If we examine single crystal dimensional changes as a function of supercooling and superheating, the rates of crystallization and melting may be similar in the vicinity of the equilibrium melting temperature. The equilibrium melting temperature can thus be experimentally identified when the crystal dimensional change through time is zero, with a dynamic equilibrium between the attachment and detachment processes of the particles.

Several examples of crystal growth and melting can be found in reports for crystalline solids of phosphorous pentoxide (Cormia *et al.*, 1963), germanium dioxide (Vergano and Uhlmann, 1970a,b), sodium disilicate (Fang and Uhlmann, 1984), and sodium (Tymczak and Ray, 1990) near their equilibrium melting temperatures. Figure 2.10 shows a schematic illustration of the crystal growth and melting rates which summarizes the behavior exhibited by simple, small molecules and oligomers. For small molecules, the crystal size increases (the crystal grows) below the melting temperature, while when slightly higher than this temperature, the crystal size decreases (the crystal melts). The growth and melting rates are similar but opposite. If the size of the molecules is increased to oligomers, however, the symmetry between crystal growth and melting rates with respect to equal amounts of

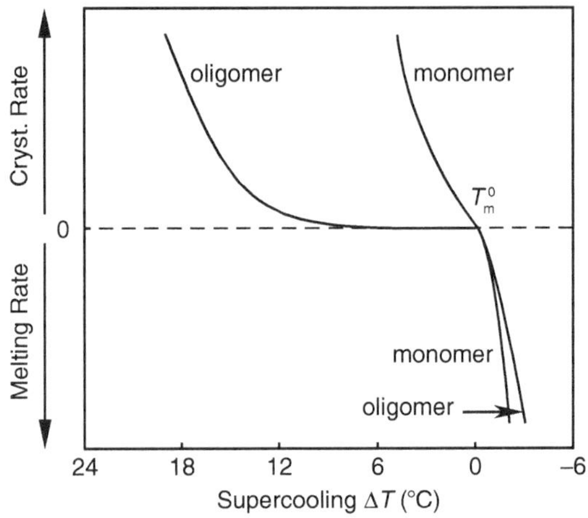

Figure 2.10 Schematic illustration of crystal growth and melting rates for crystals of simple, small molecules and oligomers. With increasing molecular size, the supercooling required to initiate crystallization increases [replotted from Wunderlich (2005), with permission].

supercooling or superheating may be lost. This is also shown in Fig. 2.10. This is due to the fact that the formation of nuclei for growth is more difficult than forming the "nuclei" needed to initiate crystal melting.

If a linear melting rate at each superheating is observed and this linear melting rate has an exponential dependency on superheating, similar to Eq. (2.9), while the single crystal habit is maintained (crystal facets and sectors), the melting process is controlled by a "heterogeneous (surface) nucleation" process. In this case of crystal melting, similar to the surface nucleation process in crystal growth, melting is limited to a two-dimensional surface, and its rate can generally be expressed as (Toda *et al.*, 2002):

$$R = R_0(T_{\text{sh}})\exp\left[\frac{-K\gamma^2 T_{\text{m}}^0}{(\Delta h_{\text{f}} k T_{\text{sh}} \Delta T_{\text{sh}})}\right] \tag{2.18}$$

where T_{sh} is the isothermal temperature above the equilibrium melting temperature, T_{m}^0. ΔT_{sh} is the superheating defined by $T_{\text{sh}} - T_{\text{m}}^0$. R_0 is the pre-exponential factor, which is weakly dependent upon the temperature T_{sh}. K is the geometric factor related to the specific crystal shape, and γ is the surface free energy. Since we use absolute temperature in Eq. (2.18), the change of temperature T_{sh} compared to the change of the superheating, ΔT_{sh}, can be neglected, and K is a constant. Therefore, R is approximately proportional to the exponential of superheating ΔT_{sh}. It then becomes:

$$\log R \propto \Delta T_{\text{sh}} \tag{2.19}$$

However, in contrast to crystal growth nuclei, which can form on the surface in a finite size at relatively large supercoolings, "liquid phase nuclei" are practically available at crystal corners, edges, and ledges. Therefore, the melting process starts at those "liquid nuclei" and proceeds from the surface to the center of the crystal. Melting may also be controlled by other rate-limiting steps, such as heat transport (conduction-limited).

When a crystal melts at temperatures much higher than the equilibrium melting temperature, "heterogeneous nucleation" can no longer dominate the process. The melting in this temperature region is called "homogenous melting." During "homogenous melting," the superheated crystal spontaneously generates a large number of spatially correlated internal local lattice instabilities. The accumulation and coalescence of these local lattice instabilities constitute the primary mechanism for "homogeneous liquid-nucleation" inside the crystal (Iwamatsu, 1999).

Several criteria can be used to characterize the superheating limit at which "homogeneous liquid-nucleation" becomes the dominant process. The first is called the "isentropic catastrophe." At a critical temperature, the entropy of a superheated crystal is identical to that of the liquid. This critical temperature has been estimated to be about 1.38 times the equilibrium melting temperature (Fecht and Johnson, 1988). Another criterion is called the "isochoric catastrophe." The entropy of a superheated crystal equals that of the amorphous solid at this critical temperature. This generates a lower critical temperature of about 1.29 times the equilibrium melting temperature (Tallon, 1989). Other older criteria include the introduction of a shear or rigidity instability when the shear modulus of a crystal goes to zero (Born, 1939) or when the amplitude of lattice vibration in the crystal reaches a critical value (Lindemann, 1910). Recent computer simulations utilizing molecular dynamics showed that melting takes place when the spatially correlated clusters generated by the superheated crystal can simultaneously meet the Lindemann and Born instability criteria (Jin *et al.*, 2001). The instability at isochoric melting was also simulated (Belonoshko *et al.*, 2006). On the other hand, the kinetic instability associated with catastrophic "homogeneous liquid-nucleation" at a critical temperature was also simulated utilizing molecular dynamics. The superheating limit for a crystal was theoretically determined via the vibration and elastic lattice instability criteria, leading to the conclusion that "homogeneous liquid-nucleation" starts at about 1.2 times the equilibrium melting temperature (Jin and Lu, 2000; Jin *et al.*, 2001; Lu and Li, 1998).

2.3. Transition kinetics involving mesophases

Most of the transitions involving mesophases are at or close to thermodynamic equilibrium. Since the structural order of the mesophases is in between that of the crystalline solid and the isotropic liquid, first-order phase transitions involving low ordered mesophases from the isotropic liquid only undergo relatively small changes of enthalpy, entropy, and volume. The phase transition kinetics involving mesophases with low structural order are usually difficult to monitor experimentally because the transition rate is fast near the equilibrium transition temperature. With increasing structural order of the mesophase, the transition kinetics from the isotropic liquid gradually slows down to rates close to crystallization, if there is a crystalline phase.

Only a few reported examples of phase transition kinetics involving low ordered mesophases can be found. The characterization of these phase transition kinetics is similar to the treatment of crystallization kinetics, which usually involves using the Avrami approach (Avrami, 1939, 1940, 1941):

$$1 - v^{c} = \exp\{-Kt^{n}\} \tag{2.20}$$

For a crystallization process, v^c in Eq. (2.20) represents the crystalline volume fraction, t is the isothermal crystallization time, and K is a collection of constants that includes the number of nuclei per unit volume, a geometric factor, and the linear growth rate. The "Avrami exponent," n, is usually associated with the dimensionality of crystallization. By taking the double log of both sides of Eq. (2.20), we can plot a linear relationship between $\log\{-\ln(1 - v^c)\}$ and $\log(t)$. The slope of this linear relationship is n, and the vertical intercept of this relationship is $\log(K)$. Using the Avrami treatment to describe the overall transition kinetics requires the assumption that the kinetics are nucleation limited. However, when Eq. (2.20) is used to describe mesophase transition kinetics, the physical meaning of these symbols needs to be changed. In a liquid crystalline transition, for example, v^c now represents the liquid crystalline volume fraction.

Several examples of transition kinetics from the isotropic liquid to a cholesteric nematic liquid crystal phase to a smectic liquid crystal phase then to a crystalline solid phase during cooling were reported for cholesteryl acetate, myristate, nonanoate, and esters. All of their transition kinetics can be fitted using the Avrami treatment of Eq. (2.20) (Jabarin and Stein, 1973; Price and Wendorff, 1971). For example, cholesteryl myristate shows two discrete transitions from the isotropic liquid to the cholesteric liquid crystalline phase. The first step is the very rapid formation of a turbid "blue" homeotropic state, and the second step is a slow transition to form a focal conic spherulitic-like texture (Jabarin and Stein, 1973). Figure 2.11 shows the

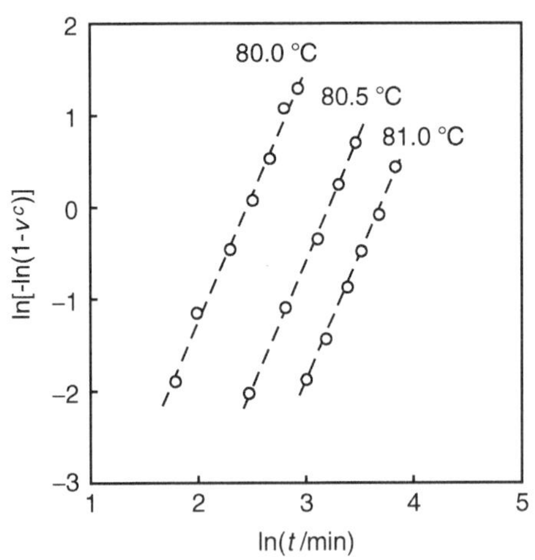

Figure 2.11 Avrami plots of cholesteryl myristate for the second step of the transition from the isotropic liquid to the cholesteric liquid crystal phase. The Avrami exponents were around three [replotted from Jabarin and Stein (1973), with permission].

transition kinetics of the second step utilizing the Avrami treatment of Eq. (2.20). This analysis indicates that these transitions are nucleation and growth processes, although the volume changes in the cholesteric nematic and smectic liquid crystal transitions are small. It also needs to be pointed out that the structural order and symmetry changes involved in these transitions are certainly different from the crystallization.

For transitions from a mesophase to a crystalline solid, the transition kinetics are less difficult to monitor. For example, the transition kinetics from cholesteryl pelargonate and caproate liquid crystals to crystalline solids can be analyzed using the Avrami treatment of Eq. (2.20) (Adamski and Czyzewski, 1978; Adamski and Klimczyk, 1978). However, some of these transitions exhibit much smaller values of the Avrami exponent, *n*. Note that this exponent is intended to represent the growth dimensionality of the ordered phase. If this value is only a fraction of unity, the physical meaning of this exponent is lost. Explanations for low exponents include a non-negligible volume fraction of pre-formed nuclei or the isothermal phase transition rate is not constant but changes with time for reasons such as the existence of a large interfacial region (Cheng, 1988; Cheng and Wunderlich, 1988).

A final caution needs to be pointed out in applying the Avrami treatment to analyze kinetic data obtained during crystallization and mesophase transitions. If we do not have morphological evidence and a microscopic model to explain how these phase transitions take place, the Avrami treatment is only a data-fitting process. No significant understanding of the phase transition mechanism can be achieved using this treatment alone.

Recently, phase transition kinetics involving the discotic columnar mesophases of 2,3,6,7,10,11-hexa(4′-octyloxybenzoyloxy)triphenylene were reported. Multiple phase transformations provided opportunities to investigate the transition kinetics between different ordered states. It is particularly interesting that this discotic liquid crystal possesses a discotic nematic phase, a rectangular columnar phase, and a highly ordered discotic orthorhombic phase upon cooling, as well as a highly ordered discotic monoclinic phase during heating (Tang *et al.*, 2001). The monoclinic phase was found to be metastable with respect to the orthorhombic phase. When the crystallization kinetics of the rectangular columnar phase to both highly ordered phases were measured, the monoclinic phase grows faster than the orthorhombic phase in a large region of relatively low temperatures. This is surprising since the former is thermodynamically metastable. At long times under isothermal conditions, the monoclinic phase transforms into the more stable orthorhombic phase (Tang *et al.*, 2003). This phenomenon is a very interesting one which will be thoroughly discussed later in this book.

3. Phases and Phase Transitions in Multiple Component Systems

3.1. Gibbs phase rule

The phase rule of heterogeneous equilibrium systems was first described by J. W. Gibbs in a series of articles published in the *Transactions of the Connecticut Academy of Arts and Sciences* between 1875 and 1878. In a single-component system, when plotting the free energy, G, versus temperature at constant pressure for a first-order phase transition, as shown in Fig. 1.5, the free energies of both phases (G_A and G_B) are equal at the equilibrium transition temperature, indicating that the stabilities of both phases are identical. The changes in the G_A and G_B, expressed on a per mole basis for a system with a constant number of moles, can be written as:

$$\left(\frac{\partial G_A}{\partial T}\right)_P \mathrm{d}T + \left(\frac{\partial G_A}{\partial P}\right)_T \mathrm{d}P = \left(\frac{\partial G_B}{\partial T}\right)_P \mathrm{d}T + \left(\frac{\partial G_B}{\partial P}\right)_T \mathrm{d}P \tag{2.21}$$

In this equation, there is only one independently adjustable variable, either temperature or pressure. In other words, when these two phases are in equilibrium with each other, only one degree of freedom exists in a single-component system. This type of system is called univariant. In a more general form, the Gibbs phase rule can be written as:

$$p + f = c + 2 \tag{2.22}$$

where p is the number of phases, f is the degrees of freedom, and c is the number of components in the system being described. The number 2 represents the two variables: pressure and temperature. Therefore, in a single-component system, if there are three coexisting phases in equilibrium with each other, there are no degrees of freedom. In such a system, there are no independently adjustable variables, so the system is invariant. This means that in a two-dimensional pressure–temperature phase diagram (which is a cross section of a three-dimensional pressure–temperature–volume phase diagram; see Fig. 2.3), a triple point represents the coexistence of the three phases at a fixed pressure and temperature. On the other hand, in a single phase there are two independently adjustable variables indicating that the system is bivariant.

When we discuss a multiple-component system, the number of independently adjustable variables starts to increase. The simplest multiple component systems are two-component binary mixtures. In the following sections, we concentrate our discussion on binary mixtures to illustrate general rules of phases and phase transitions in multiple component systems, using the Gibbs phase rule.

3.2. General thermodynamics of binary mixing

In a single-component system, all of the molecules are chemically identical and all interactions among the molecules are equivalent. In a multi-component mixture, interactions among the different types of molecules are the origin of different phase behaviors. These microscopic interactions lead to the study of partial molar properties which change with mixture concentration, such as partial molar volume, partial molar Gibbs free energy, or the chemical potential in binary and multiple component mixtures.

Let us first study the simplest case of two ideal gases mixing with n_A moles of gas A and n_B moles of gas B. Based on the Gibbs phase rule, one more degree of freedom needs to be identified. In this case, it is concentration. Since particles in ideal gases do not interact, the enthalpy of mixing is zero. This mixture is also a "permanent" gas. However, the entropy of mixing is larger than what it is for pure gases, since the mixing makes the system more random. This system generates a different Gibbs free energy of mixing:

$$\Delta G_{\text{mix}} = RT\left[n_{\text{A}} \ln\left(\frac{P_{\text{A}}}{P}\right) + n_{\text{B}} \ln\left(\frac{P_{\text{B}}}{P}\right)\right] \tag{2.23a}$$

where P_{A} and P_{B} are the partial pressures of gases A and B, respectively, and $P_{\text{A}} + P_{\text{B}} = P$, which is the pressure of the mixture. Based on Dalton's law, P_{A}/P can be replaced by molar fraction x_{A}, which is defined as n_{A}/n. Equation (2.23a) can thus be rewritten as:

$$\Delta G_{\text{mix}} = RT\{n_{\text{A}} \ln x_{\text{A}} + n_{\text{B}} \ln x_{\text{B}}\} \tag{2.23b}$$

Since the enthalpy of mixing, ΔH_{mix}, is zero, the entropy of mixing is thus:

$$\Delta S_{\text{mix}} = -R\{n_{\text{A}} \ln x_{\text{A}} + n_{\text{B}} \ln x_{\text{B}}\} \tag{2.24}$$

Since both x_{A} and x_{B} are smaller than one, the ln x_{A} and ln x_{B} in Eq. (2.24) are always negative values, making the entropy of mixing always positive, indicative of a spontaneous process.

If we have two liquids in an ideal solution, their mixing behavior is identical to ideal gas mixing, as described in Eqs. (2.23) and (2.24). Although molecules in the liquid interact with each other, the average A–B interactions in the mixture are the same as the A–A and B–B interactions in pure liquids. However, when the A–B interactions are different from the A–A and B–B interactions, we have a real solution, which possesses excess mixing properties, such as excess entropy and excess free energy. When these interactions are significantly different, the two liquids in the binary mixture become only partially miscible or even immiscible. Furthermore, miscibility changes with other parameters, such as temperature and pressure, because molecular interactions also change with respect to these parameters. For example, raising the temperature may enhance the miscibility of the two liquids. If above a certain temperature or in a certain temperature range phase separation does not occur at any concentration, then this system has an upper critical solution temperature (UCST). The UCST represents the temperature limit below which two separate phases can coexist. On the other hand, in lower critical solution temperature (LCST) systems, phase separation takes place at high temperature, and with decreasing temperature, the two liquids become miscible. Namely, the LCST is the temperature limit above which two separate phases can coexist. Some mixtures of liquids can exhibit complicated behaviors involving both upper and lower critical solution temperatures, depending upon the interaction changes with temperature and pressure.

3.3. Liquid–liquid phase separation in binary mixtures

Liquid–liquid phase separation of a binary mixture of two simple, small molecules with equal sizes is a phase transition. This phase separation relies on the difference of Gibbs free energy between the mixed and unmixed states, ΔG_{mix}. We can plot the relationship between the differences of Gibbs free energy of mixing against concentration for a specific binary mixture, as schematically shown in Fig. 2.12. Whether this mixture exhibits a liquid–liquid phase separation depends upon the temperature at a constant pressure. In case (a) of Fig. 2.12, both liquids are miscible at all concentrations. The criterion for complete miscibility is that $\partial^2\Delta G_{mix}/\partial x^2$ has to be larger than zero for all concentrations at a given temperature and pressure as shown in this figure, which is convex downwards over the entire concentration range. Case (b) of Fig. 2.12 represents the mixture at the critical point. This mixture requires that both $\partial^2\Delta G_{mix}/\partial x^2$ and $\partial^3\Delta G_{mix}/\partial x^3$ are equal to zero at this point. It further indicates that the miscibility of these two liquids has reached a point above which the sign of $\partial^2\Delta G_{mix}/\partial x^2$ becomes negative, and thus serves as the onset of liquid–liquid phase separation. In case (c) of Fig. 2.12, there are two concentrations, $x_A{}^1$ and $x_A{}^2$, at which each $\partial\Delta G_{mix}/\partial x$ is equal to zero and $\partial^2\Delta G_{mix}/\partial x^2$ is larger than zero. When the mixture has a concentration in between the concentrations $x_A{}^1$ and $x_A{}^2$, the mixture separates into two distinct liquid phases, with concentrations of $x_A{}^1$ and $x_A{}^2$ in order to decrease the ΔG_{mix} and assure the minimum Gibbs free energy of the system. Therefore, in this concentration range, liquid–liquid phase separation takes place.

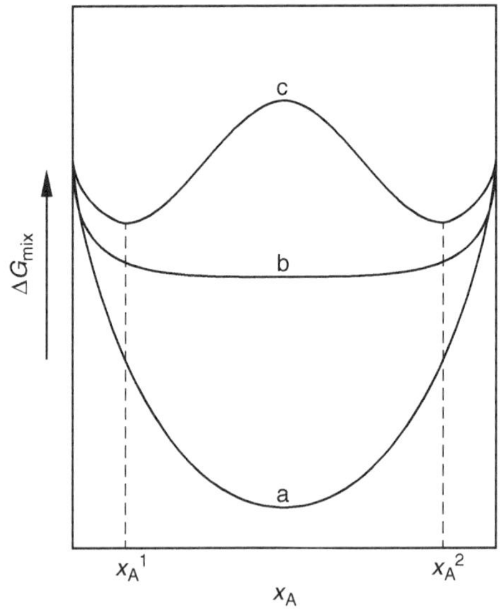

Figure 2.12 Differences of Gibbs free energies of mixing, ΔG_{mix}, with respect to concentration for a binary mixture. (a) Both liquids are completely miscible forming one liquid phase; (b) this mixture is at the critical point, and (c) liquid–liquid phase separation occurs. Note that these three cases are at three different temperatures.

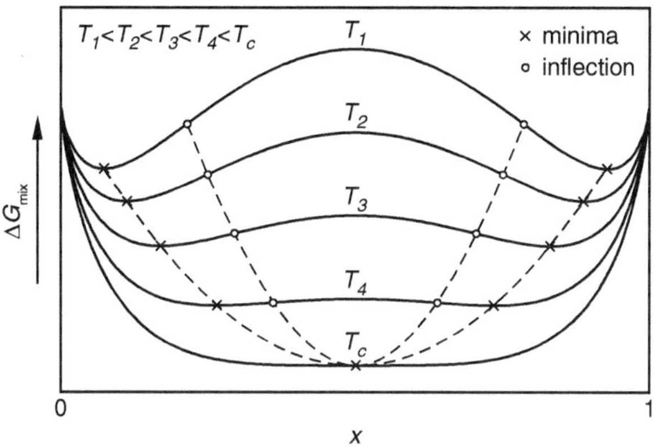

Figure 2.13 Differences of Gibbs free energy of mixing, ΔG_{mix}, with respect to concentration at different temperatures for a binary mixture with an upper critical solution temperature. The × points are the minima at which $\partial \Delta G_{mix}/\partial x$ values are equal to zero. The ○ points are the inflection points at which $\partial^2 \Delta G_{mix}/\partial x^2$ values are equal to zero. The curve formed by the minima points is the binodal curve; while, the curve created by the inflection points is the spinodal curve.

We can further investigate liquid–liquid phase separation of the binary mixture based on our knowledge of equilibrium thermodynamics. We will concentrate on case (c) of Fig. 2.12 to study its temperature dependence at a constant pressure. In Fig. 2.13, we plot a system with an upper critical solution temperature. At high temperatures above its critical point, both liquids are completely miscible and form a single liquid phase. When the temperature is slightly below the critical point, ΔG_{mix} minima appear on either side of the critical point. Further decreases in the temperature lead to a gradual increase in the concentration separation between these two minima, while the ΔG_{mix} barrier between these two minima increases. In fact, this is case (d) of Fig. 1.8 of Chapter 1. In Fig. 2.13, we know that at these two minima $\partial \Delta G_{mix}/\partial x = 0$ (identical chemical potentials). Additionally, we can also find two inflection points at which $\partial^2 \Delta G_{mix}/\partial x^2$ is equal to zero between the concentrations of these two minima. Therefore, between the minima and inflection points, $\partial^2 \Delta G_{mix}/\partial x^2$ is larger than zero. However, between the two inflection points, $\partial^2 \Delta G_{mix}/\partial x^2$ is smaller than zero.

When we connect the minima ($\partial \Delta G_{mix}/\partial x = 0$) at all temperatures, we obtain the binodal curve. Connecting all the inflection points ($\partial^2 \Delta G_{mix}/\partial x^2 = 0$) gives the spinodal curve, as shown in Fig. 2.13. We can now construct an upper critical solution temperature phase diagram for the liquid–liquid phase separation, as shown in Fig. 2.14, which plots temperature against concentration. Note that the binodal and spinodal curves in this figure are symmetric due to the fact that we only consider two simple, small molecule liquids of equal size. If the two liquids have very different molecular sizes, such as a polymer in solution with a small molecule solvent, the molar fraction needs to be changed to the volume fraction and asymmetric binodal and spinodal curves may appear. In this case, the identical chemical potentials may not be located at the minima of the mixing Gibbs free energy. Instead, the volume

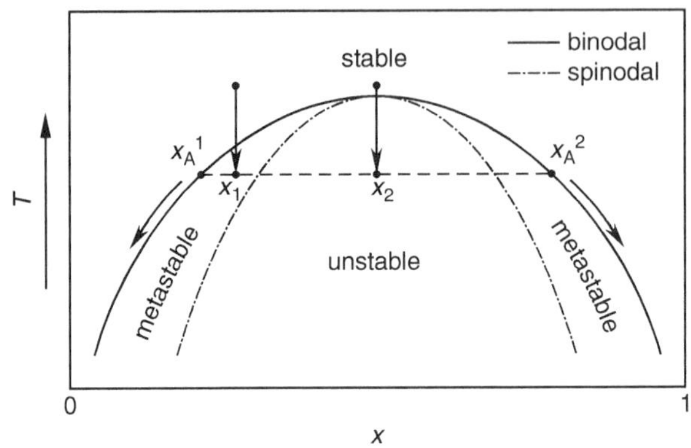

Figure 2.14 Liquid–liquid phase separation at two different concentrations (x_1 and x_2) in an upper critical solution temperature system. The solid line is the binodal curve, and the dashed line is the spinodal curve. The quench along x_1 is an off-critical quench, while that along x_2 is a critical quench.

fractions composing the two phases can be found by determining where a tangential line with a non-zero slope touches both free energy wells. In fact, at these two volume fractions where the phase separation occurs, the minimum possible Gibbs free energy of the system is achieved.

Let us return to Fig. 2.14, when we quench a binary system with a fixed concentration of x along the vertical line and the temperature enters the region between the binodal and spinodal curves, phase separation takes place immediately. For the temperature located at x_1 in this figure, the phase-separated concentrations of the two liquid phases are x_A^1 and x_A^2, as dictated by the binodal curve. As the quench temperature is lowered, the two concentrations, x_A^1 and x_A^2, correspondingly move toward pure liquid components.

The path that the quench process takes influences the mechanism of phase separation. When the quenching process brings the system in contact with the metastable region, as in the case along concentration x_1 previously discussed, it is called an off-critical quench. A critical quench occurs if we quench the system along concentration x_2, which passes through the critical point, as is also shown in Fig. 2.14. Here, the system enters the spinodal decomposition region without passing through the metastable region. Although the criterion differentiating these two kinds of quenches is not absolutely rigid, generally speaking, a critical quench requires the phase separation to take place at or around the critical concentration of the two liquids.

These phase separation processes are different depending on whether the system experiences a critical or off-critical quench from a homogeneous mixture. The off-critical quench may occur by entering the metastable region between the binodal and spinodal curves (Siggia, 1979). In the metastable region, free energy and concentration fluctuations are needed to overcome the nucleation barrier to initiate the phase separation (Cahn and Hilliard, 1959). Critical quenching places the system in the spinodal decomposition region without passing through the metastable region. This region is unstable, and the mixture spontaneously decomposes into two equilibrium concentration phases at a specific temperature (Cahn and Hilliard, 1958). Therefore, even though the final phase concentrations (x_A^1 and x_A^2), which are determined by thermodynamics, are identical in both types of quenching, the phase separation mechanisms are different, i.e., the phase morphology pathway taken to develop the ultimate equilibrium morphology.

3.4. Kinetics of liquid–liquid phase separation in binary mixtures

Considerable attention has been given to both the theoretical and experimental aspects of liquid–liquid phase separation kinetics in binary mixtures. The phase separation processes can generally be broken down into three stages: the early, intermediate, and late stages. In the early stage, the new phase is small with a diffuse boundary, while the concentration develops toward equilibrium. The late stage is defined by fixed phase concentrations and sharp interfacial boundaries with the phase morphology coarsening via the coalescence of micro-domains to form the final stable phase morphology. During this time, the total volume of each phase is conserved. The intermediate stage is simply the crossover period between the early and late stages.

At the initial stage of phase separation, depending upon the concentration of a binary mixture and the phase diagram, liquid–liquid phase separation is bounded by the binodal curve. When the system is directly quenched from a single liquid phase region at a high temperature to the metastable region in an upper critical solution temperature system like the one shown in Fig. 2.14, the phase separation is a nucleation-limited process. When the system is critically quenched into the unstable region bounded by the spinodal curve, also shown in Fig. 2.14, a spinodal decomposition process takes place. Therefore, the difference between the nucleation and the spinodal decomposition processes in an ideal "mean-field" system must be clearly understood.

From the perspective of kinetics, the nucleation-limited process (both homogeneous and heterogeneous) is an activated process in which a metastable state relaxes toward an equilibrium state. A very small droplet with one major component forms first a nucleus out of the binary mixture despite the higher free energy it generates. Therefore, a phase nucleus must grow in size to overcome the free energy barrier and stabilize itself. Classical nucleation theory attributes this barrier to the large surface area-to-volume ratio when the nucleus is small (see the description in Section 2.1). The height of this barrier depends on the system's degree of metastability, which is determined by the amount of supercooling in the isobaric case. In this nucleation-limited process, the metastable state thus relaxes by the activated growth of localized density fluctuations of large amplitude (Debenedetti, 1996).

On the other hand, the major characteristic of a spinodal decomposition process is that there is no free energy barrier. This process involves the growth of concentration fluctuations of small amplitude that exceed a critical wavelength. There is an optimal fluctuation wavelength for growth. Above this wavelength value, the fluctuations grow too slowly because the molecules need to diffuse over long distances. Below this value, the fluctuations generate large interfaces which cost too much energy. Therefore, the interfaces are not sharp, revealing that an unstable state relaxes by the spontaneous growth of long-wavelength fluctuations of small amplitude (Debenedetti, 1996).

From the phase morphological perspective, nucleation and spinodal decomposition generate completely different micro-domain structures before they reach the ultimate equilibrium state. In a nucleation-limited process, the new phase in the binary mixture takes the form of discrete small droplets imbedded in a matrix. If phase separation follows the spinodal decomposition process, a bicontinuous morphology develops. This phase morphology possesses a characteristic length scale, which is determined by the fluctuation wavelength of fastest growth. Therefore, in many practical cases, the observation of a phase morphological change is taken as evidence of a change in the phase separation mechanism.

Since the nucleation process is similar to that described in Section 2.1, our focus will only be on the spinodal decomposition process. In the early stage, the kinetics of this process can be described by the Cahn–Hilliard–Cook theory originally developed for metal alloys (Cahn, 1965; Cahn and Hilliard, 1958, 1959; Cook, 1970). After the system is critically quenched into the spinodal region, the amplitude of the concentration fluctuations is sufficiently small compared to the difference between the initial and final equilibrium concentration. The time evolution of the structure factor (S) is a function of scattering vector (q) and time (t), such that $S(q,t)$ can be written (Nakatani and Han, 1998):

$$\frac{\partial S(q,t)}{\partial t} = -2Mq^2\left\{\left[\left(\frac{\partial^2 \Delta G_{\text{mix}}}{\partial x^2}\right)_0 + 2\kappa q^2\right] S(q,t)\right\} + 2MkTq^2 \quad (2.25)$$

where M is the mobility defined as the proportionality constant in the relationship between the inter-diffusion current density and the chemical potential gradient. The inter-diffusion coefficient can be defined as $D_{\text{inter}} = M(\partial^2 \Delta G_{\text{mix}}/\partial x^2)_0$. The parameter κ is an interfacial free energy coefficient defined as the proportionality constant in the relationship between the interfacial free energy density and the square of the concentration gradient. The last term on the right side of Eq. (2.25) is the thermal noise (associated with kT), with the amplitude that is observed in fluctuation-dissipation theory. The solution of the differential Eq. (2.25) is given by Nakatani and Han (1998):

$$S(q,t) = S_\infty + [S_0(q) - S_\infty(q)]\exp\{2R(q)t\} \quad (2.26)$$

The $S_\infty(q)$ here is the virtual structure factor, and it can be written as:

$$S_\infty(q) = \frac{kT}{(D_{\text{inter}}/M + 2\kappa q^2)} \quad (2.27)$$

and the characteristic length becomes

$$R(q) = -(D_{\text{inter}}q^2 + 2\kappa M q^4) \quad (2.28)$$

Therefore, within the spinodal decomposition region, any concentration fluctuation grows exponentially at a scattering vector q with a characteristic length $R(q)$. The optimal growth wavelength of $R(q)$ sets the length scale of the phase separation of the spinodal decomposition. When we plot $R(q)/q^2$ versus q^2 based on Eq. (2.28), the intercept is the inter-diffusion coefficient, D_{inter}. This is usually called a Cahn plot. Note that this coefficient is negative when we study fluctuation growth, and it is positive when we study diffusion (Nakatani and Han, 1998).

As time passes, the amplitude of the concentration fluctuations increases, and the characteristic length grows. In scattering experiments, the spinodal peak continuously sharpens and gradually shifts to smaller scattering vectors. The system thus enters the intermediate stage, and this stage is then followed by the late stage of the spinodal decomposition process (Siggia, 1979). A universal scaling of the reduced characteristic length versus time can be observed in these stages. Experimentally, one plots the maximum-intensity reduced scattering vector, $S(q_{\text{m}},t)$, represented by Q_{max}, with respect to reduced time. Here, the reduced time is defined as:

$$\tau = \frac{t}{t_{\text{c}}}, \text{where } \frac{1}{t_{\text{c}}} = \left(D_{\text{inter}} q_{\text{m}}^2 \text{ at } t = 0\right) \tag{2.29}$$

The reduced scattering vector is thus:

$$Q_{\text{max}} = \frac{q_{\text{m}}(\tau)}{q_{\text{m}}(t = 0)} \tag{2.30}$$

A more insightful analytical solution for the mode coupling calculation is also available (Langer *et al.*, 1975) for the time-dependent scattering structure factor of binary mixtures.

Generally, the scaling law is used to describe the experimental scattering data in the late stage of liquid–liquid phase separation in binary mixtures as the system coarsens. If q_{m} represents the peak intensity scattering vector of the scattering structure factor $S(q_{\text{m}}, t)$ at a time t, the evolution of the phase-separated domains with time can be characterized by the following scaling law:

$$q_{\text{m}} \propto t^{-\alpha} \tag{2.31}$$

and

$$S(q_{\text{m}}, t) \propto t^{\beta} \tag{2.32}$$

where α and β are the scaling exponents characterizing the phase evolution with time. In the intermediate stage of spinodal decomposition, the relationship between α and β follows the inequality:

$$\beta > 3\alpha \tag{2.33}$$

since the characteristic wavelength and amplitude of the concentration fluctuations increase with time. However, in the late stage of spinodal decomposition, the amplitude of the concentration fluctuations reaches the equilibrium value determined by the coexistence curve in Fig. 2.14, and the relationship between α and β becomes:

$$\beta = 3\alpha \tag{2.34}$$

for a three-dimensional system. For the more general case, Eq. (2.34) can be written as:

$$\beta = d\alpha \tag{2.35}$$

where d represents the dimensionality of the system. For example, in a three-dimensional system ($d = 3$), Eq. (2.35) becomes Eq. (2.34). If $d < 3$, it means that the system has a less dense fractal structure. On the other hand, if $\alpha = 1$, it is an indication that the structure is coarsening with a hydrodynamic flow dominated process (Furukawa, 1998; Siggia, 1979). The phase separation can usually be monitored in real time by scattering experiments, and the peak scattering vector in these experiments should be proportional to the reciprocal of the characteristic size, $R(t)$. In the late stage of phase separation, the phase morphology continues to increase in length scale, but the morphological shape remains invariant. In this case, the scattering structure factor should also be self-similar:

$$S(q, t) = (q)^{-\mathrm{d}} f\left(\frac{q}{q_{\max}}\right) = (q)^{-\mathrm{d}} g\left(\frac{t}{t_{\max}}\right) \tag{2.36}$$

Therefore, we can plot $S(q,t)(q)^{\mathrm{d}}$ versus $q/q_{\max}$ at different times, or $S(q,t)(q)^{\mathrm{d}}$ versus $t/t_{\max}$ at different wave numbers. Doing so will result in a universal structure factor.

A well-known experimental example of spinodal decomposition for a simple binary mixture very near the critical temperature (only 0.6 microdegrees C away) is the mixture of 2,6-lutidine and water reported by Chou and Goldburg (1981). Figure 2.15A shows the light scattering experimental results for this system at times ranging from 10 to 500 s. The time period is around the late stage of the spinodal decomposition process. In this figure, scattering intensity is plotted against wave number. As the phase-separated morphology coarsens and grows, the maximum wave number, $q_{\max}$, decreases with time. However, the phase-separated morphology does not change its shape, only its length scale. The experimental data in Fig. 2.15A can thus be reduced with respect either to the maximum length scale or maximum wave number at different times to construct a universal master scattering curve for time, t, as shown in Fig. 2.15B. This data indicates that during the phase separation in the late stage of the spinodal decomposition process, the morphological change is self-similar. In these phase-separated bicontinuous domains, the equilibrium concentrations have been reached.

For the late stage of the coarsening process, a simple treatment of coarsening kinetics in phase separation was proposed by Ostwald and is called "Ostwald's ripening" (1900). If a system contains two liquid phases with sharp interfacial boundaries during phase separation, the newly formed minority liquid phase

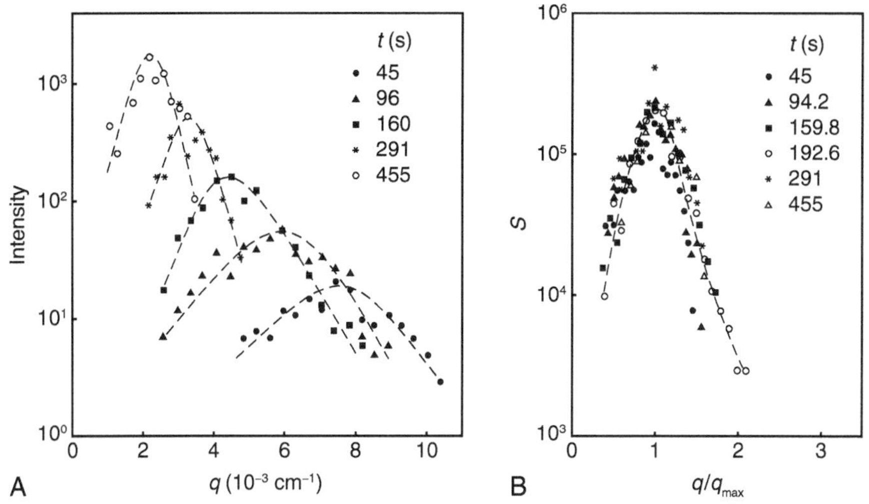

Figure 2.15 Light scattering results for liquid–liquid phase separation in a simple binary mixture of 2,6-lutidine and water: (A) data plotted for the scattering intensity versus wave number at different times and (B) scaled universal scattering curve from the experimental data of (A) after plotting the structure factor, S, versus reduced wave vector [reported after Chou and Goldburg (1981), with permission].

domains are dispersed in the majority liquid matrix. These droplets are not in thermodynamic equilibrium. The total free energy associated with the interfaces can be decreased in this liquid–liquid phase-separated system by having the droplets coalesce, thus increasing the size of the minority liquid phase domains. If this minority phase is a dispersion of droplets with a variety of sizes, statistically the large droplets will grow by absorbing the small droplets. Ostwald's ripening consists of two processes: a mass transport process between the droplets via diffusion, and attachment–detachment growth of the minority liquid phase. The mass transport process between the droplets is governed by a diffusion field in the dominant liquid phase; thus, this is a diffusion-limited process. On the other hand, the attachment–detachment process occurs at the interfacial boundaries. When the attachment–detachment process is the slowest, it becomes the rate-limiting step. The coarsening is then reaction-limited. These reaction-limited processes have characteristics similar to nucleation-limited processes.

There has been extensive effort to quantitatively describe the kinetics of diffusion-limited Ostwald's ripening. When a binary mixture is in a deep off-critical quench and the minority phase is a small fraction of the mixture, theoretical calculations indicate that, on average, the growing droplet size of the second liquid phase should obey the equation:

$$\langle R(t) \rangle = (K_1 t)^{1/3} \tag{2.37}$$

where $\langle R(t) \rangle$ represents the average droplet size and K_1 is the coarsening rate constant that depends on the diffusion constant and the surrounding environment.

This calculation assumes that the secondary liquid phase grows in size along with the droplet radius, $R(t)$, and that the volume fraction of this second phase is small, such that the phase domains are not in contact and not correlated. Furthermore, the droplet size distribution reaches a universal scaling form which is material independent. In fact, the 1/3 power in Eq. (2.37) can be obtained by general dimensional arguments, using a mean-field approximation (Lifshitz and Slyozov, 1961; Wagner, 1961). This is the well-known Lifshitz–Slyozov–Wagner law in the case of three-dimensional bulk systems.

On the other hand, in the reaction-limited process, the average droplet size follows (Schmalzried, 1981; Wagner, 1961):

$$\langle R(t) \rangle = (K_2 t)^{1/2} \tag{2.38}$$

where K_2 is the coarsening constant that now relies on the reaction constant and the surrounding environment. Detailed statistical discussions of both Eqs. (2.37) and (2.38) can be found in textbooks (see e.g., Ratke and Voorhees, 2002). Other growth laws have also been derived (see e.g., Furukawa, 1998).

It should be noted that the Lifshitz–Slyozov–Wagner law ignores the effects of the volume fraction of the droplets and their interactions. In order to understand more realistic systems with non-zero volume fractions, modifications of the Lifshitz–Slyozov–Wagner law were derived by Tsumuraya and Miyata, using a series of different kinetic coarsening interaction laws (Tsumuraya and Miyata, 1983). However, all these modifications share identical approximations; thus, they generate similar behaviors in terms of the kinetic prediction of late-stage phase morphology evolution. Further developments have been continued by studying the phase coarsening using the "effective medium" theory (Brailsford and Wynblatt, 1979). These developments included the statistical "field cell" acting on each size-class of droplets (Marsh and Glicksman, 1996), limiting the spatial extent of the diffusion field by taking into account "screening" provided by a two-phase medium comprised of the matrix and a distribution of droplets (Marqusee and Ross, 1984), extending the kinetics to finite clusters (Fradkov *et al.*, 1996) and others. From these approaches, the common prediction is that a non-zero volume fraction of droplets does not alter the temporal coarsening exponent of the Lifshitz–Slyozov–Wagner law; however, accounting for the volume fraction does change the coarsening rate [the coefficients in Eqs. (2.37) and (2.38)] and the resultant droplet size distribution. It should be mentioned that these theoretical treatments are limited to describing the kinetics of Ostwald's ripening. A comprehensive review by Baldan provides a summary of the developments in the theory of phase coarsening that have occurred in the past half-century (Baldan, 2002).

Experimental observations beyond the bulk liquid–liquid phase separation have also used the concept of Ostwald's ripening process. Some inorganic systems, such as super-alloys, can be quantitatively described by Ostwald's ripening. Many recent reports have appeared concerning the coarsening processes of two-dimensional phase-separated systems. Additionally, reaction-limited Ostwald's ripening has been used to explain the observed experimental results found for structural perfection processes. For example, the growth of crystal grains can occur in either of two ways: two or more grains merge together or one grain grows at the expenses of other grains. Other annealing processes in crystals may also be described by this ripening principle in both two- and three-dimensional systems.

3.5. Crystalline solid–liquid transitions in binary mixtures

If both liquids, A and B, in a binary mixture can crystallize at lower temperatures, we are dealing with a two-component crystalline solid–liquid system. If, in thermodynamic equilibrium, this system includes miscible liquids and immiscible solids across the entire concentration range, then this is a simple eutectic system as shown in Fig. 2.16.

Consider the phase transition behavior for a concentration x upon cooling from a high temperature along the vertical dashed line in Fig. 2.16. At x_1, both liquids are completely miscible. When the point x_2 is reached, pure crystalline solid A begins to precipitate out of the solution. Upon further decrease of the temperature, more of crystal A precipitates, and the solution concentration of B is higher than at the initial condition, x_1. This concentration change is along the x_2P line. At the point x_3, in the two-phase region, the concentrations of both components in the separated phase are dictated by the phase diagram. When point x_4 is reached, the residual solution has the eutectic concentration at point P. Further cooling leads to the simultaneous precipitation of both A and B crystals at the concentration ratio of P. Note that the eutectic point is an invariant point in this phase diagram at a particular pressure. This process occurs because in this system, there are three phases in equilibrium; therefore, based on the Gibbs phase rule, only one degree of freedom remains, which is the chosen pressure.

Another type of crystalline solid–liquid transition in binary mixtures occurs in solid solutions. Different from a eutectic system, in solid solutions, there are solid phases containing more than one component. There are two kinds of solid solutions: substitutional and interstitial. In substitutional solutions, the solvent

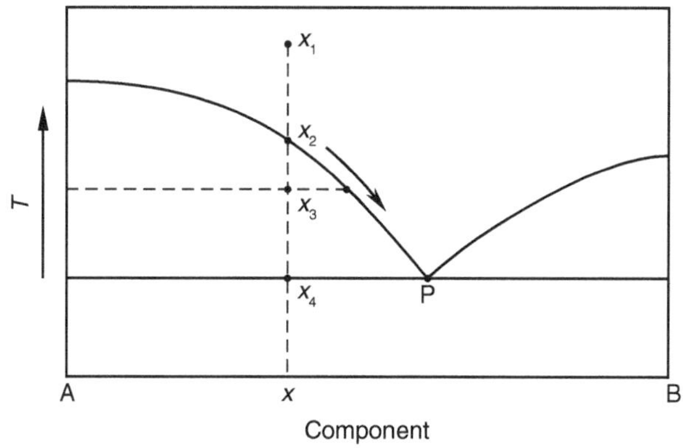

Figure 2.16 Simple temperature–concentration eutectic phase diagram for two components, A and B. These two components are completely miscible in their liquid phase but immiscible in their crystalline solid phases.

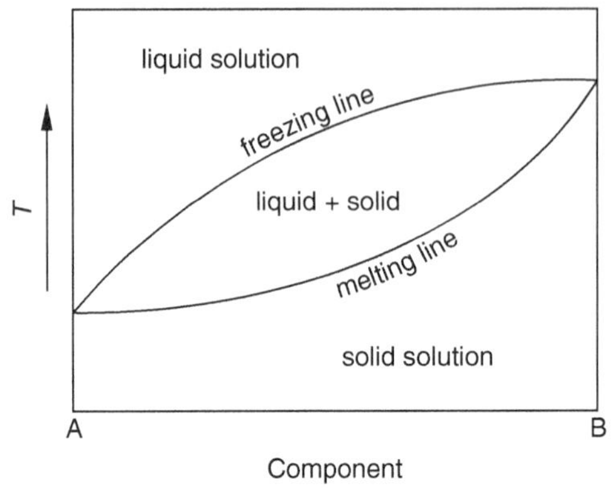

Figure 2.17 Typical temperature–concentration phase diagram for a solid solution.

building blocks (atoms, molecules, or even bigger groups) are substituted for solute blocks in the crystal structure. In interstitial solutions, the solute building blocks occupy interstices in the solvent crystal structure. Figure 2.17 shows a typical solid solution phase diagram for A and B components. Across the entire concentration range, there is a liquid solution at higher temperatures and a solid solution at the lower temperatures. The boundaries of these two areas are solidus and liquidus lines. Within the area bounded by the solidus and liquidus lines, there is a solid–liquid mixture. When we cool down a solid solution system from a high temperature, the solids precipitate as soon as the temperature reaches the solidus line. Continuous cooling leads to the solids precipitating at different concentrations of A (and B), depending on the exact temperature at which the solid precipitates. Since the Gibbs phase rule does not make a distinction between what kinds of phases are in the system, solid solutions in phase equilibrium are identical to other types of binary mixtures.

3.6. Mesophase–liquid transitions in binary mixtures

Knowledge concerning mesophase–liquid transitions in binary mixtures has developed extensively in recent years after several successful industrial discoveries involving polymers. The renewed interest in this area is also due to substantial efforts in amphiphilic and biological systems (see e.g., Livolant and Leforestier, 1996). In many cases, amphiphilic and biological systems assemble into supra-molecular structures through the micro-phase separation of two immiscible components on the nanometer scale. On the molecular scale, lyotropic liquid crystals are a basic molecular model system to study mesophase–liquid transitions. This system consists of two or more components that exhibit liquid crystalline behavior for a fixed concentration range. The phase behavior is characterized by the fluidity of the system, caused by solvent molecules surrounding the liquid crystalline compounds. Therefore, the concentration becomes another degree of freedom in the phase diagram that can be used to induce a variety of different liquid crystalline phases.

From the theoretical perspective, Onsager proposed a simple model to predict the lyotropic phase transition: the hard-rod model (Onsager, 1949). This model considers the excluded volume of one hard rod (with a certain length-to-diameter aspect ratio) when it is approached by another rod. Only a repulsive interaction between the rods is taken into account in this model, such that if two rods are oriented parallel to each other, there is little excluded volume between them. On the other hand, when one rod approaches by another at some tilt angle, there is a larger excluded volume around them. Therefore, the angular orientation decreases the net positional entropy of the approaching rod. In the parallel orientation, the rods have decreased orientational entropy but increased positional entropy. When the concentration of the rods is small, greater positional order may be favorable in terms of entropy. As the concentration of the rods increases, greater orientational order is entropically favorable. This model predicts a phase transition for rods at sufficiently high concentrations (and temperatures). The rods enter a nematic phase with long range orientational order. The rods in this model are an analog to the rigid mesogenic part of liquid crystalline molecules.

Flory extended his lattice model of flexible polymers in solution to the case of rigid molecules with a certain aspect ratio (Flory, 1956; Flory and Abe, 1978). With this model, there is a critical volume fraction of the rods above which the rods spontaneously orient to form a nematic phase. In both Onsager and Flory models, a critical order parameter, S_C, is introduced. This is an ensemble average over the angular distribution of the long axes of the rods with respect to the nematic director in a mono-domain of the liquid crystalline phase:

$$S_C = \frac{\langle 3\cos^2\theta - 1 \rangle}{2} \tag{2.39}$$

In this equation, θ represents the angle between the long axis of the rigid molecules and the orientational director (in a nematic phase) of a mono-domain. Combined, these two models predict that the lyotropic-rigid molecule phase diagram takes the form sketched in Fig. 2.18 (Khokhlov and Grosberg, 1981). Studies also show that

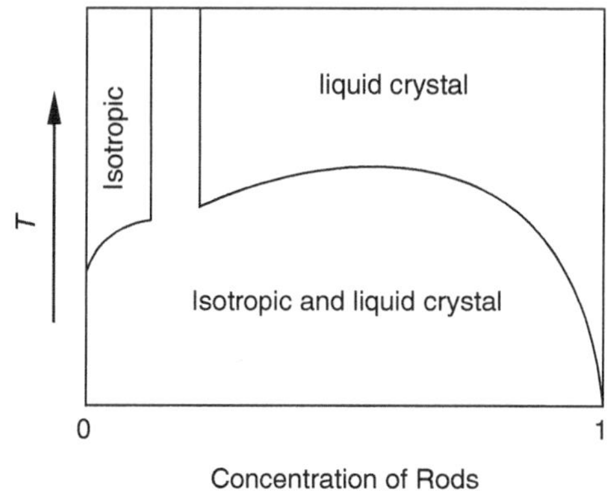

Figure 2.18 Illustration of a phase diagram for a rigid rod-diluent system.

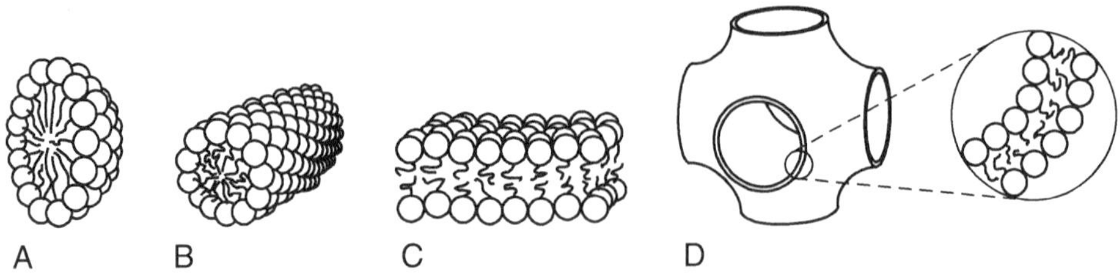

Figure 2.19 Different surfactant supra-molecular structures, including spheres (A), cylinders (B), bilayer lamellae (C), and bicontinuous structures (D).

lyotropic liquid crystals can form phase structures with greater order than the nematic phase, such as smectic phases.

There has also been extensive research into amphiphilic systems (see e.g., Evans and Wennerström, 1999). The simplest class of amphiphilic systems is surfactants, which usually consist of a hydrophilic head and a hydrophobic tail. It is known that this class of surfactants in aqueous solutions can form different micellar supra-molecular morphologies above their critical micelle concentration. These morphologies include spheres, cylinders, bilayer lamellae, vesicles, and even bicontinuous morphologies, as shown in Fig. 2.19. These supra-molecular morphologies are at thermodynamic equilibrium with a minimum Gibbs free energy at various concentrations and temperatures at a constant pressure. One of the most popular examples is common everyday soap. Soap is potassium *n*-octanoate with a polar $CO_2^-K^+$ head and a non-polar hydrocarbon tail, *n*-alkane $[CH_3(CH_2)_7^-]$. Depending on the concentration of the soap in water, different micellar morphologies can be observed.

Aggregates with different micellar morphologies as the building blocks (see Fig. 2.19) form different phases. These phases possess symmetries which resemble the symmetry of the micellar morphological building blocks. For example, the spherical micelles can closely pack into a body-centered cubic phase; while, the cylindrical micelles form a hexagonal close-packed phase. The bilayer lamellar micelles can stack together to form a lamellar phase, and the micelles constructed by saddle-splay surfaces can construct a bicontinuous cubic phase, which forms a network called a gyroid phase and is defined by threefold node points (Hamley, 2000). Figure 2.20 shows a few examples of the ordered phases formed by packing together micellar building blocks. Some of these phases can be readily observed in many micellar systems, such as in the popular example of soap in water, as shown in Fig. 2.21 (Fontell, 1981). In this figure, both the phase diagram and micellar morphologies are included. It is illustrated that when the soap concentration is low, only individual micelles are formed. With increasing soap concentration, these individual micelles aggregate and pack into the ordered phases. As the temperature is lowered in the high concentration region, the Krafft temperature, T_K, is reached. The Krafft temperature is the minimum temperature at which soap molecules form micelles. Below this temperature, the soap molecules remain in a crystalline phase. These phases are a new class of lyotropic mesophase behavior on

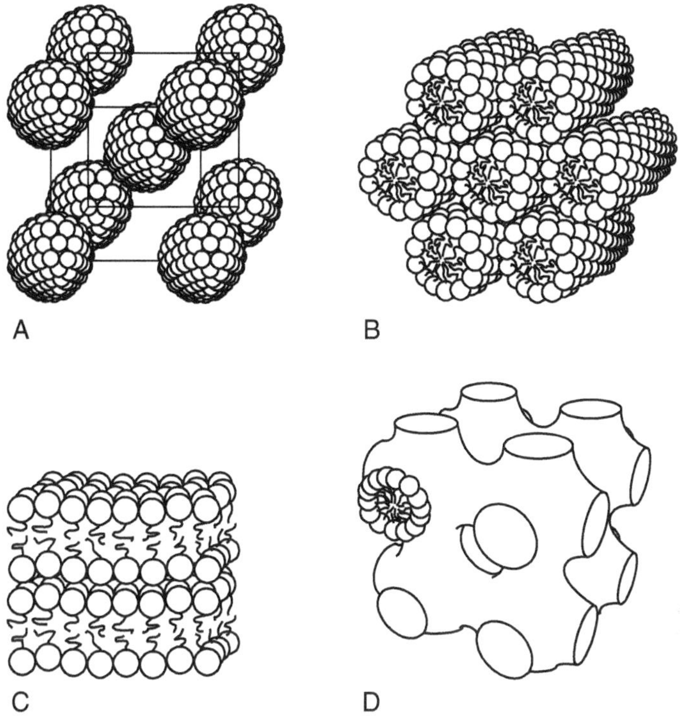

Figure 2.20 Aggregated micelle phases, including body-centered cubic (A), hexagonal cylinder (B), lamellar stacks (C), and bicontinuous cubic (gyroid) (D).

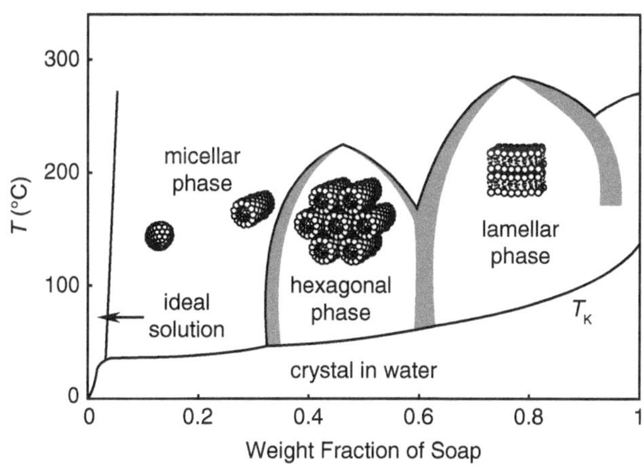

Figure 2.21 Phase diagram and corresponding micelles and aggregate micellar phases in a system of soap in water. The near vertical boundary at low weight fraction of soap is the critical micelle concentration. The Krafft temperature is represented as T_K in the figure [replotted after Fontell (1981), with permission].

the supra-molecular scale. They are critically important in understanding the thermodynamics and kinetics of phase formation in biological systems, such as cell membranes and the development of new biotechnologies, such as controlled gene therapy and targeted drug delivery.

REFERENCES AND FURTHER READING

Adamski, P., and Czyzewski, R. (1978). Activation energy and growth rate of spherulites of cholesteric liquid crystals. *Kristallografiya* **23**, 1284–1285.

Adamski, P., and Klimczyk, S. (1978). Study of the crystallization rate constant and the Avrami index for cholesterol pelargonate. *Kristallografiya* **23**, 154–160.

Atkins, P. W. (1990). "Physical Chemistry," 4th edn. Freeman and Company, San Francisco.

Avrami, M. (1939). Kinetics of phase change. I. General theory. *J. Chem. Phys.* **7**, 1103–1112.

Avrami, M. (1940). Kinetics of phase change. II. Transformation-time relations for random distribution of nuclei. *J. Chem. Phys.* **8**, 212–224.

Avrami, M. (1941). Kinetics of phase change. III. Granulation, phase change, and microstructure. *J. Chem. Phys.* **9**, 177–184.

Bahadur, B. (1976). A review on the specific volume of liquid crystals. *J. Chim. Phys. Phys.-Chim. Biol.* **73**, 255–267.

Baldan, A. (2002). Progress in Ostwald ripening theories and their applications to nickel-base superalloys—Part I: Ostwald ripening theories. *J. Mater. Sci.* **37**, 2171–2202.

Barois, P. (1998). Phase transition theories. *In* "Handbook of Liquid Crystals. Volume 1. Fundamentals" (D. Demus, J. Goodby, G. W. Gray, H.-W. Spiess, and V. Vill, eds.), Chapter VII.6.1. Wiley-VCH, Weinheim.

Barrall, E. M., and Johnson, J. F. (1974). Thermal properties of liquid crystals. *In* "Liquid Crystals and Plastic Crystals. Volume 2. Physico-Chemical Properties and Methods of Investigation" (G. W. Gray and P. A. Winsor, eds.), Chapter 10. Ellis Horwood Limited, Chichester.

Beguin, A., Billard, J. U., Bonamy, F., Buisine, J. M., Cuvelier, P., Dubois, J. C., and Le Barny, P. (1984). Sources of thermodynamic data on mesogens. *Mol. Cryst. Liq. Cryst.* **115**, 1–326.

Belonoshko, A. B., Skorodumova, N. V., Rosengren, A., and Johansson, B. (2006). Melting and critical superheating. *Phys. Rev., B* **73**, 012201.

Blumstein, R. B., and Blumstein, A. (1988). Inherently flexible thermotropic main chain polymeric liquid crystals. *Mol. Cryst. Liq. Cryst.* **165**, 361–387.

Blumstein, A., and Thomas, O. (1982). Odd-even effect in thermotropic liquid-crystalline 4,4′-dihydroxy-2,2′-dimethylazoxybenzene-alkandioic acid polymers. *Macromolecules* **15**, 1264–1267.

Born, M. (1939). Thermal dynamics of crystals and melting. *J. Chem. Phys.* **7**, 591–603.

Brailsford, A. D., and Wynblatt, P. (1979). The dependence of Ostwald ripening kinetics on particle volume fraction. *Acta Metall.* **27**, 489–497.

Cahn, J. W. (1965). Phase separation by spinodal decomposition in isotropic systems. *J. Chem. Phys.* **42**, 93–99.

Cahn, J. W., and Hilliard, J. E. (1958). Free energy of a nonuniform system. I. Interfacial free energy. *J. Chem. Phys.* **28**, 258–267.

Cahn, J. W., and Hilliard, J. E. (1959). Free energy of a nonuniform system. III. Nucleation in a two-component incompressible fluid. *J. Chem. Phys.* **31**, 688–699.

Cammidge, A. N., and Bushby, R. J. (1998). Synthesis and structural features. *In* "Handbook of Liquid Crystals. Volume 2B. Low Molecular Weight Liquid Crystals II" (D. Demus, J. Goodby, G. W. Gray, H.-W. Spiess, and V. Vill, eds.), Chapter VII. Wiley-VCH, Weinheim.

Cheng, S. Z. D. (1988). Kinetics of mesophase transitions in thermotropic copolyesters. 1. Calorimetric study. *Macromolecules* **21**, 2475–2484.

Cheng, S. Z. D., and Wunderlich, B. (1988). Modification of the Avrami treatment of crystallization to account for nucleus and interface. *Macromolecules* **21**, 3327–3328.

Cheng, J., Jin, Y., Liang, G., and Wunderlich, B. (1992). Condis crystal of small molecules. V. Solid state ^{13}C NMR and thermal properties of N,N′-bis(4-n-octyloxybenzal)-1,4-phynylenediamine (OOBPD). *Mol. Cryst. Liq. Cryst.* **213**, 237–258.

Chou, Y. C., and Goldburg, W. I. (1981). Angular distribution of light scattered from critically quenched liquid mixtures. *Phys. Rev., A* **23**, 858–864.

Cladis, P. E. (1998). Re-entrant phase transitions in liquid crystals. *In* "Handbook of Liquid Crystals. Volume 1. Fundamentals" (D. Demus, J. Goodby, G. W. Gray, H.-W. Spiess, and V. Vill, eds.), Chapter VII.6.4. Wiley-VCH, Weinheim.

Cladis, P. E., Bogardus, R. K., Daniels, W. B., and Taylor, G. N. (1977). High-pressure investigation of the reentrant nematic-bilayer-smectic-A transition. *Phys. Rev. Lett.* **39**, 720–723.

Cook, H. E. (1970). Brownian motion in spinodal decomposition. *Acta Metall.* **18**, 297–306.

Cormia, R. L., Mackenzie, J. D., and Turnbull, D. (1963). Kinetics of melting and crystallization of phosphorus pentoxide. *J. Appl. Phys.* **34**, 2239–2244.

de Gennes, P. G., and Prost, J. (1995). "The Physics of Liquid Crystals." Oxford University Press, Oxford.

Debenedetti, P. G. (1996). "Metastable Liquids: Concepts and Principles." Princeton University Press, Princeton.

Demus, D. (1998). Chemical structure and mesogenic properties. *In* "Handbook of Liquid Crystals. Volume 1. Fundamentals" (D. Demus, J. Goodby, G. W. Gray, H.-W. Spiess, and V. Vill, eds.), Chapter VI. Wiley-VCH, Weinheim.

Demus, D., Diele, S., Grande, S., and Sackmann, H. (1983). Polymorphism in thermotropic liquid crystals. *In* "Advances in Liquid Crystals" (G. H. Brown, ed.), Vol. 6. Chapter 1. Academic Press, New York.

Evans, D. F., and Wennerström, H. (1999). "The Colloidal Domain: Where Physics, Chemistry, Biology and Technology Meet." Wiley-VCH, New York.

Fang, C-Y., and Uhlman, D. R. (1984). The process of crystal melting. II. Melting of sodium disilicate. *J. Non-Crystal. Solids* **64**, 225–228.

Fecht, H. J., and Johnson, W. L. (1988). Entropy and enthalpy catastrophe as a stability limit for crystalline material. *Nature* **334**, 50–51.

Flory, P. J. (1956). Statistical thermodynamics of semi-flexible chain molecules. *Proc. R. Soc. Lond. A: Math. and Phys. Sci.* **234**, 60–73.

Flory, P. J., and Abe, A. (1978). Statistical thermodynamics of rodlike particles. 1. Theory for polydisperse systems. *Macromolecules* **11**, 1119–1122.

Fontell, K. (1981). Liquid crystallinity in lipid-water systems. *Mol. Cryst. Liq. Cryst.* **63**, 59–82.

Fradkov, V. E., Glicksman, M. E., and Marsh, S. P. (1996). Coarsening kinetics in finite clusters. *Phys. Rev. E* **53**, 3925–3932.

Frenkel, J. (1946). "Kinetic Theory of Liquids." Oxford University Press, London; reprinted by Dover, New York, 1955.

Furukawa, H. (1998). Dynamics of phase separation and its application to polymer mixture. *In* "Structure and Properties of Multiphase Polymeric Materials" (T. Araki, Q. Tran-Cong, and M. Shibayama, eds.), Chapter 2. Marcel Dekker, New York.

Garland, C. W. (1992). Calorimetric studies of liquid crystal phase transitions: AC techniques. *In* "Phase Transitions in Liquid Crystals" (S. Martellucci and A. N. Chester, eds.), Chapter 11. Plenum Press, New York.

Gibbs, J. W. (1878). On the equilibrium of heterogeneous substances. *Trans. Conn. Acad. Arts Sci.* **III**, 108–248, 343–524.

Guillon, D., and Skoulios, A. (1976a). Smectic polymorphism and melting progresses of molecules in the case of 4,4′-di(p,n-alcoxybenzylidene-amino) biphenyls. *J. Phys.* **37**, 797–800.

Guillon, D., and Skoulios, A. (1976b). Etude du polymorphisme smectique par dilatométrie et diffractométrie X. *J. Phys.* **37**, C3: 83–84.

Guillon, D., Cladis, P. E., Aadsen, D., and Daniels, W. B. (1980). X-ray investigation of the smectic A reentrant nematic transition under pressure (CBOOA). *Phys. Rev., A* **21**, 658–665.

Hamley, I. W. (2000). "Introduction to Soft Matter: Polymers, Colloids, Amphiphiles and Liquid Crystals." John Wiley & Sons, Chichester.

Hoeben, F. J. M., Jonkheijm, P., Meijer, E. W., and Schenning, A. P. H. J. (2005). About supramolecular assemblies of π-conjugated systems. *Chem. Rev.* **105**, 1491–1546.

Huang, C. C., Viner, J. M., and Novack, J. C. (1985). New experimental technique for simultaneously measuring thermal conductivity and heat capacity. *Rev. Sci. Instrum.* **56**, 1390–1393.

Hunt, J. D., and Jackson, K. A. (1966). Binary eutectic solidification. *Trans. Metall. Soc. AIME* **236**, 843–852.

Iwamatsu, M. (1999). Homogeneous nucleation for superheated crystal. *J. Phys., Condens. Matter.* **11**, L1–L5.

Jabarin, S. A., and Stein, R. S. (1973). Light scattering and microscopic investigations of mesophase transitions of cholesteryl myristate. *J. Phys. Chem.* **77**, 409–413.

Jackson, K. A, Uhlmann, D. R., and Hunt, J. D. (1967). On the nature of the crystal growth from the melt. *J. Cryst. Growth* **1**, 1–36.

Jin, Z., and Lu, K. (2000). Melting as a homogeneously nucleated process within crystals undergoing superheating. *Zeitschrift für Metallkunde* **91**, 275–279.

Jin, Z., Gumbsch, P., Lu, K., and Ma, E. (2001). Melting mechanisms at the limit of superheating. *Phys. Rev. Lett.* **87**, 055703.

Khokhlov, A. R., and Grosberg, A. Yu. (1981). Statistical theory of polymeric lyotropic liquid crystals. *Adv. Polym. Sci.* **41**, 53–97.

Langer, J. S., Bar-on, M., and Miller, H. D. (1975). New computational method in the theory of spinodal decomposition. *Phys. Rev., A* **11**, 1417–1429.

Lifshitz, I. M., and Slyozov, V. V. (1961). The kinetics of precipitation from supersaturated solid solutions. *J. Phys. Chem. Solids* **19**, 35–50.

Lindemann, F. A. (1910). Über die Berechnung molekularer Eigenfrequenzen. *Physikalische Zeitschrift* **11**, 609–612.

Livolant, F., and Leforestier, A. (1996). Condensed phases of DNA: Structures and phase transitions. *Prog. Polym. Sci.* **21**, 1115–1164.

Lu, K., and Li, Y. (1998). Homogeneous nucleation catastrophe as a kinetic stability limit for superheated crystal. *Phys. Rev. Lett.* **80**, 4474–4477.

Marqusee, J. A., and Ross, J. (1984). Theory of Ostwald ripening: Competitive growth and its dependence on volume fraction. *J. Chem. Phys.* **80**, 536–543.

Marsh, S. P., and Glicksman, M. E. (1996). Kinetics of phase coarsening in dense systems. *Acta Mater.* **44**, 3761–3771.

Moore, W. J. (1972). "Physical Chemistry." Prentice Hall, Englewood Cliffs.

Nakatani, A. I., and Han, C. C. (1998). Shear dependence of the equilibrium and kinetic behavior of multicomponent systems. *In* "Structure and Properties of Multiphase Polymeric Materials" (T. Araki, Q. Tran-Cong, and M. Shibayama, eds.), Chapter 7. Marcel Dekker, New York.

Onsager, L. (1949). The effects of shapes on the interaction of colloidal particles. *Ann. N. Y. Acad. Sci.* **51**, 627–659.

Ostwald, W. (1900). Über die vermeintliche Isomerie des roten und gelben Quecksilberoxyds und die Oberflächenspannung fester Körper. *Zeitschrift für Physikalische Chemie, Stöchiometrie und Verwandschaftslehre* **34**, 495–503.

Pisipati, V. G. K. M., Rao, N. V. S., and Alapati, P. R. (1989). DSC characterization of various phase-transitions in mesomorphic benzylidene anilines. *Cryst. Res. Technol.* **24**, 1285–1290.

Price, F. P., and Wendorff, J. H. (1971). Transitions in mesophase forming systems. I. Transformation kinetics and pretransition effects in cholesteryl myristate. *J. Phys. Chem.* **75**, 2839–2849.

Ratke, L., and Voorhees, P. W. (2002). "Growth and Coarsening: Ostwald Repining in Material Processing." Springer, Berlin.

Richards, J. W. (1897). Relations between the melting-points and the latent heats of fusion of the metals. *J. Franklin Inst.* **143**, 379–383.

Rosenberger, F. (1982). Crystal growth kinetics. *NATO ASI Ser., C Math. Phys. Sci.* **87**, 315–364.

Schantz, C. A., and Johnson, D. L. (1978). Specific heat of the nematic, smectic-A, and smectic-C phases of 4-n-pentylphenylthiol-4′-n-octaoxybenzoate: Critical behavior. *Phys. Rev. A* **17**, 1504–1512.

Schmalzried, H. (1981). "Solid State Reactions." Verlag Chemie, Weinheim.

Siggia, E. D. (1979). Late stages of spinodal decomposition in binary mixtures. *Phys. Rev. A* **20**, 595–605.

Smith, G. W. (1975). Plastic crystals, liquid crystals, and the melting phenomenon. The importance of order. *In* "Advances in Liquid Crystals" (G. H. Brown, ed.), Vol. 1, Chapter 4. Academic Press, New York.

Tallon, J. L. (1989). A hierarchy of catastrophes as a succession of stability limits for the crystalline state. *Nature* **342**, 658–660.

Tammann, G. (1903). "Kristallisieren und Schmelzen." Verlag Johann Ambrosius Barth, Leipzig.

Tammann, G., and Mehl, R. F. (1925). "The States of Aggregation." Nan Nostrand Company, New York.

Tang, B. Y., Ge, J. J., Zhang, A., Calhoun, B., Chu, P., Wang, H., Shen, Z., Harris, F. W., and Cheng, S. Z. D. (2001). Liquid crystalline and monotropic phase behaviors of 2,3,6,7,10,11-hexa (4′-octyloxybenzoyloxy)triphenylene discotic molecules. *Chem. Mater.* **13**, 78–86.

Tang, B. Y., Jing, A. J., Li, C. Y., Shen, Z., Wang, H., Harris, F. W., and Cheng, S. Z. D. (2003). Role of polymorphous metastability in crystal formation kinetics of 2,3,6,7,10,11-hexa(4′-octyloxybenzoyloxy)-triphenylene discotic molecules. *Cryst. Growth Des.* **3**, 375–382.

Toda, A., Hikosaka, M., and Yamada, K. (2002). Superheating of the melting kinetics in polymer crystals: A possible nucleation mechanism. *Polymer* **43**, 1667–1679.

Tsumuraya, K., and Miyata, Y. (1983). Coarsening models incorporating both diffusion geometry and volume fraction of particles. *Acta Metall.* **31**, 437–452.

Tsykalo, A. L. (1991). "Thermophysical Properties of Liquid Crystals." Gordon and Breach, New York.

Turnbull, D., and Fisher, J. C. (1949). Rate of nucleation in condensed systems. *J. Chem. Phys.* **17**, 71–73.

Tycko, R., Dabbagh, G., Fleming, R. M., Haddon, R. C., Makhija, A. V., and Zahurak, S. M. (1991). Molecular dynamics and the phase transition in solid C_{60}. *Phys. Rev. Lett.* **67**, 1886–1889.

Tymczak, C. J., and Ray, J. R. (1990). Asymmetric crystallization and melting kinetics in sodium: A molecular-dynamics study. *Phys. Rev. Lett.* **64**, 1278–1281.

Vergano, P. J., and Uhlmann, D. R. (1970a). Crystallization kinetics of germanium dioxide: The effect of stoichiometry on kinetics. *Phys. Chem. Glasses* **11**, 30–38.

Vergano, P. J., and Uhlmann, D. R. (1970b). Melting kinetics of germanium dioxide. *Phys. Chem. Glasses* **11**, 39–45.

Volmer, M. (1939). "Kinetik der Phasenbildung." Steinkopff, Dresden.

Volmer, M., and Schmidt, O. (1937). Über den Schmelzvorgang. *Zeitschrift für physikalische Chemie, Abteilung B: Chemie der Elementarprozesse, Aufbau der Materie* **35**, 467–480.

Wagner, C. Z. (1961). Theorir der Alterung von Niederschlägen durch Umlösen (Ostwald-reifung). *Zeitschrift für Elektrochemie* **65**, 581–591.

Walden, P. (1908). Über die Schmelzwärme, spezifische Kohäsion und Molekulargrösse bei der Schelztemperatur. *Zeitschrift für Elektrochemie* **14**, 713–724.

Wunderlich, B. (1980). "Macromolecular Physics. Volume III. Crystal Melting." Academic Press, New York.

Wunderlich, B. (2005). "Thermal Analysis of Polymeric Materials." Springer, Berlin.

Wunderlich, B., and Grebowicz, J. (1984). Thermotropic mesophases and mesophase transitions of linear, flexible macromolecules. *Adv. Polym. Sci.* **60**, 1–59.

Wunderlich, B., and Jin, Y. M. (1993). The thermal properties of four allotropes of carbon. *Thermochim. Acta* **226**, 169–176.

Yandrasits, M. A., Cheng, S. Z. D., Zhang, A., Cheng, J., Wunderlich, B., and Percec, V. (1992). Mesophase behavior in thermotropic polyethers based on the semiflexible mesogen 1-(4-hydroxyphenyl)-2-(2-methyl-4-hydroxyphenyl)ethane. *Macromolecules* **25**, 2112–2121.

Yoon, Y., Zhang, A., Ho, R.-M., Cheng, S. Z. D., Percec, V., and Chu, P. (1996a). Phase identification in a series of liquid crystalline TPP polyethers and copolyethers having highly ordered mesophase structures. 1. Phase diagrams of odd-numbered polyethers. *Macromolecules* **29**, 294–305.

Yoon, Y., Ho, R.-M., Moon, B., Kim, D., McCreight, K. W., Li, F., Harris, F. W., Cheng, S. Z. D., Percec, V., and Chu, P. (1996b). Mesophase identifications in a series of liquid crystalline TPP polyethers and copolyethers having highly ordered mesophase structures 2. Phase diagrams of even-numbered polyethers. *Macromolecules* **29**, 3421–3431.

CHAPTER THREE

Concepts of Metastable States

Contents

This chapter introduces our current understanding of the metastable states. The idea of metastable states can be traced back, over 100 years, to Ostwald. It is now recognized that metastable states are thermodynamically defined, and the appearance of a metastable state can either be associated with the supercooling or superheating of a phase before a phase transition has time to occur or because of the competing kinetics caused by the existence of multiple free energy pathways to reach the equilibrium phase. Examples for both appearances of metastable states are provided. The current concept of metastable states is limited by disregarding the effects of size on phase stability. In order to explain various experimental observations, it is also necessary to extend the concept of metastable states to multiple length scales. Furthermore, the degree of metastability needs to be quantitatively measured to determine how far a metastable state is from its equilibrium state.

1. Ostwald's Stage Rule and Definition of a Metastable State

Since metastable states frequently occur during phase transitions, they were identified soon after the study of phase transition behaviors started. The existence of metastable states was first suggested by the experimental observation that phase transitions do not take place at the equilibrium transition temperature. Since these supercooled or superheated nonequilibrium states are not at the lowest free energy for a specific temperature and pressure, they are thermodynamically defined as metastable and only exist for finite periods of time.

In addition to supercooling and superheating, metastable states can also develop because of the kinetic pathway taken during the phase transition. Ostwald first focused on phase transition phenomena over 100 years ago when formulating his stage rule (Ostwald, 1897). As cited by Schmelzer *et al.* (1998), "... in the course of transformation of an unstable (or metastable) state into a stable one the system does not go directly to the most stable conformation (corresponding to the modification with the lowest free energy) but prefers to reach intermediate stages (corresponding to other metastable modifications) having the closest free energy to the initial state." This implies that phase transitions proceed via a series of metastable states of increasing stability. This rule is supported by an abundance of experimental data. If the transition from a metastable state to another metastable state, or even the ultimately stable state, is slow enough, the metastable state can be observed. However, the rule does not explain why the system chooses to pass through a series of intermediary steps instead of transferring straight to the equilibrium phase.

By definition, a metastable state is thermodynamically stable, yet is not the lowest free energy, stable equilibrium state. In other words, it is a stable state with a local free energy minimum. In a plot of free energy (G) versus an order parameter (Φ), as shown in Fig. 3.1, both the metastable state and ultimately stable equilibrium state have a first derivative of the free energy with respect to the order parameter that is equal to zero, $\mathrm{d}G/\mathrm{d}\Phi = 0$ and a second derivative of the free energy with respect to the order parameter that is positive, $\mathrm{d}^2G/\mathrm{d}\Phi^2 > 0$ (also see Fig. 1.8b). The difference between the two free energy minima, ΔG in this figure, reflects the stability difference between the metastable and equilibrium states and, thus, represents the driving force for the particle's relaxation from the metastable state to the equilibrium state. The free energy height ΔG^* is the activation energy barrier a particle must overcome during the transition to reach the equilibrium state. Relaxation of the metastable state toward a more stable one (or the equilibrium state) thus requires activation to overcome a barrier.

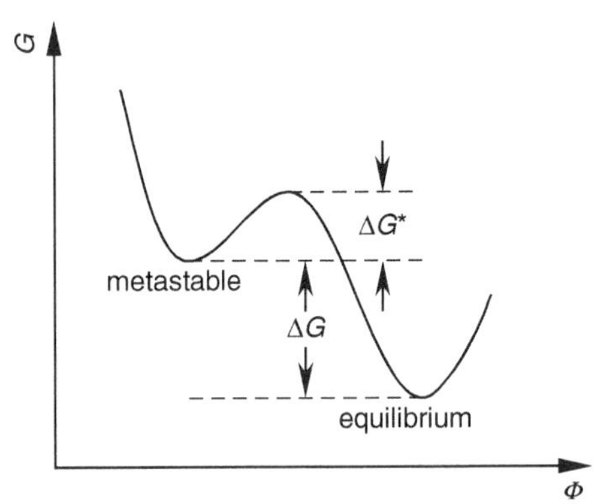

Figure 3.1 Representation of a metastable state using a plot of free energy (G) versus order parameter (Φ). Note that the metastable state possesses a local stable minimum free energy, while the equilibrium state has a global stable minimum free energy. A relaxation of the metastable state toward the equilibrium state needs to overcome the barrier (ΔG^*); while, the driving force of the relaxation from the metastable state to the equilibrium state is the difference between the two minima of the free energies (ΔG).

Metastable states appear for two reasons. The first is determined by the thermodynamics of phase transitions. When the limit of metastability is absolute and the transition barrier is high, the metastable state can exist long enough to be detected experimentally. Examples of this type of metastable state include the region between the binodal and spinodal curves of a liquid–gas transition, supercooled amorphous liquids, and superheated crystalline solids. These examples will be briefly discussed in Sections 2.1 and 2.2 of this chapter. The key feature of this type of metastable state is that it is associated with the supercooling or superheating of a phase when only one transition barrier exists between the two phases. At a particular pressure and temperature that is not an equilibrium transition pressure and temperature, only one phase is stable, and another phase is metastable.

We are more interested in the phase transitions involving multiple free energy pathways with several different local free energy minima and free energy barriers of varying heights that cause the growth of a particular metastable state. This state is a structural intermediary of the initial and final states with a local free energy minimum. The question becomes: How can one clearly understand the process of a system being trapped in a local free energy minimum? A general explanation is that the transition from the initial state to the final equilibrium stable state must have more than one free energy pathway available. The pathway through the metastable states must have a lower free energy barrier than the direct transition to the equilibrium stable state. Furthermore, experimental observation of the metastable state requires that this state exists on a time scale that is longer than the time resolution of the instrument used (Debenedetti, 1996). Namely, the transition kinetics of the metastable state to the next more stable (or equilibrium) state must be sufficiently slow to detect the metastable phase.

Using non-scientific language to illustrate the microscopic process, the particles have to make a choice among several energy barriers, but they are unable to predict, or are "blind" to, the phase stability and free energy landscape behind these energy barriers. Since the ability to overcome a barrier is determined by thermal (density) fluctuations, which are associated with thermal energy, kT, and the particle interactions, most particles choose the pathway with the lowest free energy barrier, regardless of the phase stability behind this barrier. The particles can then be trapped in a local free energy minimum (a metastable state), and as long as it is deep enough to prevent particles from escaping this phase within experimentally observable time frames, a metastable phase can be found.

2. Examples of Metastable States in Phase Transitions

2.1. Metastable states in liquid–gas transitions

A classical example of a metastable state is the supercooling or superheating involved in the condensation of a gas phase or evaporation of a liquid phase in the region between the binodal and spinodal curves, as shown in Fig. 3.2. In fact, the binodal

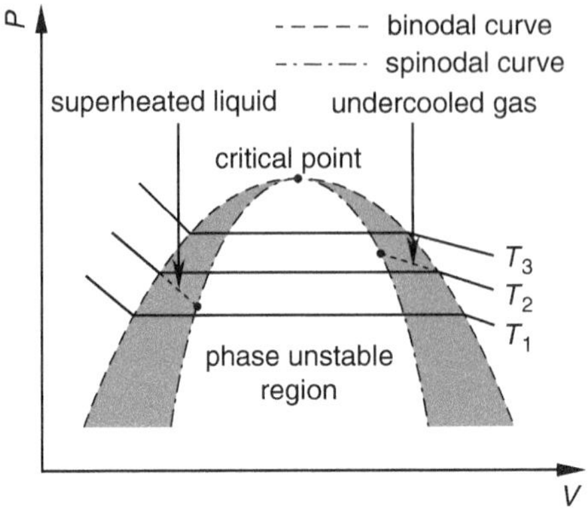

Figure 3.2 Illustration of the phase transition between the liquid and gas phases in a pressure–volume phase diagram. The binodal and spinodal curves are obtained from $\partial G/\partial \rho = 0$, and $\partial^2 G/\partial \rho^2 = 0$. In this figure, both the superheated liquid and supercooled gas regions are shadowed.

curve of the pressure–volume phase diagram is defined as when the first derivative of the Gibbs free energy with respect to density (the order parameter) is equal to zero, $\partial G/\partial \rho = 0$; the spinodal curve is defined as when the second derivative of Gibbs free energy with respect to density is equal to zero, $\partial^2 G/\partial \rho^2 = 0$. Therefore, the binodal curve represents a first-order transition within the limits of absolute thermodynamic stability, and the spinodal curve is the limit of metastability. Note that the spinodal and binodal curves discussed here deal with density changes for a gas to liquid transition in a single-component system, while the curves discussed in Section 3.3 of Chapter 2 are for a binary mixture and, therefore, refer to concentration changes. If one starts from the gas phase and decreases the temperature (or increases the pressure) to make the isobaric (or isothermal) system reach the critical point (Fig. 2.2A), liquid and gas phases will not be distinguishable because there is no symmetry difference between these two phases, as described in Section 1.2 of Chapter 1.

In the pressure–volume regions bounded between the binodal and spinodal curves, metastable superheated liquid or supercooled gas phases exist. However, as soon as the spinodal curve is reached, as shown in Fig. 3.2, the metastable liquid or gas can no longer exist. Therefore, if we illustrate the first-order liquid–gas phase transition by plotting free energy with respect to temperature at a constant pressure (isobaric) or with pressure at a constant temperature (isothermal), as shown in Fig. 3.3A and B, the superheated liquid and supercooled gas, or superexpanded liquid and supercompressed gas, will reach the limit of metastability. The system will thus reach the end points of the metastable region, as shown in these two figures as filled circles. They are represented on the spinodal curve in Fig. 3.2. Beyond these two end points, metastable states are thermodynamically forbidden. On the other hand, the phase transition temperature (or pressure) in these figures is defined by the binodal curves.

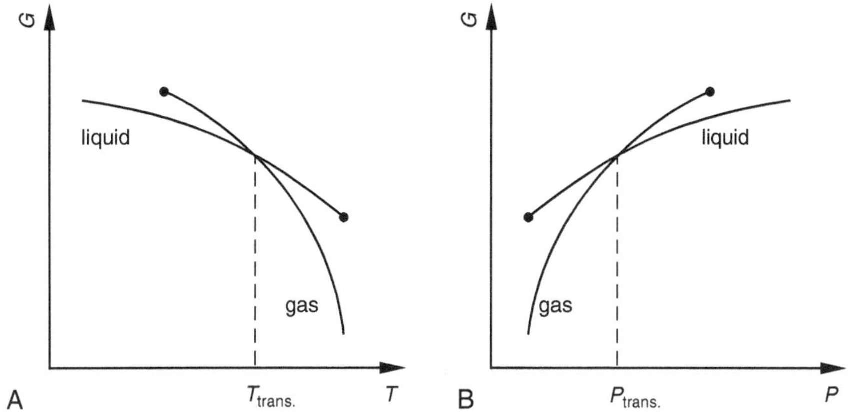

Figure 3.3 Relationship between free energy and temperature at constant pressure (A) and between free energy and pressure at constant temperature (B). For the liquid–gas transition, the limit of metastability of the liquid and gas can be defined by the spinodal curve, and therefore, the extent of metastability in both the gas and liquid phases are fixed (the filled circles in these figures).

2.2. Metastable states in crystalline solid–liquid transitions

An equally important example of a metastable state occurs in the vicinity of the first-order transition temperature for a crystalline solid–liquid transition. In the case of an isobaric transition, a superheated crystal phase at temperatures higher than the equilibrium melting temperature, or a supercooled liquid phase at temperatures lower than the equilibrium melting temperature, may exist. This can be seen in plots of the free energy versus temperature, as shown in Fig. 3.4A, in the vicinity of

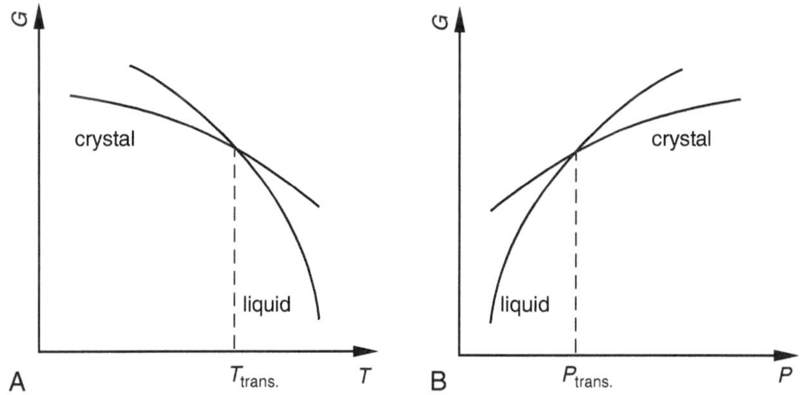

Figure 3.4 Relationship between free energy and temperature at constant pressure (A) and between free energy and pressure at constant temperature for a crystalline solid–liquid transition (B). Note that there is not a limit of metastability in these two figures, compared with Fig. 3.3A and B, because a critical point in the crystalline solid–liquid transition does not exist.

an isobaric first-order transition between a liquid and a crystalline solid. In comparing this figure with Fig. 3.3A, however, one notices that there are no end points for the free energy lines associated with the supercooled liquid and the superheated crystal. Since phase symmetry is broken in the crystalline solid–liquid transition, no critical point exists. At constant pressure, the free energy line of the crystal phase can extend to higher temperatures, and the free energy line of the liquid phase can extend to lower temperatures on both sides of the equilibrium melting temperature (Fig. 3.4A). In the isothermal case, we can plot the free energy versus pressure, as shown in Fig. 3.4B. Again, there is no end point for either the supercompressed liquid or superexpanded crystal lines. This means that there is always an activation barrier in the classical nucleation theory for a crystalline solid–liquid transition. To restate this concept differently, no unstable states exist. It is certain that these free energy lines associated with metastable states extend indefinitely with respect to temperature or pressure. They can only intersect another phase (a polymorph) with a different thermodynamic stability, as shown in the phase diagrams in Fig. 2.4A and B.

An alternative possibility was envisaged by Tammann over 100 years ago (Tammann, 1903; Tammann and Mehl, 1925). In a two-dimensional temperature–pressure phase diagram, the crystalline solid–liquid phase curve could also form a closed circle or terminate at zero on the temperature or pressure axis. This process would require that the phase curve bend back on itself, thus implying the existence of a re-entrant liquid phase, as shown in Fig. 3.5A (if the temperature is low enough below the glass transition temperature, it is also possible to enter an amorphous solid phase). In this case, two melting temperatures may exist at a given pressure, and two pressures may possess the same melting temperature (Rastogi *et al.*, 1993).

This back bending of the melting temperature with increasing pressure is caused by a change in the sign of the slope of the relationship between the melting

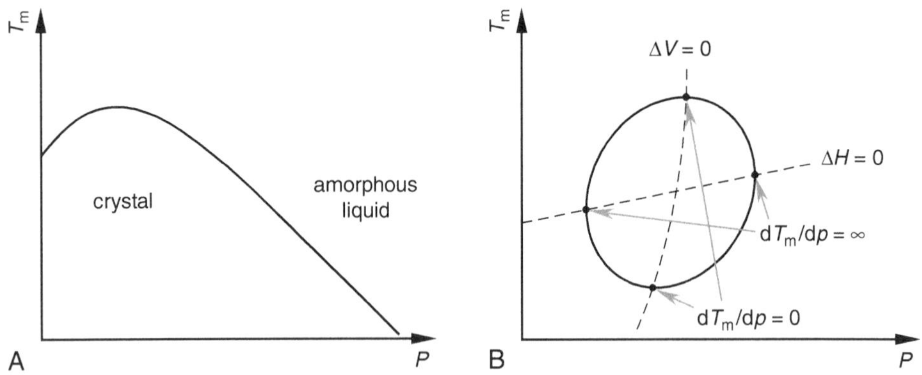

Figure 3.5 Two-dimensional temperature–pressure phase diagrams demonstrating how the slope of dT_m/dp changes its sign from positive to negative leading to amorphization with increasing pressure (A); and a closed loop phase curve with both $dT_m/dp = 0$ and $dT_m/dp = \infty$ at the point where the slopes changes its sign (B) [replotted from Rastogi *et al.* (1993), with permission].

temperature and pressure. Note that this relationship is given by the Clausius–Clapeyron equation:

$$dT_m/dp = T\Delta V/\Delta H \quad (3.1)$$

where ΔV represents the change in volume and ΔH is the change in heat content. When there is increasing pressure, the density of the crystal is higher than the density of the liquid. Here, Eq. (3.1) has a positive slope, and the melting temperature rises. However, if the density of the crystal becomes lower than that of the isotropic liquid, the slope of this equation becomes negative, causing the melting temperature to decrease with increasing pressure (Rastogi *et al.*, 1991; van Ruth and Rastogi, 2004). This change leads to amorphization with increasing pressure, as shown in Fig. 3.5A.

To analytically summarize the discussion above, if there is a point at which the first derivative of the melting temperature with respect to pressure is equal to zero, $dT_m/dp = 0$ (which has been experimentally observed) and a point at which the first derivative of the melting temperature with respect to pressure approaches infinity, $dT_m/dp = \infty$ (which has not yet been experimentally proven) in a two-dimensional temperature–pressure phase diagram, then at both points, the slope of Eq. (3.1) changes its sign (Rastogi *et al.*, 2006). Figure 3.5B schematically illustrates the closed circle phase diagram where no other phase intervenes (Rastogi *et al.*, 1993; Tamman, 1903; Tamman and Mehl, 1925).

3. Appearance of Metastable States Controlled by Competing Kinetics

In many practical cases, a single-component system may possess more than one crystal structure with different crystallographic symmetries, yet all of the crystal forms have long-range positional, bond orientational, and molecular orientational order as described in Section 1.2 of Chapter 1. In other cases, multiple phases may exist because one or more types of order lose their long range, step-wise correlation decay, resulting in mesophases. Since the purpose of this section is to illustrate the development of metastable states in the presence of several competing kinetic pathways, we will only focus on the stability between the metastable phases and describe their development despite the fact that it is not the most stable state. These principles are applicable to all types of coexisting ordered phases. We will temporarily avoid distinguishing between the phase structures (crystals or mesophases) and call this type of phase behavior "polymorphs."

Polymorphs, different variants of ordered phases, exhibit multiple transitions between the crystalline solid and liquid phases. We use similar isobaric free energy–temperature and isothermal free energy–pressure relationships to describe the thermodynamic properties of each of these phases and their phase transitions. In these cases, however, the free energy lines represent different ordered and liquid phases instead of only crystalline solid and liquid phases. Unless we are at the intersection of the two free energy lines, all but one of the possible polymorphs correspond to individual metastable states, such that at a specified fixed pressure and temperature (or volume), only one of the polymorphs is thermodynamically stable. The rest of the phases are defined as thermodynamically metastable.

With regard to the phase transitions in polymorphs, two types of behavior can be identified. The first type is called "enantiotropic." Vorländer wrote of enantiotropic phases: "... the transition temperatures are reached both by heating the solid crystalline form and by cooling ... the amorphous molten mass" (Vorländer, 1923). For a single-component system which contains more than one ordered phase, enantiotropic phase behavior describes a thermodynamic stability sequence in the temperature region where the phase transitions take place isobarically. The more-ordered phase melts first, followed by the disordering of the less-ordered phase while heating. During cooling from the isotropic liquid, the reverse transition sequence is observed in the formation of the less-ordered phase, first at higher temperatures, then the more-ordered phase forms at a lower temperature. From a purely thermodynamic point of view, one can readily understand the enantiotropic behavior with an isobaric plot of the free energy versus temperature for different phases, as shown in Fig. 3.6A. In this figure, each phase possesses its own thermodynamic free energy, and within each particular temperature region, one phase has the lowest free energy, making it the most stable phase. Figure 3.6B, which only takes thermodynamics into account, schematically illustrates the transition sequence as observed in calorimetric experiments. During cooling from the isotropic liquid phase down to above the disordering temperature of the less-ordered phase, the liquid is most stable. Between the equilibrium melting temperature of the less-ordered phase and the disordering temperature of the more-ordered phase, the less-ordered phase is most stable. Below the equilibrium disordering temperature, the more-ordered phase is most stable. This transition sequence is reversible during subsequent heating, as illustrated in Fig. 3.6B; thus, it is enantiotropic. In this figure, we do not consider the kinetic aspects of the phase transitions, such as overcoming

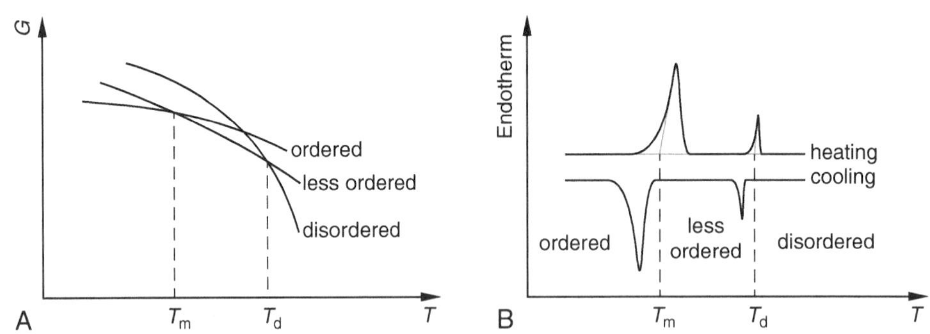

Figure 3.6 Illustration of the free energy versus temperature at constant pressure for "enantiotropic" phase transition behavior (A); and a schematically drawn calorimetric experimental result for the corresponding transition sequences during cooling and subsequent heating (B). The disordering temperature is denoted as T_d and the melting temperature as T_m.

the free energy barrier, which requires supercooling and melting of each ordered phase, which may involve superheating.

On the other hand, "monotropic" phase behavior occurs when the less-ordered phase is metastable throughout the entire temperature range studied. The recognition of monotropic behavior can be traced back to 1877 (Lehmann, 1877). Vorländer defined a monotropic phase as: "... said phase only appears when the amorphous molten substances are cooled, and not when the solid crystalline forms are melted" (Vorländer, 1923). Let us consider the monotropic transition behavior of a single-component system, including one liquid, one metastable (less-ordered), and one stable (more-ordered) phase to illustrate how kinetics enables the development of the metastable phase. It is assumed that there is a smaller barrier for the formation of the metastable phase than for the formation of the more-ordered phase, which does not necessarily imply that the transition from the liquid to the metastable phase possesses a faster kinetic pathway at equivalent supercoolings, but means that the metastable state has sufficient supercooling to achieve faster growth rates than the stable phase at higher temperatures.

In this monotropic phase transition, Fig. 3.7A illustrates an isobaric plot of the free energy versus temperature for the different phases. Figure 3.7B is a schematic illustration of the calorimetric behavior of the monotropic transition. It is evident in Fig. 3.7A that the metastable phase is not the most stable phase at any temperature. This phase may be experimentally accessible if two conditions are met. First, the thermodynamic stability of the metastable phase (the free energy line) cannot be too far from the neighboring stable phase. Second, the stable phase must require a larger supercooling to form compared to the metastable phase. During fast cooling from the isotropic liquid phase, the stable phase formation can be bypassed because it has a higher free energy barrier than the metastable phase, and thus requires larger supercooling. As a result, the transition from the isotropic liquid to the metastable phase

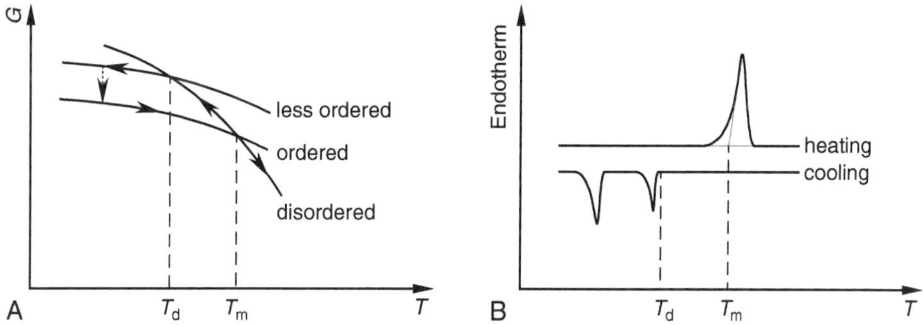

Figure 3.7 Illustration of the free energy versus temperature at constant pressure for "monotropic" phase transition behavior (A). Note that in the monotropic case, the metastable phase is not the lowest free-energy phase for the entire temperature region studied. Therefore, the appearance of this metastable phase relies entirely on the slow nucleation kinetics of the stable phase. A schematic drawing of the calorimetric experimental results for the corresponding phase transition sequences during cooling and subsequent heating (B). The disordering temperature is denoted as T_d and the melting temperature as T_m.

occurs first at high temperatures near the melting temperature of this less-ordered phase (the first exothermic process in Fig. 3.7B). After the metastable phase is formed, it is assumed that the transformation to the stable phase takes place with further cooling, as indicated by the second exothermic process in Fig. 3.7B. Therefore, during subsequent heating only one endothermic process can be seen, which is the melting of the stable phase to the isotropic liquid.

So far, the largest body of research on metastable states with monotropic behavior is in the area of liquid crystals. Of the over 100 liquid crystal molecules reviewed more than 80 years ago (1923), Vorländer listed at least 37 that were identified as monotropic with metastable liquid crystalline phases. In the past 20 years, the monotropic liquid crystalline behavior has continued to be reported (Andrews *et al.*, 1985; Carr and Gray, 1985). Recent reports in the area of monotropic discotic phase formation have also appeared (Tang *et al.*, 2001, 2003).

The kinetic origin of the monotropic phase transition behavior will now be discussed. To make a comparison on an absolute temperature scale, we need to note that the metastable phase has a lower melting temperature than the melting temperature of the stable crystal phase. Figure 3.8 is a schematic of the overall kinetic phase transformation rates across the temperature range of interest. Since the melting temperature of the metastable phase, $(T_m)_{meta}$, is lower than the melting temperature of the stable phase, $(T_m)_{st}$, the phase transition always goes directly from the isotropic liquid to the stable phase when between these two temperatures in region I of

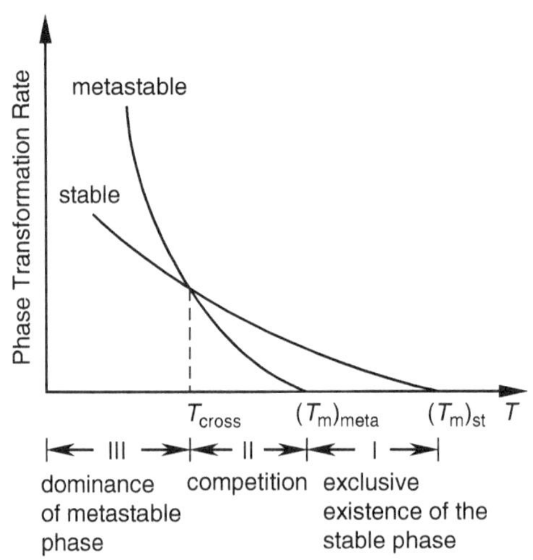

Figure 3.8 Schematic of the overall phase transformation rates (kinetics). There are two phase transition rates: one for formation of the metastable phase and another for the formation of the stable phase. Note that in this particular case, the $(T_m)_{meta} < (T_m)_{st}$. Three temperature regions are therefore classified: region I where $(T_m)_{meta} < T < (T_m)_{st}$; region II where $T_{cross} < T < (T_m)_{meta}$; and region III where $T < T_{cross}$. T_{cross} is the temperature where the two phase transformation rates cross, as shown in the figure [replotted from Keller and Cheng (1998), with permission].

Fig. 3.8, $(T_m)_{meta} < T < (T_m)_{st}$. When the temperature is lowered below the melting temperature of the metastable phase, $T < (T_m)_{meta}$, the rate of the metastable phase formation increases. Furthermore, the metastable phase growth rate increases faster with decreasing temperature than the stable phase growth rate. This increase results in a crossover of the two growth rate curves as shown in Fig. 3.8. At temperatures below the crossover temperature, T_{cross}, in region III of Fig. 3.8, the growth of the metastable phase becomes fastest and forms before the stable phase (Keller and Cheng, 1998).

This analysis left out the most interesting question: What happens to the phase growth process between the melting temperature of the metastable phase, $(T_m)_{meta}$, and the crossover temperature, T_{cross} (region II in Fig. 3.8)? In this region, the growth rates of the metastable and the stable phases are of the same order of magnitude. It is currently not certain what will occur. We could observe two distinct phase growth processes, each at its own rate, indicating a mixed phase containing both structures, or the metastable phase could become a precursor to the formation of the stable phase. A recent study on the metastable phase transition kinetics in a polymer has at least initiated the effort to determine the answer to this question (Jing *et al.*, 2002).

In practice, when a liquid is being cooled, a metastable phase can usually be reached before the ultimately stable equilibrium phase has time to form. The question is: Upon further cooling of the metastable phase, can the metastable phase transition into the stable phase? This is entirely dependent upon the kinetics of the transition, as determined by the free energy barrier, between the metastable and stable phases. When this transition growth rate is extremely slow, indicating a very high free energy barrier for this transition, the metastable phase may be "permanent," with a lifetime longer than the experimental time frame. As such, attempts to justify Ostwald's stage rule are based on kinetics.

4. What Are the Limitations of the Current Understanding of Metastable States?

So far, all of the cases described illustrating the observation of metastable states have either been attributed to the free energy barrier delaying the transition from the initial to the final phase and causing superheating/supercooling or the growth rate competition between different kinetic pathways opened up by local free energy minima embedded in the free energy landscape. These metastable states are "classical" in that they exist on a single length scale (usually on the atomic or molecular scale) with no effect of phase size on the stability of these states taken into account. This traditional understanding causes us to deal with one metastable state at a time with no size effect on or interaction between metastable states on different hierarchical length scales. A variety of recent experimental observations and theoretical simulations have shown intriguing and potentially far-reaching examples of size and interaction effects on metastable states. We use these results to illustrate the limits of the metastable state concept described above.

The first case illustrates that the stability of a metastable state is size dependent. The simplest type of metastable state arises from phases of limited size because the ultimate stability of a phase assumes the phase dimensions are large enough so that the effect of surfaces on the stability is negligible. In practice, the lower size limit where surfaces start having an effect, although dependent upon the particular system, is generally on the micrometer length scale. However, a phase domain limited to such small dimensions, like dispersed droplets in a liquid phase or the fine grain structure in a crystalline solid, is by definition metastable. These smaller domains coarsen toward equilibrium stability, causing coalescence of the phase texture. This "Ostwald's ripening" process (Ostwald, 1900) has been discussed in Section 3.4 of Chapter 2. If the metastable state is formed in a confined environment where the confinement is on the nanometer scale, the aging processes can only occur within the confined space.

In addition to phase size, phase domain habit, or shape, will also affect phase stability. For example, in a simple, free (unsupported) liquid, thermodynamic stability is maximized with a spherical phase shape, which minimizes the surface area-to-volume ratio. In a crystal, the most thermodynamically stable shape is the Wulff surface (the overall surface free energy is minimized). All other phase domain shapes correspond to metastable states. This is specifically pertinent for crystal phases in which the habit may be determined by the kinetics of crystal growth (Section 2.1 of Chapter 2), rather than by the ultimate thermodynamic stability.

Increasing experimental and theoretical evidence indicates that phase stability changes with the phase size. Extensive reports exist on the decreasing melting temperature of small inorganic and metal crystals with decreasing size (Bachels *et al.*, 2000; Dick *et al.*, 2001). This observation can be traced back to a report by Pawlow (1909). Melting point depression has been attributed to an increasingly large surface area-to-volume ratio with decreasing crystal size. Physically speaking, the microscopic particles (atoms or molecules) at the surface of a crystal have fewer physical interactions with their neighbors and, thus, are more easily excited by thermal motion. A linear relationship between melting temperature and inverse crystal size (radius) has been found in inorganic and metal small crystals (Buffat and Borel, 1976; Couchman, 1979; Lai *et al.*, 1996), as long as the crystal size is larger than several nanometers. Compared to perfect crystals of infinite size with an equilibrium melting temperature, these small crystals of different sizes and shapes are metastable.

Furthermore, when polymorphs exist in a one-component system, it is possible that not only are the stabilities of these phases determined by their phase size, but the relative phase stability between the polymorphs may also switch with phase size. One example involves the phase behavior of uniform colloidal particles in suspension. This system serves as a model for atomic and molecular phase behavior, including the role of metastability. It is known that uniform spherical colloid particles with sub-micron diameters can form three-dimensionally ordered arrays corresponding to colloidal "crystal" lattices. Here, the "crystal" forms through a first-order phase transition with the concentration of the suspended particles as a variable. Recently, it was shown that all three states of matter (gas, liquid, and crystal) can be represented by such a system, in which the transitions often pass through metastable states and/or become locked into such states, as defined in terms of bi-component phase diagrams (Evans *et al.*, 1997; Poon *et al.*, 1995). Note that for an ideal hard sphere system, the driving force is purely entropic. However, through recent developments, enthalpic interactions can also be introduced in a controlled manner, which enables the colloids to model a wider array of phase transitions. This approach to study ideal systems has given us several insights into metastability. In particular, the origin of metastable phases is now considered to exist because they kinetically grow faster with a lower free energy barrier with respect to the concentration order parameter (Evans and Cates, 1997; Evans and Poon, 1997).

Our interest is in the closely packed "crystal" phase of these colloidal systems. They have been the focus of a number of thorough computational studies (Moroni *et al.*, 2005; Oxtoby, 2003; Oxtoby and Shen, 1996; ten Wolde *et al.*, 1996). Theoretical computations and experimental observations have shown that the hexagonal close-packed lattice is metastable with respect to the face-centered cubic lattice, although the free energy difference is small (Bolhuis *et al.*, 1997). The metastable hexagonal close-packed phase has been found to be the favored polymorph, as compared with the ultimate stability of the face-centered cubic structure. This condition occurs because the kinetic barrier to form the hexagonal close-packed lattice is smaller than that of the face-centered cubic lattice. In fact, an important connection between phase size and phase structure is demonstrated in this type of system. When the system size is sufficiently small, the hexagonal close-packed polymorph becomes the stable one. An additional consequence arising from these computations is that even colloidal crystals of sufficient size to have the face-centered cubic lattice should have an outer shell of finite thickness with a hexagonal close-packed structure. Note that at infinite size, this hexagonal close-packed lattice is metastable to the face-centered cubic phase.

It is evident that the above example possesses metastable states on two length scales, which must meet the necessary conditions, as follows: two different ordered structures can grow at the same conditions and their phase stabilities are associated with size. The interplay between the two stability levels of these metastable states generates a metastability inversion with respect to phase size. More specifically, the stabilities of these two phases and the free energy barrier heights associated with forming the two phases must change with phase size during structural formation at constant temperature and pressure.

If this computational result concerning a closely packed, spherical colloid system is generally applicable to atomic and molecular crystallization, then during crystal formation, the growth front may be a different polymorph that is more stable and grows faster on small length scales, as compared with the larger thermodynamically stable phase being formed. Hence, the crystal would grow from a phase structure that is different from the one within the crystal interior. The consequences of this far-reaching deduction could lead to new understandings in crystal growth.

5. Concept of Metastability

In order to quantitatively describe metastable states, we need to define metastability. The reference state is always the equilibrium state at infinite size. Therefore, metastability is the stability difference between the metastable state and the equilibrium state at infinite size. This metastability is analogous to the development of the concept of crystallinity more than half a century ago. Crystallinity was introduced initially to explain the difference between experimentally measured densities and those calculated based on the unit cell lattice dimensions, but it quickly became a quantitative characterization of semi-crystalline polymeric materials and is associated with mechanical and other material properties. The crystallinity can be experimentally determined via calorimetric, diffraction, and spectroscopic methods. Since these three methods will generate different crystallinities for one semi-crystalline sample, it is more meaningful to utilize a single method to compare the crystallinities of a set of samples obtained from one type of polymer having different thermo-mechanical histories. The question then becomes: Can we identify measurable quantities which represent how far a metastable state is from the equilibrium state (thermodynamics) and how fast this metastable state can transfer to the equilibrium state (or to the closest more stable metastable state)? In other words, what is the lifetime of this metastable state before its metastability changes?

The metastabilities of the thermodynamically defined metastable states can be represented by the Gibbs free energy difference between the metastable state and its corresponding equilibrium state, as long as these metastabilities are phase-size independent. If we can isobarically measure the heat capacity with calorimetric methods, these thermodynamic properties can be calculated based on the description in Section 1.1 of Chapter 1. However, this has been proven to be a very difficult task because every metastable state has a limited lifetime and is only stable with respect to infinitesimal fluctuations. Therefore, the metastability of a phase may shift toward the equilibrium stable state with time, as well as with environmental changes involving temperature and pressure. As shown in Fig. 3.9, in which we plot an isobaric relationship of free energy versus temperature for a crystalline solid-to-liquid transition in a single-component system, the metastable states possess higher free energy levels compared to the equilibrium state in both the crystal and the liquid. With increasing temperatures, the crystalline solid in a metastable state is often continuously annealed, thereby decreasing its free energy level toward the equilibrium value. As such, during heating, the crystal solid changes its entropy level.

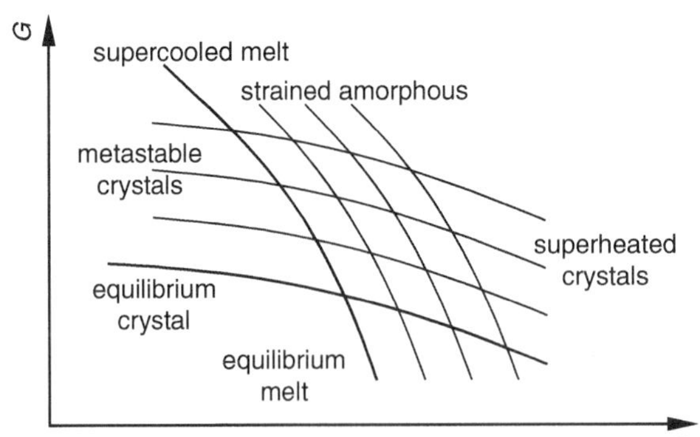

Figure 3.9 Schematic illustration of the free energies of the equilibrium state and various metastable states with respect to temperature at a constant pressure for a crystalline solid–liquid transition [replotted from Wunderlich (1980), with permission].

This process is called entropy production and has to be described by nonequilibrium thermodynamics. Therefore, people are often hesitant to use equilibrium thermodynamics to describe metastable states and their phase transitions. This hesitation is fully understood since a metastable state is not at thermodynamic equilibrium.

However, if we assume that the metastability of this crystalline solid does not change in the pressure–temperature phase diagram during heating, its free energy would cross the free energy line of the equilibrium liquid, causing the metastable phase to melt, as shown in Fig. 3.9. This melting temperature, given that during heating the metastable crystal does not change its level of entropy, is characteristic of that metastable state of the crystal. This process is called "zero-entropy production" (Wunderlich, 1980). With zero-entropy production, equilibrium thermodynamics can be used to describe the behavior of a metastable state. Furthermore, as shown in Fig. 3.9, the metastable states not only exist in the crystal, but also in the liquid. By decreasing the entropy of the liquid with stretching or deformation, the free energy line can be pushed toward higher temperatures. This process results in the superheating of the crystalline solid, which is also a metastable state. Experimentally, keeping the metastability of a phase constant during heating is a difficult but necessary issue to solve.

In practice, the stability difference needs to correspond to an experimentally measurable quantity. In different systems, the measurable quantity is different. Finding an appropriate structural parameter by which to measure metastability is difficult since metastable states cannot be readily unified in a formalized manner. There are distinct and definitive structural parameters in some systems which can be precisely measured and used to represent metastability. These systems offer the opportunity for quantitative discussions concerning a more general understanding of phase behavior in polymers. Throughout the remainder of this book, we will focus on the analysis of these systems and use them to construct a platform connecting different research areas in polymer physics.

REFERENCES AND FURTHER READING

Andrews, B. M., Gray, G. W., and Bradshaw, M. J. (1985). The preparation and liquid crystal behavior of pyrimidines and dioxans incorporating a dimethylene linking group. *Mol. Cryst. Liq. Cryst.* **123**, 257–269.

Bachels, T., Güntherodt, H.-J., and Schäfer, R. (2000). Melting of isolated tin nanoparticles. *Phys. Rev. Lett.* **85**, 1250–1253.

Bolhuis, P. G., Frenkel, D., Mau, S.-C., and Huse, D. A. (1997). Entropy difference between crystal phases. *Nature* **388**, 235–237.

Buffat, P. h., and Borel, J.-P. (1976). Size effect on the melting temperature of gold particles. *Phys. Rev. A* **13**, 2287–2298.

Carr, N., and Gray, G. W. (1985). The properties of liquid crystal materials incorporating the $-CH_2O-$ inter-ring linkage. *Mol. Cryst. Liq. Cryst.* **124**, 27–43.

Couchman, P. R. (1979). The Lindemann hypothesis and the size dependence of melting temperatures II. *Philos. Mag. A* **40**, 637–643.

Debenedetti, P. G. (1996). "Metastable Liquids, Concepts and Principles." Princeton University Press, New Jersey.

Dick, K., Dhanasekaran, T., Zhang, Z., and Meisel, D. (2001). Size-dependent melting of silica-encapsulated gold nanoparticles. *J. Am. Chem. Soc.* **124**, 2312–2317.

Evans, R. M. L., and Cates, M. E. (1997). Diffusive evolution of stable and metastable phases.1. Local dynamics of interfaces. *Phys. Rev. E* **56**, 5738–5747.

Evans, R. M. L., and Poon, W. C. K. (1997). Diffusive evolution of stable and metastable phases.2. Theory of nonequilibrium behavior in colloid-polymer mixtures. *Phys. Rev. E* **56**, 5748–5758.
Evans, R. M. L., Poon, W. C. K., and Cates, M. E. (1997). Role of metastable states in phase ordering dynamics. *Europhys. Lett.* **38**, 595–600.
Jing, A. J., Taikum, O., Li, C. Y., Harris, F. W., and Cheng, S. Z. D. (2002). Phase identifications and monotropic transition behaviors in a thermotropic main-chain liquid crystalline polyether. *Polymer* **43**, 3431–3440.
Keller, A., and Cheng, S. Z. D. (1998). The role of metastability in polymer phase transitions. *Polymer* **39**, 4461–4487.
Lai, S. L., Guo, J. Y., Petrova, V., Ramanath, G., and Allen, L. H. (1996). Size-dependent melting properties of small tin particles: Nanocalorimetric measurements. *Phys. Rev. Lett.* **77**, 99–102.
Lehmann, O. (1877). "Über Physikallische Isomerie." Dissertation, Strassburg. From Kelker, H. (1973). History of liquid crystals. *Mol. Cryst. Liq. Cryst.* **21**, 1–48.
Moroni, D., Ten Wolde, P. R., and Bolhuis, P. G. (2005). Interplay between structure and size in a critical crystal nucleus. *Phys. Rev. Lett.* **94**, 235703.
Ostwald, W. (1897). Studien über die Bildung und Umwandlung fester Körper. *Z. Phys. Chem. Stöchiometrie und Verwandschaftslehre* **22**, 289–300.
Ostwald, W. (1900). Über die vermeintliche Isomerie des roten und gelben Quecksilberoxyds und die Oberflächenspannung fester Körper. *Z. Phys. Chem. Stöchiometrie und Verwandschaftslehre* **34**, 495–503.
Oxtoby, D. W. (2003). Crystal nucleation in simple and complex fluids. *Philos. Transact. Math. Phys. Eng. Sci.* **361**, 419–428.
Oxtoby, D. W., and Shen, Y. C. (1996). Density functional approaches to the dynamics of phase transitions. *J. Phys. Condens. Matter* **8**, 9657–9661.
Pawlow, P. (1909). Über die Abhängigkeit des Schmelzpunktes von der Oberflächenenergie eines festen Körpers. *Z. Phys. Chem. Stöchiometrie und Verwandschaftslehre* **65**, 1–35.
Poon, W. C. K., Pirie, A. D., and Pusey, P. N. (1995). Gelation in colloid-polymer mixtures. *Faraday Discuss.* **101**, 65–76.
Rastogi, S., Newman, M., and Keller, A. (1991). Pressure-induced amorphization and disordering on cooling in a crystalline polymer. *Nature* **353**, 55–57.
Rastogi, S., Newman, M., and Keller, A. (1993). Unusual pressure-induced behavior in crystalline poly-4-methyl-pentene-1. *J. Polym. Sci. Polym. Phys. Ed.* **31**, 125–139.
Rastogi, S., Vega, J. F., van Ruth, N. J. L., and Terry, A. E. (2006). Non-linear changes in the specific volume of the amorphous phase of poly(4-methyl-1-pentene); Kausmann curves, inverse melting, fragility. *Polymer* **47**, 5555–5565.
Schmelzer, J., Möller, J., and Gutzow, I. (1998). Ostwald's rule of stages: The effect of elastic strains and external pressure. *Z. Phys. Chem. Stöchiometrie und Verwandschaftslehre* **204**, 171–181.
Tammann, G. (1903). "Kristallisieren und Schmelzen." Verlag Johann Ambrosius Barth, Leipzig.
Tammann, G., and Mehl, R. F. (1925). "The States of Aggregation." Nan Nostrand Company, New York.
Tang, B. Y., Ge, J. J., Zhang, A., Calhoun, B., Chu, P., Wang, H., Shen, Z., Harris, F. W., and Cheng, S. Z. D. (2001). Monotropic phase behaviors of 2,3,6,7,10,11-hexa (4′-octyloxy benzoyloxy) triphenylene discotic molecules. *Chem. Master.* **13**, 78–87.
Tang, B. Y., Jing, A. J., Li, C. Y., Shen, Z, Wang, H., Harris, F. W., and Cheng, S. Z. D. (2003). Role of polymorphous metastability in crystal formation kinetics of 2,3,6,7,10,11-hexa (4′-octyloxy benzoyloxy) triphenylene discotic molecules. *Crystal Growth and Design* **3**, 375–382.
ten Wolde, P. R., Ruiz-Montera, M. J., and Frenkel, D. (1996). Numerical calculation of the rate of crystal nucleation in a Lennard-Jones system at moderate undercooling. *J. Chem. Phys.* **104**, 9932–9947.
van Ruth, N. J. L., and Rastogi, S. (2004). Nonlinear changes in specific volume, A route to resolve an entropy crisis. *Macromolecules* **37**, 8191–8194.
Vorländer, D. (1923). Die Erforschung der molekularen Gestalt mit Hilfe der kristallinischen Flüssigkeiten. *Z. Phys. Chem. Stöchiometrie und Verwandschaftslehre* **105**, 211–254.
Wunderlich, B. (1980). "Macromolecular Physics. Volume III. Crystal Melting." Academic Press, New York.

CHAPTER FOUR

Metastable States in Phase Transitions of Polymers

Contents

From this chapter onward, all discussions focus on polymer metastable states and phase transitions. In the first part of this chapter, the classical understanding of metastable states is introduced in terms of the supercooling of the isotropic melt and crystallization, as well as the superheating of the crystal and melting in the bulk of semicrystalline homopolymers. A general review is provided summarizing the development of polymer crystallization theories advanced so far and current issues remaining in this area. A new definition of the nucleation barrier, consisting of both enthalpic and entropic components, is proposed. The observation of lower melting temperatures for the metastable polymer crystals, compared with the corresponding equilibrium crystals, leads to the recognition of the phase size dependence of crystal metastability. However, the most critical issue in experimentally determining a crystal's metastability is how to avoid changing its metastability during heating. Furthermore, crystal melting kinetics has not been well established, yet limited observations indicate that the melting process may be controlled by different mechanisms.

The second part of this chapter deals with metastable states in phase-separated polymer blends and copolymers. In particular, the concept of morphologically metastable states is introduced to illustrate the different stages of the morphological evolution observed in phase-separated polymer blends. These morphologies are on the kinetic pathway to the two-layer equilibrium, phase-separated morphology. Specifically, the liquid–liquid phase separation kinetics of nucleation-limited and spinodal decomposition mechanisms are described, and the different pathways for the metastable phase morphological formation are discussed. On the other hand, the nanophase separation of block copolymers generates phase morphologies that are almost always the thermodynamic equilibrium morphology due to the chemical connectivity between the two chemically dissimilar blocks. There is, however, still a special hexagonally perforated layer metastable phase that is induced by mechanical shearing. Furthermore, when crystallization of one block takes place within the nanoscale-separated phase morphology, we obtain hierarchical structures on different length scales, in which at least one phase structure is metastable.

1. Supercooled Liquids and Crystallization

1.1. Supercooled liquids

Compared to crystallization in simple, small molecules, as described in Section 2.1 of Chapter 2, polymer crystallization occurs even further away from thermodynamic equilibrium. At a constant pressure, polymer crystallization always takes place from a supercooled liquid. Below the equilibrium melting temperature, the Gibbs free energy of the isotropic liquid is always higher than that of the crystalline solid. Therefore, the supercooled liquid is, by definition, thermodynamically metastable. The crystallization in a supercooled liquid is initiated by a nucleation event. Traditionally, it has been understood that the nucleus is small in volume and large in surface area. At this initial stage, the positive surface free energy overwhelms the negative volume free energy, thus constructing a free energy barrier. The question that then arises is: How do the thermal (density) fluctuations in the metastable, supercooled polymer liquid trigger primary nucleation? Even after primary nucleation, polymer crystal growth is generally dominated by a surface nucleation process. A similar question can also be asked: How do the thermal (density) fluctuations in the supercooled liquid affect surface nucleation? To answer these questions, we need to understand the microscopic structure and dynamics of the supercooled liquid, as well as their temperature and pressure dependences. At this stage, the term "supercooled liquid" refers to a liquid that is below its equilibrium melting temperature and will form a glass below its glass transition temperature, such that crystallization or other ordering processes are suppressed.

Although we currently do not completely understand the microscopic structure and dynamics of supercooled liquids, the temperature dependence of important macroscopic properties of supercooled liquids, such as viscosity, have been experimentally observed. It has been determined that the viscosity, and thus the structural relaxation times, drastically increase with decreasing temperature.

The strength of this temperature dependence is a critical characteristic of the liquid. A fragile liquid, which has a strong relaxation time–temperature dependence, is thought to have weak physical intermolecular interactions, like van der Waals interactions; whereas, strong liquids, which have a weaker relaxation time–temperature dependence, are thought to have stronger, "network-like" interactions between the molecules (Angell, 1985). The vast majority of supercooled linear polymeric melts are fragile.

For a fragile liquid at low levels of supercooling, the relaxation dynamics of the supercooled liquid are similar to the equilibrium liquid where the relaxation time follows an Arrhenius relationship. The microscopic relaxation requires the cooperative rearrangement of molecules. The characteristic length scale of these cooperatively rearranging regions is estimated to be around 2 to 3 nm (Cicerone and Ediger, 1995; Schmidt-Rohr and Spiess, 1991; Tracht *et al.*, 1998). However, when the supercooled liquid is cooled even further, the cooperative relaxation behavior becomes increasingly non-Arrhenius with a nonexponential temperature dependency. This behavior can be empirically described with a stretched-exponential function $\exp\{-(t/\tau)^{\beta}\}$, where t represents the experimental time and τ is the material's relaxation time. The $\beta(>1)$ exponent stretches this exponential function. Furthermore, in this temperature region, liquids produce excess small angle light scattering beyond predictions, indicating the existence of a long-range density fluctuation with a correlation length of up to 300 nm in polymeric (Debye and Bueche, 1949; Fischer, 1993) and small molecule liquids (Fischer, 1993). The characteristic length scale of local cooperative motion has been found to be nearly independent of the long-range fluctuation correlation. Therefore, two types of relaxation processes with very different length and time scales have been experimentally observed.

A detailed microscopic description of liquid motion using hydrodynamic theories indicates that rotational and translational motion are decoupled; namely, the product of the translational diffusion constant and the reorientation time should be essentially independent of temperature in a supercooled liquid. However, this temperature independence breaks down in highly supercooled liquids. The reason for this breakdown may be the result of the existence of structurally transient domains in the supercooled liquid. It is postulated that hydrodynamic theories are only valid within one domain, and averaging over all of the domains favors larger diffusion constants and slower reorientation rates (see, e.g., Fourkas *et al.*, 1997; Murry *et al.*, 1999).

The "Fischer's cluster" concept was introduced to explain the long-range correlation of density fluctuations in a single component supercooled liquid (see, e.g., Fischer *et al.*, 2002). A theory of the structure and dynamics of a Fischer's cluster was also put forward (Bakai and Fischer, 2004). This theory is based on heterophase fluctuations in the supercooled liquids that consist of solid-like and liquid-like transient associations of molecules. These two types of clusters possess different short-range order yet are randomly distributed in the supercooled liquid. Therefore, in the heterophase liquids, one can discuss phase aggregation and equilibrium between these two types of clusters and their critical phenomena.

When we discuss supercooled liquids in the bulk, one must deal with the concept of the glass transition temperature. The vitrification or devitrification process will be extensively discussed in the coming chapters. However, a detailed review of the glass transition temperature is beyond the scope of this volume. Only a very brief description is given here. As the temperature is decreased below the glass transition temperature, the relaxation time becomes so large that the system can no longer maintain the equilibrium structure of the supercooled liquid. Thus, it appears that the global thermal and density fluctuations stop because the structural relaxation time and the experimental time scale intersect at the glass transition temperature. Unlike the equilibrium crystal melting temperature, which is defined thermodynamically, the temperature at which the relaxation time crosses the experimental time scale varies according to the experimental setup. So, the glass transition temperature is not a fixed temperature. Instead, for any particular experimental setup, the glass transition temperature is defined empirically by using a threshold physical reference, such as the viscosity reaching a certain value, for example, 10^{13} poise. The glass transition temperature can be experimentally measured utilizing calorimetric, volumetric, mechanical, dynamic, dielectric, and other methods.

Glass transition phenomena have been extensively studied in noncrystallizable molecules. For crystallizable molecules, the amorphous glass can be reached by very fast quenching. In most cases, maintaining a deeply supercooled liquid without crystallization, for systems such as water, is a difficult task. Because semicrystalline polymeric melts can be highly supercooled (up to 100 °C) for long periods of time, they give us the opportunity to study the structures and dynamics of deeply supercooled liquids. However, a difficulty in dealing with homopolymers is that they have been treated as if they are single-component systems, but because of the polydisperse nature of homopolymer systems, the original concepts applicable to single-component simple, small molecule systems have to be relaxed. However, if this simplifying approach was not adopted, every polymer system would have to be treated as a multicomponent system with an infinite number of components. This is practically impossible to accomplish.

Understanding crystallization from polymer melts with increasing supercooling at a constant pressure (or increasing pressure at a constant temperature) is essential from both the perspectives of fundamental science and practical application. Qualitatively, the temperature at which crystallization takes place with respect to the equilibrium melting temperature is dependent on internal and external parameters, such as molecular structure, stereoregularity, molecular weight, polydisperity, and cooling rate. Generally speaking, the faster the cooling rate, the more irregular the chemical structure, and the longer the chain molecule, the greater the level of supercooling that can be reached. This process can also be observed in multiple component polymer systems such as polymer crystals grown in solutions and in blends. The lifetime of the metastable supercooled liquid state, which is able to be experimentally detected, is governed by the nucleation process in the initial stage of polymer crystallization. The degree of supercooling is defined as the difference between the equilibrium melting temperature and the crystallization temperature at constant pressure.

1.2. Comments on polymer crystallization theories

Modern polymer physicists have exerted significant effort to understand polymer crystallization since the discovery of chain-folded, lamellar single crystals of semicrystalline polymers (Fischer, 1957; Jaccodine, 1955; Keller, 1957; Till, 1957). There are basically two major ways to understand polymer crystallization. The first is from the structural and morphological perspective. Although we cannot monitor every molecular trajectory to "see" the molecular pathways during crystal growth, the resulting polymer crystal structure and morphology encompass information about this process and thus, provide insight into what happened during the nucleation and crystal growth processes. This approach uses structure and morphology as probes from which explanations concerning crystal growth on the subnanometer to nanometer scale can be deduced. It does, however, rely on the assumption that the final observed structure and morphology are truly representative of the crystallization process. The other approach is based on the results observed via scattering experiments. Because the scattering results can be conducted *in situ* in real time, they represent an overall average of the density fluctuations in a supercooled liquid during crystallization. This approach requires detailed microscopic models assisted by theoretical descriptions to explain the experimental scattering data and must be supported with direct morphological observations.

Almost all theories so far have been kinetic in nature and describe the crystal growth behavior on a preexisting surface. Both the lateral and normal crystal growth models, which were originally utilized to describe crystallization in simple, small molecules, as discussed in Section 2.1 of Chapter 2, have been used in describing polymer crystallization. A successful theory must have two important features. First, it must explain major experimental results. Second, it must successfully predict the crystallization behavior of polymers. All polymer crystallization theories proposed so far only accomplish the first goal. Therefore, let us begin with a brief summary of the important experimental results that are critical to understanding polymer crystallization. They can be categorized into the following five areas.

A number of isothermal crystal growth experiments with semicrystalline polymers in the melt at different crystallization temperatures (or supercoolings) showed that crystal growth rate, R, is linear with time at a constant temperature, as shown in Fig. 4.1A (Cheng and Lotz, 2005, and the references cited therein). As long as the spherulitic radial growth direction always follows a specific crystallographic plane normal and the growth is not affected by the local environment, the measured growth rates are representative of the growth of lamellar single crystals in this specific direction. However, the experimentally ideal case is to directly measure the growth rate of lamellar single crystals. This measurement can be achieved by transmission electron microscopy, *in situ* differential interference contrast optical microscopy, and *in situ* atomic force microscopy, as observed in polyethylene (Toda, 1992; Toda and Keller, 1993) and syndiotactic polypropylene (Zhou *et al.*, 2000; Zhu *et al.*, 2007) or by using a self-decoration method on single crystals under polarized light microscopy for poly(ethylene oxide), originally developed by Kovacs *et al.* (Cheng and Chen, 1991; Kovacs and Gonthier, 1972; Kovacs and Straupe, 1979, 1980; Kovacs *et al.*, 1975, 1977). An earlier survey of linear crystal growth rates can be found in

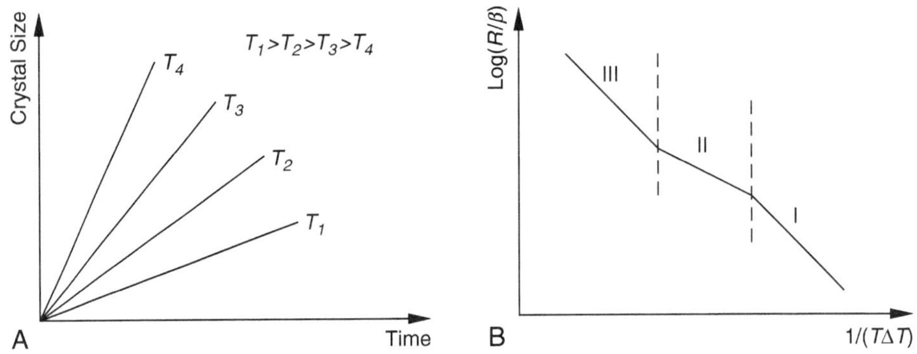

Figure 4.1 Schematic illustration of the general features of polymer crystal growth: (A) linear relationships between the crystal sizes and temperature with time under isothermal crystal growth in the temperature region that is near and below the melting temperature. If the temperature region is near and above the glass transition temperature, the trend will be reversed. (B) relationship between logarithmic crystal growth rate (R) and $1/(T\Delta T)$, where β is the prefactor, including an activation term, T is the crystallization temperature, and ΔT is the supercooling. Growth regimes I, II, and III are based on experimental observations of polyethylene [replotted from Hoffman and Miller (1997), with permission].

Lovinger *et al.* (1985). Furthermore, at different crystallization temperatures, the growth rates exhibit an exponential relationship with respect to the reciprocal of the product of the crystallization temperature and supercooling, $1/(T\Delta T)$, as shown in Fig. 4.1B. Note that when the lamellar thickness is a constant, as in the case of growing extended chain crystals with oligomers, the growth rate decreases linearly with decreased supercooling (Cheng *et al.*, 1992; Leung *et al.*, 1985).

Second, the lamellar thickness measured after crystallization is linearly proportional to the reciprocal supercooling for crystals grown from an isotropic liquid (both from the melt and solution), as shown in Fig. 4.2 (see, e.g., Wunderlich, 1973). In some oligomers, quantized lamellar thickness increases with decreased supercooling have

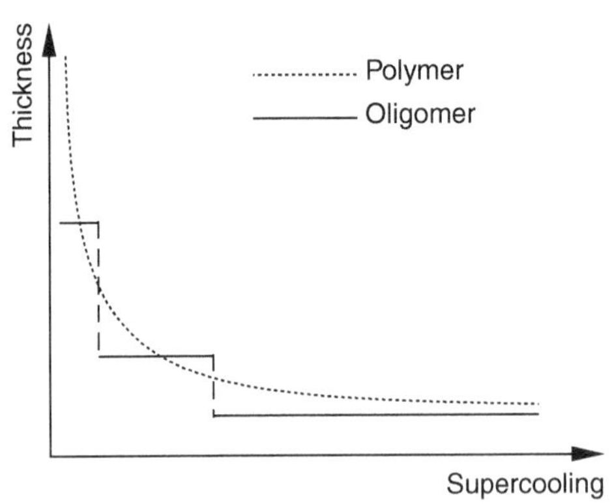

Figure 4.2 Schematic illustration showing that the lamellar thickness formed during polymer crystallization is inversely proportional to supercooling. In oligomers, quantized changes in the lamellar thickness can be observed.

been observed in monodisperse *n*-alkanes (Ungar *et al.*, 1985) and poly(ethylene oxide) fractions with narrow molecular weight distributions (Arlie *et al.*, 1965, 1966, 1967; Kovacs and Gonthier, 1972; Kovacs and Straupe, 1979, 1980; Kovacs *et al.*, 1975, 1977). Figure 4.2 schematically illustrates this quantized change of lamellar thickness in oligomers. These changes are attributed to the formation of integral-folded chain crystals and extended chain crystals with all the chain ends located on the crystal lamellar basal surface. The lamellar thickness is usually detected by small angle x-ray scattering for bulk samples, transmission electron microscopy (see, e.g., Faraday Discussions of the Chemical Society, 1979; NATO Advanced Science Institutes Series C, 1993), and atomic force microscopy for single crystals (Zheng *et al.*, 2006; Zhu *et al.*, 2007). In a few polymers, such as polyethylene and poly(ethylene oxide), the lamellar thicknesses can also be deduced using longitudinal acoustic mode Raman spectroscopy (Kim and Krimm, 1996; Ungar and Keller, 1986).

Third, the growth rates are specific to the crystal plane associated with that particular growth direction, so growth rates in different directions may be different. Polymer single crystals possess a variety of habits, ranging from elongated ribbon-like to square- or hexagonal-shaped when grown from either the melt or dilute solution (Bassett, 1981; Cheng and Li, 2002; Cheng and Lotz, 2003, 2005; Geil, 1963; Keller, 1968; Khoury and Passaglia, 1976; Wunderlich, 1973). The anisotropic single crystal habits are always associated with unit cells having lower rotational symmetry with respect to the *c* axis, while the more isotropic single crystal habits have unit cells with higher rotational symmetry. Figure 4.3 schematically illustrates that the kinetically anisotropic single crystal habit is attributed to different growth rates at one crystallization temperature along different crystallographic planes of the different growth fronts, while the kinetically isotropic single crystal habit is caused by the identical growth rates at the crystallization temperature due to the fact that the growth fronts possess the identical crystallographic plane.

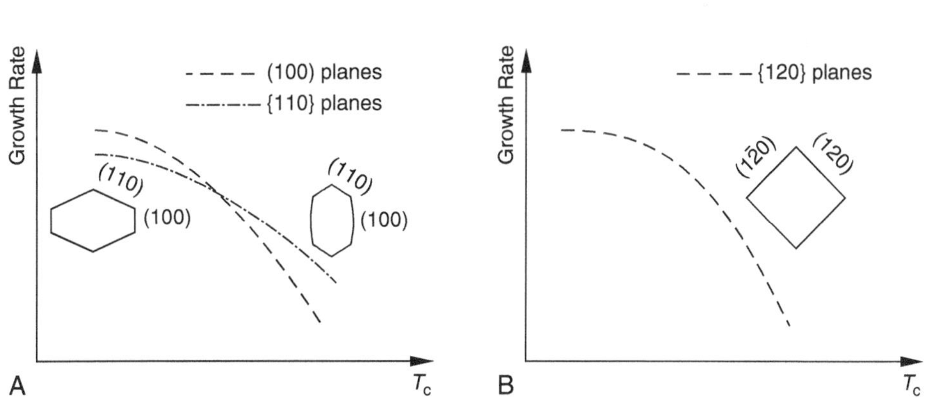

Figure 4.3 Schematic illustration of relationships between growth rates and crystallization temperatures along different crystallographic planes for a kinetically anisotropic single crystal habit such as in polyethylene (A). On the other hand, only one relationship between growth rates and crystallization temperatures can be seen for a kinetically isotropic single crystal habit such as in poly(ethylene oxide) (B).

Fourth, experimental observations reveal that in many semicrystalline polymers, the crystal growth rates depend upon molecular weight (see, e.g., Cheng and Wunderlich, 1986b, 1988; Ergoz *et al.*, 1972; Hoffman *et al.*, 1975; Lauritzen and Hoffman, 1973; Magill, 1964, 1967, 1969; Pérez *et al.*, 1984; Umemoto and Okui, 2005; Vasanthakumari and Pennings, 1983). Generally speaking, the linear growth rate decreases with increasing molecular weight. Furthermore, the supercooling required for nucleation also increases with increasing molecular weight. Figure 4.4 shows a schematic illustration of the relationship between growth rate and molecular weight. The molecular weight also remarkably affects the crystallinity and crystal morphology of polymers. Above a critical molecular weight, the crystallinity starts to decrease, while the spherulitic morphology gradually deteriorates due to the loss of specific chain orientation and cooperative lamellar stacking. Ultimately, aggregates of small crystals are dominant for polymers with ultrahigh molecular weights, such as in the case of ultrahigh molecular weight polyethylene.

Fifth, when a polymer sample has a broad molecular weight distribution or is a mixture of low and high molecular weight fractions having identical chemical repeating units (called "bimodal"), molecular segregation of the low molecular weight species takes place during crystallization (Cheng and Wunderlich, 1986a,b; Cheng *et al.*, 1988; Keith and Padden, 1964a,b; Wunderlich, 1976). It seems that the molecules can "feel" their size during crystallization until they are long enough so that intrachain segments can independently crystallize without affecting each other. For example, in a blend composed of a high and a low molecular weight poly(ethylene oxide) mixed in a miscible molecular weight range at high supercoolings, both species may cocrystallize. At low supercoolings, however, the high molecular weight species crystallizes first, and the low molecular weight chains segregate microscopically between lamellae and/or macroscopically between the spherulites (Cheng *et al.*, 1988). This segregation process takes place at lower temperatures compared with the values predicted by theoretical calculations corresponding to the

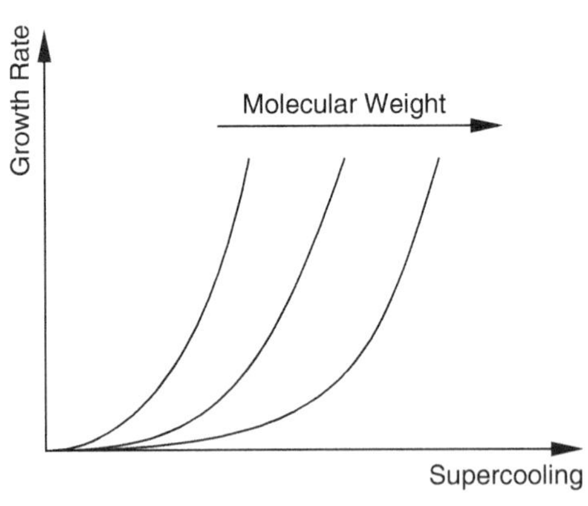

Figure 4.4 Schematic illustration of the relationship between molecular weight and growth rate at different supercoolings. With decreasing molecular weight, the supercooling required for nucleation is reduced [replotted from Armitstead *et al.* (1992), with permission].

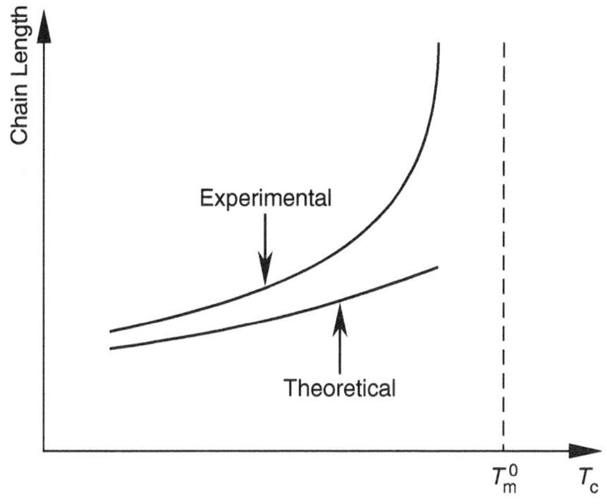

Figure 4.5 Schematic illustration of a segregation process for a bimodal polymer blend with two different molecular weights but with identical chemical structure in the miscible region. Molecular segregation takes place due to the crystallization kinetics. The theoretical line is calculated based on the equilibrium fractionation process [replotted from Armitstead *et al.* (1992), with permission].

equilibrium fractionation process, as illustrated in Fig. 4.5, and substantially decreases the crystal growth rates. This phenomenon has been observed up to a molecular weight of around 20 kg/mol, indicating that only chains of sufficient length (above ~ 40 kg/mol) were able to eliminate long-range intrasegmental recognition along the entire chain. Therefore, when the chain is long enough, one polymer chain can crystallize into two separate neighboring crystals, forming tie molecules (Keith *et al.*, 1965, 1966a,b, 1971).

The crystallization process converts chain molecules that start out in a three-dimensional random coil conformation into predominately chain-folded lamellar crystals. The specific trajectory of an individual chain molecule during crystallization may be very different from others. Yet, we lack a technique to monitor each chain molecule to "see" how it crystallizes. For this reason, an analytical theory always takes a "mean-field" approach by simplifying the individual trajectories to an average form. Therefore, the analytical theory sacrifices these molecular details.

The first analytical theory was put forth more than 40 years ago by Hoffman and coworkers (Hoffman and Lauritzen, 1961; Hoffman *et al.*, 1975, 1976; Lauritzen and Hoffman, 1960, 1973), as well as Frank and Tosi (1961). They utilized a lateral growth, surface-nucleation-limited process to describe the growth rates of polymer lamellar crystals. The molecular picture of surface nucleation is clear. An existing crystal with a defined, atomically smooth, crystallographic surface provides a growth front. Chain molecules deposit onto the growth plane and start to crystallize into the lattice, one stem at a time, to form lamellae. The crystal growth rate perpendicular to the growth front is linear at a constant crystallization temperature. This kinetic model contains four parameters to describe the nucleation-limited process: the surface nucleation rate, i; the growth rate parallel to the growth plane that covers

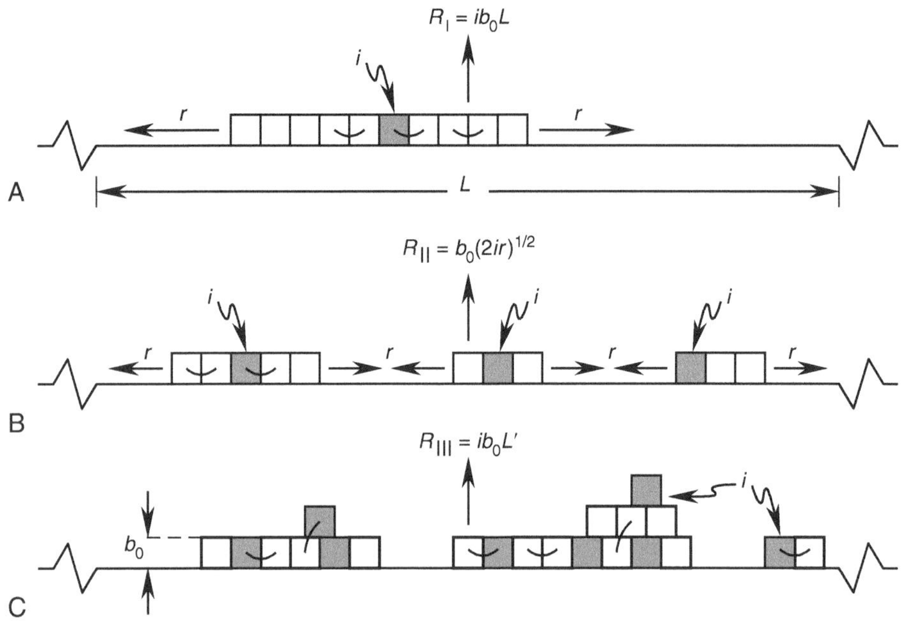

Figure 4.6 Schematic illustrations of the three growth regimes in polyethylene crystal growth based on the Hoffman–Lauritzen theory: (A) at low supercoolings, a mono-nucleation event is expected in regime I; (B) at intermediate supercoolings, multiple nucleation events occur in regime II; and (C) at high supercoolings, the growth enters regime III where the niche separation, L', is the dominant factor [redrawn from Hoffman and Miller (1997), with permission].

the growth front after surface nucleation, which is called the lateral covering rate, *r*; the width of the growth front (the substrate length), which the nuclei and growth covers, *L*; and the growth rate normal to the growth plane, *R*. The Hoffman–Lauritzen theory predicts three growth regimes, which are shown in Fig. 4.6.

In region I at low supercoolings (high crystallization temperatures), the growth from a single nucleus covers the entire growth substrate, *L*, as shown in Fig. 4.6A. The physical picture of this process is as follows: Given an atomically smooth growth front, the rate-limiting surface nucleation occurs first, then, the rest of the growth front is quickly covered by lateral growth to generate a new, atomically smooth growth front that then waits for the formation of the next nucleus on the new front. The analytical expression for the growth rate in regime I is given by the equation (Frank and Tosi, 1961; Hoffman, 1982; Hoffman and Lauritzen, 1961; Hoffman *et al.*, 1975, 1976; Lauritzen and Hoffman, 1973):

$$\mathrm{R_I} = ib_0L \tag{4.1}$$

Here b_0 is the thickness of the molecular layer crystallized on the substrate and *L* is the substrate length that is covered by one surface nucleus, under the condition that the ratio between the lateral covering rate and the substrate length, r/L, is much smaller than the product of the surface nucleation rate and the substrate length, iL (Frank, 1974).

With increased supercooling, the system enters regime II. Figure 4.6B illustrates the physical picture of regime II growth. It is evident that on a substrate of width *L*, more than one nucleus forms. Therefore, the growth rates are now associated with the surface nucleation rate, *i*, and the lateral covering rate, *r*. The critical factor in this regime is the niche separation between the two neighboring nuclei. At the higher end of crystallization temperatures in regime II, the niche separation is large. As the nucleation rate increases with increased supercooling, the niche separation distance decreases. The analytical expression of the growth rate in regime II given by (Sanchez and DiMarzio, 1972):

$$\mathrm{R_{II}} = b_0(2ir)^{1/2} \tag{4.2}$$

is independent of the substrate length, *L* (Hoffman, 1982; Hoffman *et al.*, 1979; Mansfield, 1988). Frank also derived Eq. (4.2) using a differential equation with defined boundary conditions (1974). This equation possesses analytical solutions with moving boundary conditions that predict the different morphological habits of polyethylene single crystals, from the lozenge to the lenticular habit (Mansfield, 1988; Toda, 1991). An analysis of the morphological shape change from the lozenge to the truncated lozenge habits was earlier reported by Passaglia and Khoury (1984).

On further increasing the supercooling, one passes into the lower crystallization temperature end of regime II, resulting in changes in the crystal growth. When one finally enters regime III, as shown in Fig. 4.6C, the niche separation distance is on the same order of magnitude as the stem width a_0. Therefore, the lateral covering rate, *r*, is no longer a dominant factor, so the analytical expression for the growth rate returns to:

$$\mathrm{R_{III}} = ib_0L' \tag{4.3}$$

Here L' is the width between two neighboring niches, which is about one to three stem widths (Hoffman, 1983; Hoffman and Miller, 1997).

Because the surface-nucleation rate, *i*, also takes the Turnbull and Fisher (1949) form of Eq. (2.9) in Section 2.1 of Chapter 2, the linear growth rate, *R*, always has some type of exponential relationship with respect to the free energies of the activation and nucleation barriers. The free energy of activation at high crystallization temperatures is almost constant. Significant attention has thus been focused on the free energy of the nucleation barrier. Based on the classical treatment of nucleation theory, which was also adopted by the Hoffman–Lauritzen theory, the nucleation barrier for polymer crystal growth can be described using Eq. (2.10) in Section 2.1 of Chapter 2. It reads:

$$\Delta G = -V\Delta g_{\mathrm{f}} + \mathrm{A}\gamma + B\gamma_{\mathrm{e}} \tag{4.4}$$

The barrier is caused by the lateral and folded surface free energies—$A\gamma$ and $B\gamma_e$, where γ and γ_e are the lateral and fold surface free energy densities and A and B are the lateral and fold surface areas. These terms overwhelm the bulk free energy term of the crystal when it is small. Here, the bulk free energy is defined as $V\Delta g_f$, where Δg_f is bulk free energy density and V is the volume of the crystal. A detailed analytical construction of this nucleation barrier depends on how the chain molecules place themselves into the crystalline lattice. The Hoffman–Lauritzen theory assumes that the stems attach onto the crystal growth front one at a time as the averaging step in its "mean-field" approach. Other approaches use a few segments at a time, as suggested by Point (1979a,b), or a few stems at a time, as proposed by Phillips (1990, 2003). These detailed differences may change the four structural parameters during crystal growth, but the overall exponential dependence of the growth rate, *R*, with respect to the nucleation barrier is not altered.

Of the four parameters in the theory (the surface nucleation rate, *i*; the lateral covering rate, *r*; the width of the growth front, or the so-called substrate length, *L*; and the growth rate perpendicular to the growth plane, *R*), only the linear growth rate, *R*, can be experimentally measured. In the specific case of measuring the single crystal growth of polyethylene, the lateral covering rate, *r*, can then also be determined (Toda, 1993). Using dark field imaging in transmission electron microscopy, the substrate length, *L*, can be revealed for the specific case of isotactic poly(vinyl cyclohexane). The results fit well with the Hoffman–Lauritzen theoretical prediction and the length *L* decreases with increased supercooling (Alcazar *et al.*, 2006). This substrate length can also be detected indirectly using supplemental experimental methods, such as the wide angle x-ray diffraction, which can measure the crystallite size that corresponds to the substrate length, *L*, of crystals growing in the melt (Hoffman and Miller, 1989).

From the time the Hoffman–Lauritzen theory was originally proposed, it underwent continuous improvements and modifications to accommodate new experimental findings and theoretical understandings. These improvements and modifications are simple reflections of the flexibility of this theory and the ability to manipulate structural parameters based on improved understandings of their physical significance. The first modification was in response to the so-called "δl catastrophe" (Hoffman and Lauritzen, 1961; Hoffman *et al.*, 1969; Lauritzen and Hoffman, 1973; Point *et al.*, 1986). Based on the first version of the Hoffman–Lauritzen theory, one expects that the lamellar thickness unrealistically goes to infinity at $\Delta g_f = 2\gamma/a_0$. This unrealistic thickness surge can be overcome by introducing a parameter Ψ into the Hoffman–Lauritzen theory that apportions a fraction of the free energy of crystallization to the free energy associated with stem attachment, while the remainder is released during the subsequent rearrangement of the attached stem.

The second modification dealt with the substrate length, *L*, which initially was speculated to be on the order of micrometers. However, later experimental results based on an experiment designed to measure the crystal growth rate of polyethylene single crystals in solution with a well-controlled temperature-jumping technique (the isochronous decoration method) showed that the value of the substrate length, *L*, should be much smaller (Point *et al.*, 1986). The Hoffman–Lauritzen theory predicts that the crystal growth rate has to increase as long as the crystal lateral size is smaller than *L*, where polyethylene single crystal growth rates exhibit a linear behavior down to at least the micrometer resolution of the experiment (Point *et al.*, 1986). The isochronous decoration method results imply that *L*, in the Hoffman–Lauritzen theory, must be smaller than 1 μm. Currently, it is thought that the value of substrate length should be several hundred nanometers, which corresponds to the crystallite size measured via wide angle x-ray diffraction experiments (Hoffman, 1985a,b; Hoffman and Miller, 1989; Lauritzen and Passaglia, 1967). This estimated substrate length has also very recently been confirmed via direct visualization, using dark field transmission electron microscopy in at least one semicrystalline polymer (Alcazar *et al.*, 2006).

The third modification was made to account for the growth rate minimum found in *n*-alkanes in both the melt and the solution around the crystallization temperature where the once-folded, integral-folded chain [IF($n=1$)] crystals convert to extended chain [IF($n=0$)] crystals (for details see below, Section 1.5 of this Chapter) (Organ *et al.*, 1989; Ungar and Keller, 1986). The introduction of an entropic component to the lateral surface free energy, γ, was attributed to a transient layer of "kinetic ciliation." This manipulation allows the Hoffman–Lauritzen theory to reproduce the experimentally observed rate minimum (Hoffman, 1991). Furthermore, the introduction of the C_∞ term (the characteristic ratio of the polymers in the melt) into the presentation of the lateral surface free energy, γ, was also suggested (Hoffman, 1992; Hoffman *et al.*, 1992).

The fourth modification explains how the Hoffman–Lauritzen theory can be applied to curved growth fronts, in particular, to the curved (200) planes in polyethylene single crystals that have been both theoretically calculated (Hoffman *et al.*, 1979; Mansfield, 1988) and experimentally observed (Toda, 1992; Toda and Keller, 1993). In this case, a lattice strain in the {200} sectors was argued for and introduced into the Hoffman–Lauritzen theory by means of an independently justified surface free energy parameter. Molecularly "serrated" (200) planes were also taken into account. Along the {110} planes, the crystal growth followed the standard Hoffman–Lauritzen theory for a smooth growth front (Hoffman and Miller, 1989; Miller and Hoffman, 1991). These last two modifications are still being debated because of newly reported experimental observations.

Although the I and II crystal growth regimes in polyethylene spherulites and axialites from the melt were experimentally observed in the mid-1970s, it took almost 30 years to quantitatively verify the existence of regime III growth in polyethylene, as shown in Fig. 4.7 (Armistead and Hoffman, 2002). Most of the experimental observations concerning polymer crystal growth rates can be explained by the Hoffman–Lauritzen theory.

There are other proposed theories that specifically describe some characteristics of polymer crystal growth. A multiple path approach was proposed (Point, 1979a,b) to overcome the so-called "$\delta\ell$ catastrophe." In order to explain molecular segregation in polymers with broad molecular weight distributions, a molecular nucleation concept was proposed by Wunderlich (1976). A two-dimensional nucleation process was suggested by Hikosaka (1987, 1990) to describe both lateral crystal growth and molecular sliding diffusion along the chain direction in the polyethylene hexagonal lattice at elevated pressures. Sadler proposed an approach describing rough surface growth in polymer crystals that parallels the continuous crystal growth mechanism of small molecules, as described in Section 2.1 of Chapter 2 (Sadler, 1986, 1987a,b,c,d; Sadler and Gilmer, 1984, 1986, 1988). Although this approach initially ignored the chain connectivity in the computer simulations, an entropic

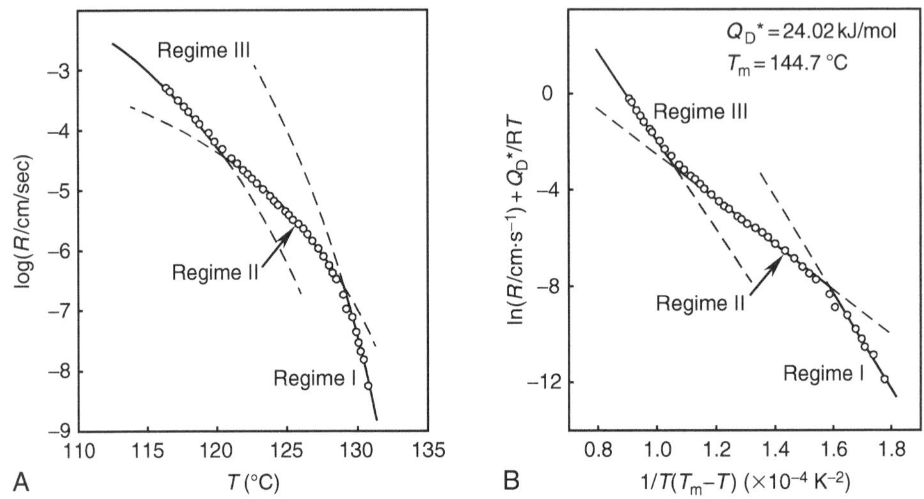

Figure 4.7 Linear crystal growth rates of a polyethylene fraction crystallized in the melt: (A) the experimentally measured growth rate data and (B) three regimes after the experimental data were treated using the Hoffman–Lauritzen theory [replotted from Armistead and Hoffman (2002), with permission].

barrier to polymer crystal growth was introduced, which leads to the development of the important concept of a "poisoning" mechanism and was suggested to be based on the experimental observation of crystallizing *n*-alkanes (Ungar and Keller, 1986). For a critical description of polymer crystal growth, an excellent and extensive review was published by Armitstead *et al.* (1992); this study remains one of the most thorough and important documents in this area.

All these theories have been sufficiently fundamental in their assumptions to describe crystallization phenomena for semicrystalline polymers, although some have "overlooked" issues that still exist in this area (Geil, 2000). Among them, the Hoffman–Lauritzen theory has been more widely utilized to quantitatively fit experimental results. The most important issue for this theory is that it cannot predict all of the polymer crystal growth behavior.

Recently, the origin of the nucleation barrier has been extended to incorporate both enthalpic and entropic contributions (Cheng and Lotz, 2003, 2005). Analytically, the entropic barrier concept is more difficult to express, but it contains some molecular details. Statistical mechanics is thus necessary to connect these microscopic contributions to the nucleation barrier, which is commonly described by classical thermodynamics. Computer simulation has become possible with the rapid increase in calculation speeds and capacity. The advantage of this technique is that it provides trajectories of individual molecules during crystallization and, therefore, it is able to average all the trajectories in the ensemble to construct the nucleation barrier. Recent results from computer simulation concerning the very initial stage of nucleation reveal the entropic origin of the barrier (Liu and Muthukumar, 1998; Muthukumar, 2000, 2003).

In addition to the nucleation-limited process in polymer crystallization, other rate-limiting processes may also appear. Under some specific conditions and environments, the crystal growth may be controlled by the transport of growth material. The most recent example of this mechanism change utilizes the crystal growth of pseudo-dewetted, thin layers of low molecular weight poly(ethylene oxide) with different end groups on a hydrophilic mica surface. Depending on the end-group chemistry and crystallization temperature, both the nucleation-limited and diffusion-limited mechanisms were found based on whether the crystal lateral size or the crystal volume was linearly proportional to crystallization time (Zhu *et al.*, 2007). On the other hand, the heat of crystallization released at the liquid–crystal interface can generate temperature gradients, resulting in a heat transport-limited process controlling crystal growth. This phenomenon can often be found in injection molding where *trans*-crystallization takes place.

Although great efforts have been taken to further understand theoretical aspects of polymer nucleation and growth, there are still unsolved questions that remain. In the following sections, some of these issues will be described. As stated previously, the views described here are mainly based on the structure and morphology observed experimentally with the assumption that these contain embedded pieces of microscopic information about the nucleation and growth processes. Note that the microscopy and diffraction based analyses rely almost exclusively on solid state observations. Determining the structure of supercooled liquids before and during polymer crystallization would rely critically on scattering techniques.

1.3. Primary nucleation process in polymer crystallization

The traditional approach to observe homogeneous (primary) nucleation is the droplet method initially devised by Vonnegut (1948). Because the droplets are sufficiently small (~3 μm or smaller in diameter), most of them will not contain heterogeneous nuclei, so homogeneous nucleation and crystallization in the droplet can be followed by polarized light microscopy. At a large supercooling, the crystal growth rate is very fast and leads to a single nucleus crystallizing the whole droplet. In general, it is estimated that when homogeneous nucleation takes place at such supercoolings, the nuclei size is around 10 nm^3. For a chain molecule with a molecular weight between 50 and 500 kg/mol, the typical volume ranges from 100 to 1000 nm^3 (Wunderlich, 1976). Therefore, a primary nucleus only requires part of one chain molecule. Reports on single-molecule, single-crystals of poly (ethylene oxide) (Bu *et al.*, 1991), isotactic polystyrene (Bu *et al.*, 1998b), and polytetrafluroethylene (Geil *et al.*, 2005) showed that a single polymer chain can crystallize into a small, facetted single crystal. Transmission electron microscopy and electron diffraction observations have established the presence of crystallographic relationships at the very initial stage of polymer crystallization.

To understand the structure and dynamics of metastable supercooled liquids, experimental work focusing on slowly crystallizing polymers such as isotactic polystyrene has been important. Samples of isotactic polystyrene were rapidly quenched to below their glass transition temperature bypassing crystallization. This quenching process freezes the thermal (density) fluctuations of the polymer liquid. The samples

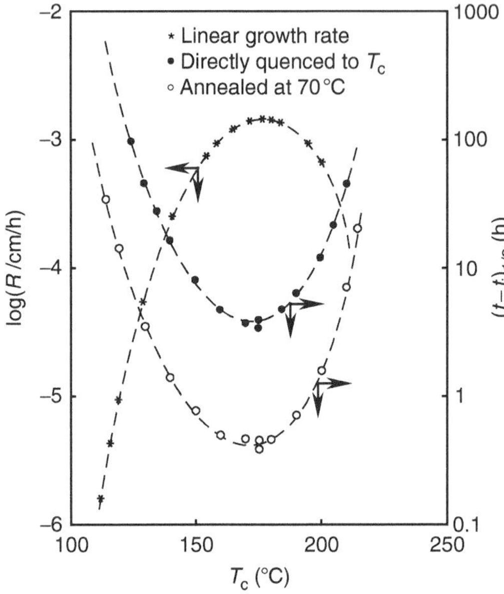

Figure 4.8 Relationships between the linear crystal growth rates (observed using polarized light microscopy) and the overall crystallization half-time (observed in differential scanning calorimetry) of isotactic polystyrene samples with different crystallization temperatures and two different thermal histories. One shows samples directly quenched from the isotropic melt at 260 °C to different crystallization temperatures. The second shows samples that were first quenched to 70 °C (30 °C below the glass transition temperature of isotactic polystyrene) and annealed there for 6 min. The samples with both thermal histories had their crystal growth rates and overall crystallization rates measured at different crystallization temperatures [replotted from Cheng and Lotz (2005), with permission].

were then aged below the glass transition temperature. The effect of physical aging on the subsequent overall crystallization process was investigated by then increasing the sample temperatures to between the glass transition temperature and the crystal melting temperature. Figure 4.8 shows that the linear growth rate of isotactic polystyrene spherulites is barely affected, or not at all, by whether they are directly quenched from the melt or after physical aging below the glass transition temperature. On the other hand, the system's overall crystallization half-time was reduced by one order of magnitude after the samples were physically aged, compared with the sample directly quenched from the melt. The increase in the overall crystallization rate results mainly from a nearly three order of magnitude increase in primary nuclei density. Part of the increase may be due to "homogeneous" nucleation taking place slightly above the glass transition temperature during the quenching process. Furthermore, extending the aging time below the glass transition temperature from 6 to 15 min resulted in a tenfold increase in the nucleation density (Cheng and Lotz, 2005). The latter observation indicates that although the large-scale thermal (density) fluctuations are frozen below the glass transition temperature, local densification contributes to the formation of primary nuclei. In essence, the thermal aging at different temperatures and times may generate local regions in the supercooled liquid with different metastabilities. Several questions remain including: How does the local densification during aging affect large-scale thermal (density) fluctuations in the supercooled liquid after the samples were brought back above the glass transition temperature? What role does denser packing in the supercooled liquid play in enhancing the primary nucleation? Are the "Fischer's clusters" described in Section 1.1 of this Chapter associated with primary nucleus formation?

An interesting investigation into primary nuclei density changes in polyethylene utilizes the melting of extended chain crystals at a constant temperature and different resident melting times. Because extended chain crystals of polyethylene are disentangled in the crystalline state, crystal melting creates a disentangled melt with a certain chain orientation. Entanglement develops after crystal melting with residence time and is accompanied by a loss of chain orientation. It was found that the nucleation density decreases with increasing residence time, implying that entanglement and chain disorienting, which increase with time, hamper the formation of primary nuclei (Yamazaki *et al.*, 2002, 2006). Furthermore, they observed a two-step decrease of the nucleation density in the polyethylene melt that followed two different exponential functions with respect to the residence time. This result suggests that there is a two-step process in forming entanglements from the disentangled melt. The significance of this observation is that during and after extended chain crystal melting, the chains reentangle and lose orientation with increasing residence time, causing the decrease in primary nucleation density.

Another approach to determine the effect of entanglement on primary nucleation is to create an isotropic disentangled melt. This can be achieved by collecting nanoparticles of single or several chains made by freeze-drying very dilute polymer solutions (Bu *et al.*, 1991, 1998b). After heating the collected nanoparticles up to their melting temperature, a disentangled melt forms. At a sufficiently long residence time in the melt, entanglements develop. In this case, the primary nucleation density change with residence time in the melt can only be attributed to the disentanglement effect eliminating the effect of chain orientation (Bu *et al.*, 1998a).

Similar to the traditional droplet experiments, the formation of primary nuclei in small droplets can also be investigated using pseudo-dewetted semicrystalline polymer thin films on an incompatible substrate (Reiter and Sommer, 1998, 2000) because this method can isolate the semicrystalline polymers into small droplets ranging in diameter size from submicrometer to a few micrometers. A recent report on primary nucleation in poly(ethylene oxide) droplets generated by dewetting poly(ethylene oxide) films on a hydrophilic amorphous polystyrene substrate was an excellent example of this approach, and the dependence of the homogeneous primary nucleation on supercooling was studied using real-time atomic force microscopy (Massa and Dalnoki-Veress, 2004; Massa *et al.*, 2003). The results showed that the nucleation rate scales with the volume of the crystallizable droplets, indicating that homogeneous primary nucleation in the bulk of these droplets was achieved. The supercooling dependence of primary nucleation using this approach fits well with data from more conventional experiments analyzed with classical nucleation theory.

The most recent result giving insight into this topic was from the crystallization of polyethylene within microemulsion droplets (Weber *et al.*, 2007). The experiments showed that polyethylene, even at a large supercooling of more than 100 °C, forms single crystals with sizes that are uniformly small, with a lateral dimension of 25 nm and a thickness of 9 nm, where the crystalline part has a thickness of 6.3 nm. The nanocrystals contained about 14 chains (equivalent to about three million molecular weight for a single chain). Additionally, the facetted nano-single crystals are a clear indication of a nucleation-limited growth mechanism in this system (Cheng, 2007).

An attempt to understand the initiation of homogeneous nucleation came from the idea of spinodal-assisted primary nucleation (Olmsted *et al.*, 1998). Different versions of this idea were also reported by other research groups (Imai *et al.*, 1993; Matsuba *et al.*, 2000). A classical spinodal decomposition process describes how an unstable system relaxes by the spontaneous growth of long wavelength density fluctuations of small amplitude, as indicated in Section 3.4 of Chapter 2. This process does not require an energy barrier, and the density perturbations in single component systems and composition perturbations in mixtures are thought to be large in extent but small in intensity. The crystallization of polymer chains with identical chemical repeat units and nearly equal length is essentially crystallization from a single component system. It was proposed that in the case of polyethylene, spinodal decomposition could form aggregates of *trans* and *gauche* conformations. The time scale of the *trans–gauche* conformational transition is at least several orders of magnitude, if not more than 10 orders of magnitude, faster than the primary nucleation rate. This mismatch of the dynamics between these two processes does not support the idea that conformational aggregation is a precursor to primary nucleation. Yet, it reflects the efforts made to understand both the structure and dynamics of supercooled liquids and the initial stage of polymer crystallization.

1.4. Structure of the interfacial liquid near the crystal growth front

For polymer crystal growth, which is a surface-nucleation-limited process, the structure of the interfacial zone near the growth front in the supercooled melt is not yet clearly understood. Recent efforts by Strobl (2000) suggest that the melt near the crystal growth front is in a mesomorphic phase with stem orientational order. It was proposed that polymer lamellar crystals form via the build up of granular crystals, then these granular crystals coalesce with each other to construct a lamellar crystal. Figure 4.9 is the schematic model provided by Strobl (2000) to illustrate this proposed polymer crystallization pathway. A few atomic force

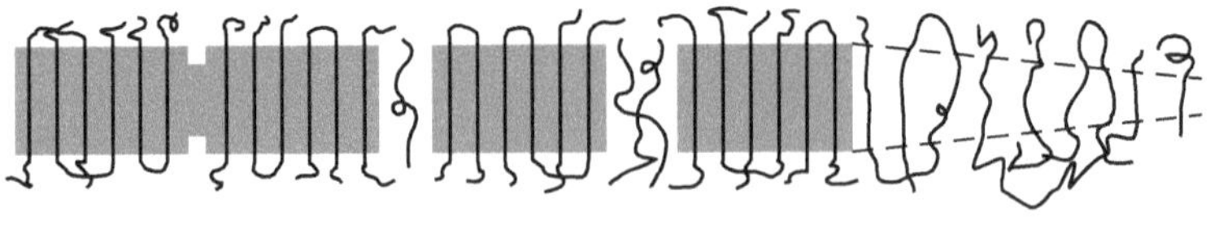

Figure 4.9 Schematic model of polymer crystallization suggested by Strobl [redrawn from Strobl (2000), with permission].

microscopy micrographs were shown to support this view. There are clear differences between this model and the classical crystallization model shown in Fig. 4.6. The existence of two proposed crystallization pathways reflects the differences in attempting to understand polymer crystal growth from the structural and the scattering perspectives (Cheng *et al.*, 2000; Lotz, 2000; Muthukumar, 2000; Strobl, 2000).

To directly detect the existence of this mesomorphic phase in the supercooled melt during polymer crystallization is experimentally difficult. Strobl's model requires two thermodynamic transitions: one from the supercooled isotropic melt to the mesomorphic phase and one from the mesomorphic phase to the crystal. This idea differs from the traditional view of a single transition process from the isotropic melt to the crystal. For Strobl's model to be considered accurate, direct or indirect experimental evidence is needed. First, if there is a mesomorphic phase in front of the crystal growth plane, the crystal growth rate data reported over the last 40 years then represents the kinetics associated with two phase transition processes going from the supercooled liquid to the crystal. Polymers that can crystallize from both the supercooled isotropic melt and a mesomorphic phase, such as a low-ordered liquid crystalline phase with only chain orientational order, can serve as a test of Strobl's model. This test requires polymers that possess monotropic phase transition behavior, as described in Section 3.3 of Chapter 3. Because in this case the crystal melting temperature is higher than the liquid crystal isotropization temperature, the crystallization kinetics can be monitored directly from both the supercooled melt and the liquid crystalline phase. For a meaningful comparison in this study, the crystal structures formed from the melt and from the liquid crystalline phase must be identical.

It was observed that both the primary nucleation and crystal growth rates in the liquid crystal to crystal transition are several times to one order of magnitude greater than those crystallized directly from the supercooled melt (Jing *et al.*, 2002; Pardey *et al.*, 1992, 1993, 1994; for details, see Section 3.1 of Chapter 5). The implication is that polymer crystallization from the supercooled isotropic melt and liquid crystalline phase (which possesses chain and thus stem orientational order) is very different. Therefore, even if the supercooled melt in front of the growth plane exists with some degree of stem orientational order, this order is short-ranged confined to a thin interfacial layer, rather than an independent mesomorphic phase.

Another approach would be to grow polymer crystals in an anisotropic, oriented melt. A study on polyethylene crystal growth from a melt generated by melting extended chain crystals shows that the linear growth rates of polyethylene are dependent upon the residence time of this melt up to about 30 min. With increasing resident melting time, the linear growth rates decrease. This has been attributed to the low chain entanglement of melts from polyethylene extended chain crystals at short residence times (Psarski *et al.*, 2000). Because the increase in entanglement is accompanied by a loss of chain orientation in the melt, the question can then be asked: Does this set of data also imply an effect of disorientation on linear crystal growth? This experimental observation at least indicates that the residual chain (and thus the stem) orientation significantly affects the linear growth rate. Therefore, the growth rates measured in quiescent polyethylene crystallization, which are much slower than the rates observed by Psarski *et al.* (2000), must only possess weak and short-ranged stem orientational order at the crystal growth front.

An example that is well-known but difficult to study is crystallization in fiber spinning. Crystallization in fiber spinning is called "orientation-induced polymer crystallization." When the chain molecules are highly oriented, the crystallization rates are several orders of magnitude faster than the rates obtained from quiescent crystallization. The crystallization rates critically depend upon the degree of chain orientation. Attempts to use nucleation theory to describe this crystallization process suggest that almost every segment of the chain molecule has to be a nucleation site, a practically impossible event because the crystallinity of the fibers does not reach 100%. Putting aside the crystallization mechanism in fiber spinning, which is another important and separate research topic, the implication of these observations is that at the growth front in the quiescent crystallization, the chain and stem orientation in the supercooled melt, at most, has only minor orientational order and short-range lateral order.

On another research front, recent experimental results from nanocalorimetric experiments via a chip calorimeter with ultrafast cooling and heating rates exceeding several thousand degrees per second showed that the recrystallization process of polymer crystals during heating, after melting the originally formed crystals, is substantially faster than those crystals grown from the supercooled isotropic melt (Adamovsky and Schick, 2004; Adamovsky *et al.*, 2003; Minakov *et al.*, 2004, 2006). Do these results indicate that shortly after melting, the global chain conformation is still far from a completely random coil, and thus recrystallization becomes a process of crystallization from a phase that retains some degree of stem orientational order? It would be very interesting if this study could be extended to semicrystalline polymers with a racemic helical chain conformation in the crystal because these chain conformations have to follow handedness selection rules to pack into the crystal lattice. Compared with the crystal growth rates measured when directly quenched from the isotropic melt, these results might indicate that preexisting stem orientational order in the interfacial liquid zone near the crystal growth front must significantly affect the crystal growth kinetics. Again, this would indicate that the stem orientation in the liquid zone is very weak, and this zone is not an independent mesomorphic phase.

1.5. What is the nucleation barrier?

In every kinetic theory, polymer crystal growth is always associated with the concept of a nucleation barrier. Traditionally, it was solely understood to be an enthalpic consequence of the competition between the bulk free energy and surface free energy when the crystal size is small, as described in Section 1.2 of this Chapter. In the last 20 years, entropic contributions to the nucleation barrier have also gradually emerged. They have been experimentally detected and supported by computer simulation. Therefore, we will next analyze the experimental observations on different length and time scales to elucidate the enthalpic and entropic origins of the nucleation barrier (Cheng and Lotz, 2003, 2005).

1.5.1. Barriers affected by chemical defects

It is the chemical and physical periodicities in crystals that provide the shortest length scale (a fraction of a nanometer) by which the nature of the nucleation barrier can be probed. When a polymer contains sizable chemical defects, these defects must be rejected from the crystals. Only small defects with similar chemical structures may be included in the crystals. Thermodynamic descriptions of defect exclusion (Flory, 1955) and inclusion (Sanchez and Eby, 1973, 1975) in polymer crystals were proposed (see also Sanchez, 1977). A technologically important example of chain defects is metallocene-synthesized polyethylene containing short-chain branching, also called linear low-density polyethylene. Generally speaking, when the short-chain branches are methyl groups, the defects may be accommodated in the polyethylene orthorhombic crystal lattice. When the methyl group content, as measured by branches per one hundred carbons, reaches about 18%, the chains gradually lose their ability to crystallize. If the short-chain branch size on the polymer backbone increases to 1-butane, 1-hexane, or 1-octane, they are excluded from the crystal. Accordingly, their crystallinity as a function of branch content decreases more drastically than methyl-branched polyethylene. When the short-chain branch content reaches about 10%, as shown in Fig. 4.10, the polyethylene loses its crystallinity (Wunderlich, 1980), and at higher concentrations of short-chain branches, the materials behave like elastomers.

Current experimental techniques cannot probe the real-time exclusion or inclusion of each defect during crystallization, but information concerning the average concentration of defects can be obtained. Because any defect in a chain molecule included in the crystal leads to lattice expansion and any chain defect deposited onto the crystal growth front hampers the growth rate, we can monitor the crystal lattice expansion and crystal growth rates to get information about defect exclusion and inclusion. For example, polyethylene samples with short-chain branches exhibit crystal lattice expansion. This lattice expansion can be attributed to three factors: an accumulation of short-chain branch defects at the interface of the lamellar crystal fold surface (Chiu *et al.*, 2000), a decrease of the lamellar thickness (Davis *et al.*, 1974),

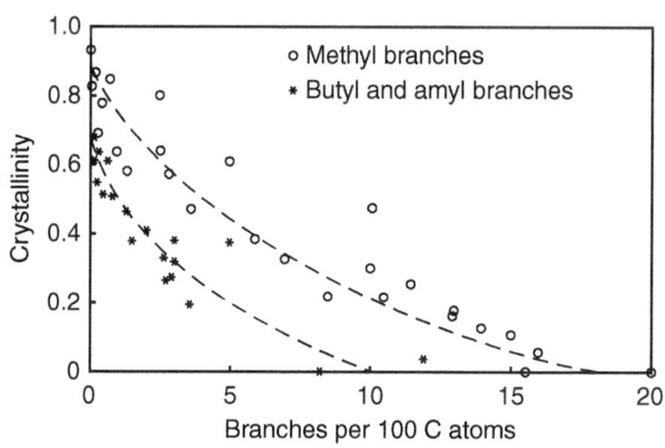

Figure 4.10 Relationship between short-chain branch content and crystallinity for polyethylene samples [replotted from Wunderlich (1980), with permission].

and the inclusion of defects into the crystals. Among these, the defect inclusion causes the largest lattice expansion. Furthermore, the lattice expands along the *a* axis several times more than along the *b* axis (Chiu *et al.*, 2000; Wunderlich, 1973). On the other hand, the slowing of growth rates occurs because crystal growth requires the defect to be removed from the crystal front (exclusion) or to form a noncrystalline point defect on the growth surface (inclusion). Both of these cases contribute to the nucleation barrier. A number of examples of defect-exclusion have been reported. In all cases, the linear crystal growth rates decrease with increasing short-chain branch content. One example of defect-exclusion during crystal growth is the slowing of linear growth rates for octane-branched polyethylene samples with increased branch content, as shown in Fig. 4.11A (Wagner and Phillips, 2001).

Another series of interesting examples includes isotactic polypropylene copolymers with short-chain branches formed by comonomers of 1-butane, 1-hexane, or 1-octane (Alamo *et al.*, 2003; Dai *et al.*, 2003; Hosier *et al.*, 2003, 2004). The existence of defects and short-chain branches provides the ability for these polymers to develop the γ-form. The overall crystallization kinetics for both the phases are defect type and concentration dependent.

Examples of defect inclusion during crystal growth include a series of isotactic polypropylene polymers with different isotacticities called "stereocopolymers" (Cheng *et al.*, 1991; Janimak *et al.*, 1991, 1992). Although the defects in these samples are not completely randomly distributed within the chains and between chains, the trend of significantly decreased crystal growth rates with decreasing isotacticity for this series of isotactic polypropylene can be observed in Fig. 4.11B. Structural analysis

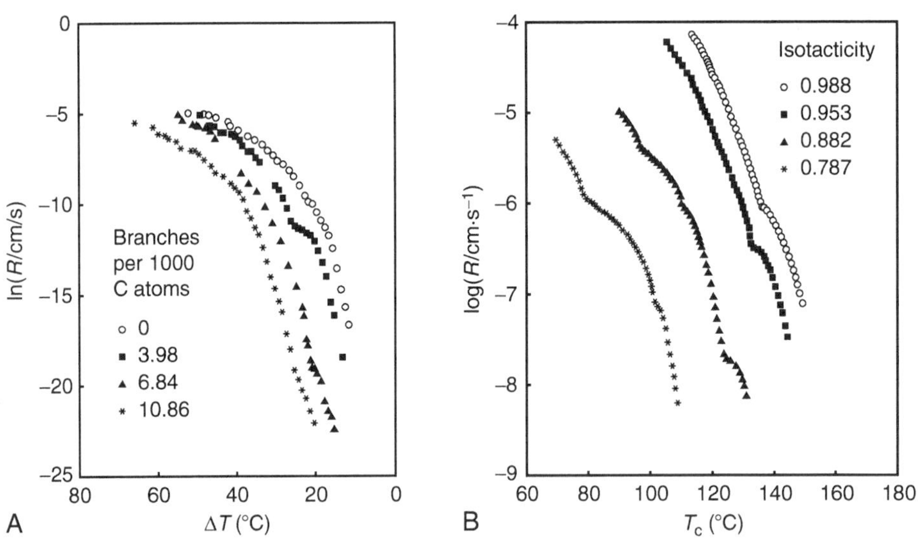

Figure 4.11 Linear growth rates of octane-branched polyethylene samples with different branching densities (A) [replotted from Wagner and Phillips (2001), with permission]; and isotactic polypropylene samples with different isotacticities (B) [replotted from Janimak *et al.* (1991), with permission].

has indicated that the stereodefects are included in the crystals; therefore, the defects significantly affect the growth rates and the nucleation barrier.

Another report was done for a series of highly regioregular isotactic polypropylenes that contained only *rr* stereoregular defects at controlled concentrations. The overall crystallization rates, which consist of the α-form initially and the γ-phase in the later stages, increase with increasing *rr* defect concentration and decreasing molecular weight. This example clearly shows that defects significantly affect the formation of the α- and γ-forms and their crystallization kinetics (De Rosa *et al.*, 2005).

1.5.2. Barriers affected by stem conformations

The next length scale important to polymer crystal growth is that of the stem conformation, which is on the length scale of a few nanometers. For example, vinyl polymer chains in crystals have helical stem conformations, which are chiral but racemic. The helical conformations can be either right- or left-handed. Additionally, the substituent group in the vinyl polymers is usually tilted with respect to the main chain axis, rather than normal to it. This defines the "up" and "down" orientations of the helices. As a result, a crystalline vinyl polymer, such as isotactic polypropylene, has four 3_1 helical conformations with identical rotational energy, calculated based on the rotational isomeric model of a single chain (Flory, 1969). The question is: How do stems with these four conformational senses pack into the crystal lattice during crystal growth? Note that because of chain folding in polymer lamellar crystals, the chain direction, including the "up" and "down" conformational sense, alternates, but the helical handedness does not necessarily change.

Polymer crystal structures with helical chain conformations determined by diffraction experiments indicate that the helical chain packing requires the precise arrangement of stems into either antichiral or isochiral packing schemes. In antichiral packing, a right-handed helical stem alternates with left-handed helical stems. On the other hand, isochiral packing requires that all the stems possess the identical handedness in a crystal structure. The unit cell lattice of the isotactic polypropylene α-phase chain packing model is monoclinic, as shown in Fig. 4.12A, and the stem handedness alternates based on the layer along the *b* axis of the unit cell, rather than on individual chains. This crystal packing thus has a coordination number of five. The arrangement of the "up" and "down" orientations of the methyl groups leads to two variants of the α-form, the α_1- and α_2-subforms (Fig. 4.12B and C). Although the packing handedness of these two subforms is identical, the α_1-phase has a random "up" and "down" arrangement, while the α_2-form arrangement is strictly alternating. This arrangement leads to two different space groups, $C2/c$ and $P2_1/c$, for these two subforms (Auriemma *et al.*, 2000; De Rosa *et al.*, 1984; Hikosaka and Seto, 1973; Napolitano *et al.*, 1990; Petraccone *et al.*, 1984, 1985; Turner-Jones *et al.*, 1964).

When a part of the helical stem with the wrong handedness is deposited onto the growth front, one of two things could occur. It could undergo a conformational transition to correct its handedness or it could be rejected back into the isotropic liquid (Lotz, 2000, 2005; Lotz *et al.*, 1991). This represents another selection process on a different length scale from the chemical defects previously described. Specifically for isotactic polypropylene, a third possibility exists but only on the lateral *ac* growth front. It could rotate 100° (or 80°) to initiate "lamellar branching" of the α-phase or

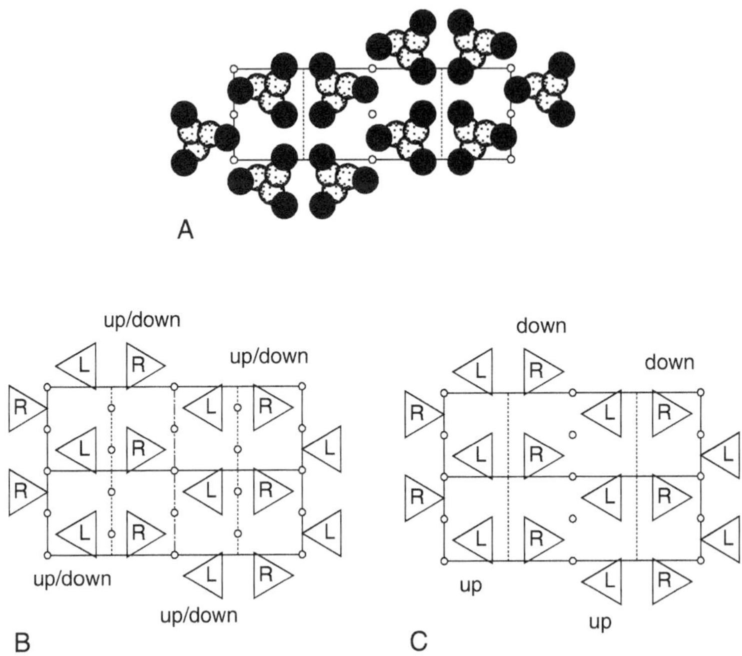

Figure 4.12 A molecular packing model of the isotactic polypropylene α-phase (monoclinic structure) with the unit cell dimensions of $a = 0.666$ nm, $b = 2.078$ nm, $c = 0.646$ nm, and $\beta = 99.62°$. In this drawing, only the "up" methyl groups are shown for simplification (A) [redrawn from Wunderlich (1973), with permission]; the α_1-phase with the random arrangement of "up"and "down"conformations (B); the α_2-phase with an alternating arrangement of "up"and "down"conformations (C). Note the different space groups in the α_1- and α_2-forms [redrawn from Cheng and Lotz(2005), with permission].

initiate the γ-phase (Brückner and Meille, 1989). Stems with the wrong "up" or "down" arrangement cannot be corrected in the solid state. Therefore, there is no solid–solid transition between the α_1- and α_2-phases. Because the selection and inclusion of the "up" or "down" arrangement requires time, the α_2-phase can only grow at low supercoolings with slow growth rates from the isotropic melt.

When crystal growth takes place at high supercoolings, where the free energy of the nucleation barrier is not a dominant factor, polymer crystal growth rates are generally fast. As a result, parts of the stems with the wrong conformation deposited onto the growth front may not have enough time to correct themselves or be rejected. This transformation causes those inevitable "mistakes" to be included in the crystal as long as the lattice can accommodate them. Although a single carbon–carbon conformational transformation takes place on the order of picoseconds, a series of sequential transformations in the solid state to correct the handedness may take much longer. When these sequential transformations cannot keep up with the crystal growth rate, as in the case of isotactic polypropylene, a "smectic" (Natta *et al.*, 1959) or "paracrystalline" phase forms in the melt (Zannetti *et al.*, 1969). In order to correct these conformational handedness defects, this "smectic" phase needs more than 18 months at room temperature to carry out the necessary cooperative conformational transformations to form the α-phase (Miller, 1960).

On the other hand, isotactic polystyrene crystals grown from solution form small, ordered clusters of gel (Atkins *et al.*, 1980; Keller, 1995). Computer simulations indicate that the correction of handedness mistakes may be possible in isotactic polypropylene crystals, but not in isotactic polystyrene crystals because the substituent phenyl group of isotactic polystyrene is much bigger (Brückner *et al.*, 2002). The inability of isotactic polystyrene to carry out conformational transitions in crystals was considered to be part of the reason for its very slow growth rates (Alemán *et al.*, 2001). Thermal correction processes for stem conformational rearrangement have been extensively reported, such as metastable crystal annealing, premelting, and/or recrystallization in semicrystalline polymers. However, almost all of the discussions are qualitative due to the difficulties of maintaining a constant level of crystal metastability. Therefore, identifying each of these annealing and reorganization processes is a difficult task.

One of the most convincing pieces of evidence for the selection process of stem conformation during crystal growth is the case of biodegradable poly(L-lactic acid) and poly(D-lactic acid). The handedness of their helical conformation is fixed by a chiral atomic center. When pure poly(L-lactic acid) or poly(D-lactic acid) crystallizes, both form the identical orthorhombic lattice with a $P2_12_12_1$ space group and an isochiral packing scheme with a lozenge single crystal habit (Sasaki and Asakura, 2003; Alemán *et al.*, 2001). However, more stable stereocomplexes can be formed using a poly (L-lactic acid)/poly(D-lactic acid) mixture with an equal composition ratio. The resulting antichiral crystal lattice is trigonal with a space group of either $R3c$ for isocline helices or $R\bar{3}c$ for a statistical "up" or "down" orientation of the helices (Cartier *et al.*, 1997). The single crystal morphology is a hexagonal habit. This example clearly illustrates that the single crystal habit embeds information concerning the unit cell symmetry, which is determined by the stem symmetry and handedness during crystal growth. The helical handedness selection process to create the lattice packing scheme is absolutely precise. Any mistakes made during stem deposition need to be corrected, and this correction process is a part of the nucleation barrier in crystal growth.

1.5.3. Barriers affected by lamellar thickness and global conformation

On the order of tens of nanometers, lamellar single crystals are the building blocks of further crystal aggregates, and lamellar thickness is the characteristic that uniquely determines crystal stability (see below for detailed discussion in Section 2.3 of this Chapter). For semicrystalline polymers, lamellar thickness has been quantitatively understood to be proportional to the reciprocal of the supercooling. Thus, the question is: Why is lamellar thickness supercooling specific? Here, we follow the explanation proposed by Armitstead *et al.* (1992). Kinetic theories presume that the growth front would have a range of possible lamellar thicknesses, each of which possesses a corresponding growth rate. The observed thickness for the crystal is, however, the thickness that allows the crystal to grow the fastest and, thus, is kinetically preferred. This is the fundamental kinetic origin of the model. All kinetic theories must possess two balancing factors: the "driving force" for polymer crystal growth and the "barrier" as shown in Fig. 4.13 (Armitstead *et al.*, 1992).

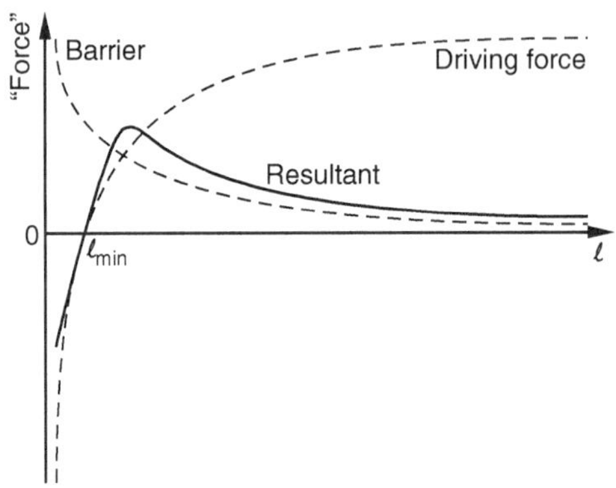

Figure 4.13 The resultant growth rate is determined by a compromise between the "driving force" and the nucleation barrier terms at a constant supercooling [replotted from Armitstead *et al.* (1992), with permission].

The "driving force" is determined by the amount of supercooling. At a given supercooling, this "driving force" also depends on the lamellar thickness, ℓ. When the lamellar thickness is smaller than the required minimum thickness, $\ell < \ell_{min}$, crystal growth will not take place because the surface free energies generated during the growth overwhelm the bulk free energy of the crystal promoted by growth. When the lamellar thickness is greater than the minimum thickness, $\ell > \ell_{min}$, the bulk free energy starts to dominate, causing the lamellae to grow thicker because the fold-surface free energy of a polymer lamellar crystal is always larger than the corresponding lateral-surface free energy. This process accelerates as the thicker lamellae provide a stronger "driving force." Therefore, if only the "driving force" determined the crystal growth, the fastest crystal growth rate would correspond to crystals with infinite thickness. However, there is another factor that needs to be taken into account.

The "barrier" must be overcome by means of a random fluctuation for a molecule to deposit onto the crystal growth front. This "barrier" increases with lamellar thickness, thus inhibiting the formation of thicker crystals. In fact, the origin of this "barrier" serves as the starting point for various kinetic theories (Armitstead *et al.*, 1992). The compromise between the driving force and barrier determines the growth rate (Fig. 4.13). The thickness that corresponds to the maximum growth rate is slightly above the minimum required thickness, ℓ_{min}, and the minimum thickness has been experimentally shown to be inversely proportional to the supercooling.

The cases of *n*-alkanes (Ungar and Keller, 1987; Ungar *et al.*, 1985) and low molecular weight poly(ethylene oxide) fractions (Arlie *et al.*, 1965, 1966, 1967, Kovacs and Gonthier, 1972; Kovacs and Straupe, 1979, 1980; Kovacs *et al.*, 1975, 1977) provide experimental evidence to illustrate the effect of global conformation on crystallization kinetics. Integral-folded chain lamellar crystals exhibit a quantized increase in lamellar thickness with decreasing supercooling. It is known that *n*-alkanes crystallize into integral-folded chain crystals at low supercoolings. In a narrow crystallization temperature region where the transition between the growth of crystals with a fold number *n* and a fold number of $(n-1)$ occurs [e.g., at the transition between once-folded chains with two stems per molecule $(n=1)$ and extended chains with one stem per molecule $(n-1=0)$], both crystal nucleation and growth rate exhibit a minimum for *n*-alkanes between $C_{162}H_{326}$ and $C_{390}H_{782}$ for crystallization from the melt and solution (Boda *et al.*, 1997; Hobbs *et al.*, 2001; Hosier *et al.*, 2000; Organ *et al.*, 1989, 1997; Morgan *et al.*, 1998; Putra and Ungar, 2003; Sutton *et al.*, 1996; Ungar and Keller, 1986, 1987; Ungar and Zeng, 2001; Ungar *et al.*, 2000, 2005). A weak minimum in the single crystal growth rates of low molecular weight poly(ethylene oxide) fractions with methoxy end groups in the melt was also reported around supercoolings where the crystal growth changes from once-folded to extended chain crystals (Cheng and Chen, 1991). The explanation for the "rate minimum" is when a chain molecule with an *n*-folded conformation deposits onto a crystal growth front that prefers to grow the $(n-1)$-folded chain crystal, this location becomes "poisoned" and hampers further growth of the $(n-1)$-folded chain crystal. Further crystal growth requires the correction of this "mistake" by either extending the folded conformation or removing this chain. Therefore, both the surface-nucleation and growth rates decrease. This is called "self-poisoning" because the mistake was made by the chain molecules themselves (Organ *et al.*, 1989; Putra and Ungar, 2003; Ungar and Keller, 1986, 1987; Ungar and Zeng, 2001; Ungar *et al.*, 2005).

Blends of two *n*-alkanes having different molecular weights showed that the growth of longer *n*-alkanes was poisoned by the deposition of shorter molecules onto the growth front, causing a rate minimum (Hosier *et al.*, 2000; Ungar and Zeng, 2001). We can extend this concept to the crystallization behavior of homopolymers composed of two different molecular weight fractions. As long as the molecular weight of one fraction is sufficiently low enough to not exactly fit on the crystal growth front, the deposition of this low molecular weight polymer onto the growth front "poisons" further growth and slows down the surface nucleation and growth rates. This process was observed in the case of a binary mixture of poly (ethylene oxide) fractions with very different molecular weights (Cheng and Wunderlich, 1986a,b, 1988). This type of "poisoning" may not generate an observable macroscopic "rate minimum," yet the molecular selection in the growth process is always present.

One can also find many experimental reports regarding the selection process in the crystallization of miscible polymer blends with one crystallizable component (Alfonso and Russell, 1986; Di Lorenzo, 2003; Mareau and Prud'homme, 2003). In this case, the phase separation is induced by the crystallization of the crystalline component. The probability concerning which component deposits onto the growth front at any particular time depends on the local concentration of that component. The selection process takes place at the growth front, and only the crystallizable component can be used for crystal growth. This selection process significantly decreases or sometimes completely stops the growth. The noncrystallizable component is now "poisoning" the growth front and reducing the crystal growth rate. Similar cases can also be found in the crystallization of crystalline–amorphous diblock copolymers in solution (Chen *et al.*, 2004a).

1.5.4. Barriers affected by the physical environment

During polymer crystal growth, it has been observed that enhanced growth rates can be achieved at the reentrant corners of twin crystals (Frank, 1949; Stranski, 1949), in particular in *n*-alkanes (Dawson, 1952; Khoury, 1963) and polyethylene (Bassett and Keller, 1962; Khoury and Padden, 1960). Although published in a journal with limited readership, a comprehensive investigation of polyethylene twin-crystal growth was carried out by Wittmann and Kovacs (1970). When polyethylene single crystals grow in dilute solution, {110} twinning often occurs. As shown in Fig. 4.14A, for a 100 kg/mol molecular weight sample crystallized at 111.8 °C from a tetradecanol solution, the {110} twin results in a 112.6° reentrant corner. This corner is bounded by two (200) planes when the single crystal has a truncated lozenge shape, which is bounded by four (110) and two (200) planes. This reentrant corner enhances polyethylene crystal growth [note that during growth, the angle gradually increases due to the curvature of the (200) planes] (Sadler *et al.*, 1986). This enhanced crystal growth occurs because the corner provides a less than 180° growth front, thus reducing the lateral-surface free energy, or enhancing the growth rate, when the polyethylene chain deposits at the corner. Enhanced polyethylene growth rates at the reentrant corners of twinned crystals were also reported by Colet *et al.* (1986) as well as Toda and Kiho (1989). The growth enhancement was found to become increasingly pronounced when the single crystals formed twins, which were only bounded by four (110) planes. The growth is now along the (200) normal direction. This enhanced growth is due to the 134.8° reentrant angle between the two (110) planes, which is larger than the reentrant angle formed by two (200) planes, as shown in Fig. 4.14B (Sadler *et al.*, 1986; Wittmann and Kovacs, 1970). Based on calculations of the Hoffman–Lauritzen theory that only take the reduced

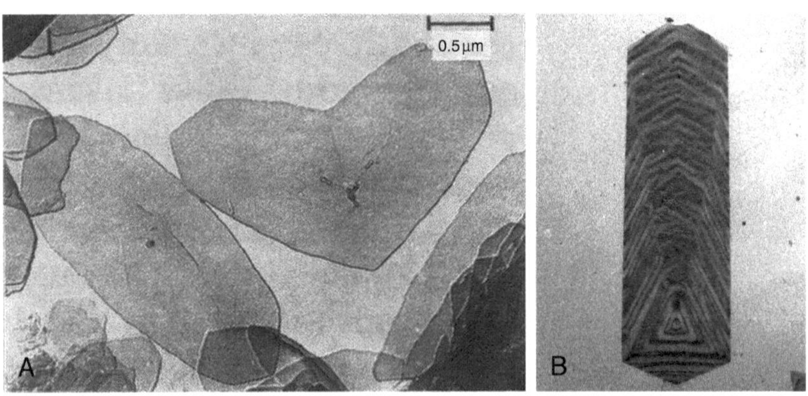

Figure 4.14 Bright field transmission electron microscope images of polyethylene crystal growth at a reentrant corner, constructed by a {110} twin in tetradecanol solution with a molecular weight of 100 kg/mol at 111.8 °C (A) [reprinted from Sadler *et al.* (1986), with permission]; and a bright field image at a reentrant corner constructed by a {110} twin in tetrachloroethylene solution (0.5%) for a polyethylene with a molecular weight of 5 kg/mol, after self-seeding at 92.5 °C and crystallization at 63.5 °C (the initial crystallization temperature was at 50 °C). The growth along the *a* axis is much faster than that of the (110) normal direction resulting in the elongated twin morphology (B) [reprinted from Wittmann and Kovacs (1970), with permission].

lateral surface free energy into account, the growth rate at the corner should increase by several orders of magnitude. However, experimental data show that the growth rate enhancement was much less than that predicted by the theory (Sadler *et al.*, 1986). This reduced enhancement maybe due to the constraint of limited space for molecular motion in front of the corner and the requirement of lattice matching at the corner. These studies clearly indicate that the geometric confinement plays a significant role in crystal growth.

When polyethylene crystallizes in the melt at low supercoolings, single crystals can grow, as shown in Fig. 4.15A (Bassett *et al.*, 1988). In this figure, the large mother single crystals developed first with a symmetrical lenticular-shape bounded by two long, curved (200) growth fronts. Smaller daughter single crystal terraces grew on the mother crystal substrate because of screw dislocations on the top and bottom of the substrate. The daughter crystals, therefore, have slower growth rates than the mother crystal due to confinement by the substrate. It is worth noting further that the shape of these daughter crystals is asymmetric along the two opposite (200) normal directions, indicating that the growth rates along both the (200) normal directions are not equal. This inequality is due to the fact that the polyethylene chains in the single crystals are usually tilted between 20° and 35° toward the (200) normal direction. The highest tilting angle reported has been 45° (see, e.g., Lotz and Cheng, 2005). This tilting angle generates two growth fronts with one inclined less than 90° (acute) and the other greater than 90° (obtuse) with respect to the substrate. Because the chains can more easily access the obtuse growth front, the growth rates along the obtuse (200) normal direction are faster than along the confined acute (200) normal direction (Bassett *et al.*, 1988). The inequality of the crystal growth rates along the (200) normal direction can also be clearly seen from the asymmetric single crystal habit formed by polyethylene in the thin film melt, as shown in Fig. 4.15B

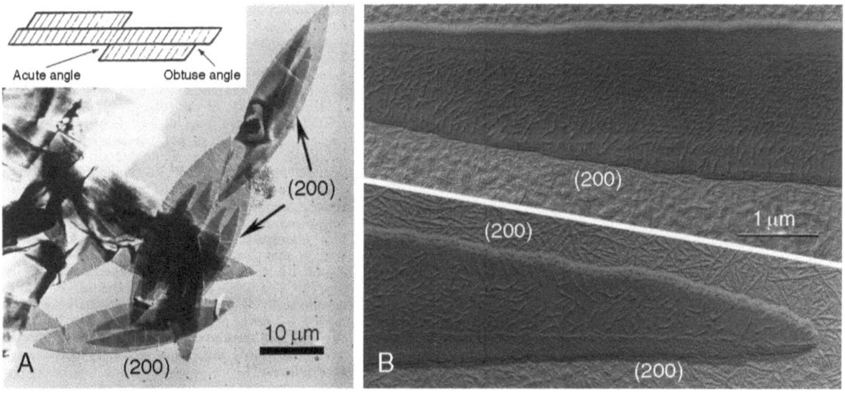

Figure 4.15 Anisotropic crystal growth along the [200] direction of polyethylene single crystals formed by screw dislocations on their mother polyethylene single crystals in the melt. The insert illustrates the chain tilting effect on the crystal growth on both sides of the curved (200) planes (A) [reprinted from Bassett *et al.* (1988), with permission]; and asymmetry of the single crystal habit grown in the polyethylene thin film melt (B) [reprinted from Keith *et al.* (1989), with permission].

(Keith *et al.*, 1989). In this case, the substrate causes significant confinement of the acute polyethylene crystal growth front. The asymmetric growth along the two (200) normal directions is even more clearly identified after lamellar decoration (detailed description of this method will be given in Section 4.1 of Chapter 5).

From a broader perspective, a new look at the nucleation barrier has been presented. Multiple selection processes on different length and time scales that take place during crystallization affect the growth kinetics. We have grouped them together and referred to them as the nucleation barrier. Various structural and morphological observations indicate the selection processes' effects on the nucleation barrier. The selection rules arise from sorting out variations in chemical structure, molecular conformation, molecular weight, and morphology. The greater the variation among these parameters, the stronger the effect of the sorting processes on the growth rate. The combination of these sorting processes creates a nucleation barrier that includes both enthalpic and entropic contributions. It is qualitatively expected that the traditional free energy enthalpic contribution dominates short range interactions at smaller length scales and that the entropic sorting processes are at larger length scales. Although our understanding of solid state structures has progressed significantly, issues remain concerning the structure and dynamics of the supercooled melt in the bulk, as well as near the interface, and their effects on the nucleation and growth processes.

2. Superheated Crystals and Crystal Melting

2.1. Superheated crystals

Thermodynamically, when the temperature is above the equilibrium melting point at a constant pressure, the Gibbs free energy of the crystal is higher than that of the liquid, resulting in a driving force for melting the crystal to the liquid. However, if we heat the crystal above its equilibrium melting temperature and it survives for a certain period of time, this crystal is now superheated. The melting rates and growth rates of a crystal are not equivalent for an equal amount of superheating or supercooling, respectively. As described in Section 2.2 of Chapter 2, when the crystal melting occurs near and above the equilibrium melting temperature, it starts as a surface (heterogeneous) melting process. In this case, the "surface nucleation" sites are generally at the corners and ledges of the crystal. After passing the superheating limit, homogeneous "nucleation" takes place when the crystals are strongly superheated. If the crystals are of infinite size (surface-free crystals), the melting occurs when the superheated crystal spontaneously generates a sufficiently large number of spatially correlated destabilized lattice locations. The accumulation and coalescence of these internal, local lattice instabilities constitute the primary mechanism for homogeneous "melt-nucleation."

Polymer crystal melting with superheating usually means that a metastable polymer crystal instead of an equilibrium polymer crystal is melting; thus, it is experimentally more difficult to monitor. The reason is that polymer folded chain crystals are thin. They generally melt below the equilibrium melting temperature, and the associated kinetics are very fast. Moreover, the original metastability of the crystal must remain unchanged up to the onset of melting. Investigating polymer crystal melting under superheating thus requires systems with significant crystal stability. Major experimental approaches have focused on generating large single (surface-free) crystals or constructing solid-confined environments around small crystals reducing their susceptibility to melting.

How do we grow crystals with thicknesses on the length scale of micrometers without chain folding? Two general approaches are used: either crystallizing long chain polymers to form extended chain crystals or crystallizing them during polymerization to avoid generating chain folding. The most well-known example of the first approach is polyethylene crystals formed at pressures above 4.5 kbar at elevated temperatures (Geil *et al.*, 1964; Wunderlich and Arakawa, 1964) (see Section 2.2 of Chapter 5). A few other polymers can also form extended chain crystals under either elevated pressures or high-temperature annealing, such as polytetrafluoroethylene (Bunn *et al.*, 1958; Melillo and Wunderlich, 1972), polychlorotrifluoroethylene (Miyamoto *et al.*, 1972), and *trans*-1,4-polybutadiene (Rastogi and Ungar, 1992). The lamellar thickness of these extended chain crystals can be over 1 μm. The formation mechanism of the extended chain crystals in these polymers is critically associated with their phase structure during crystallization. A columnar phase that has relatively perfect long-range hexagonal lateral packing, yet only short-range order along the chain direction is necessary (see Fig. 2.8B of Chapter 2). As a result, large mobility along the chain direction can be achieved to unfold the crystals and increase the lamellar thickness. A detailed discussion of this phenomenon will be given in Section 2.2 of Chapter 5.

The second approach is crystallization during polymerization. This approach requires relatively large mobility of monomers, such as in the gaseous or liquid state. A few solid state examples have also been reported. These monomers deposit onto the crystal lattice with a specific molecular orientation, and polymerization takes place to form large crystals without folds (Wunderlich, 1968a,b). The most well-known examples are poly(diacetylene) (Wegner, 1969, 1972, 1979), polyoxymethylene (Iguchi, 1973; Iguchi *et al.*, 1969; Mateva *et al.*, 1973), and poly-*para*-xylylene (Kubo and Wunderlich, 1971, 1972). In recent years, a series of aromatic polyesters (Liu *et al.*, 1996b, 1998), poly(*p*-phenylene terephthalamide) (Liu *et al.*, 1996a), and polyimides (Liu *et al.*, 1994a,b,c) have also been reported to crystallize during polymerization from the monomer melt. However, obtaining large, perfect crystals by crystallization during polymerization is critically dependent upon monomer concentration and on nucleation and growth at different temperatures and pressures. It also requires careful control of the nucleation step to prevent irregular overgrowth or the development of polycrystalline morphologies.

Most semicrystalline polymers, however, cannot form extended chain crystals using either one of the approaches described above. The question then becomes: Can we generate superheated polymer metastable crystals at small crystal sizes? The answer can be realized by providing confined solid environments to limit the volume expansion of these small crystals at the melting temperature. The first example of this approach used the crystallization of low molecular weight poly(ethylene oxide) fractions with narrow molecular weight distributions. These poly(ethylene oxide) fractions form integral-folded-chain crystals in the bulk as described in Section 1.5 of this Chapter (Arlie *et al.*,

1965, 1966, 1967; Kovacs and Gonthier, 1972; Kovacs *et al.*, 1975, 1977; Kovacs and Straupe, 1979, 1980). In the specific case of a poly(ethylene oxide) with a molecular weight of 6 kg/mol, a two-step crystal growth was carried out. The fraction was first crystallized in a highly supercooled melt to form single crystals with twice-folded chains. (The fold number was two with three stems per chain molecule). In the second step, the supercooling of the liquid was decreased by increasing the crystallization temperature. Once-folded chain lamellar growth then ensues at the periphery of the twice-folded chain single crystals. The resulting lamellar single crystals had the twice-folded chain crystals surrounded by the once-folded chain crystal, as shown in Fig. 4.16. The composite single crystal was thinner in the center (thus, having a lower melting temperature) and thicker around the edge (with a higher melting temperature). When the temperature was increased to the melting temperature of the twice-folded chain crystal, the thinner center part remained unaffected and did not melt. Only after the temperature was further increased to completely melt the once-folded chain crystal did the twice-folded chain crystal relax into the isotropic melt. This process is clearly illustrated by the sequential decrease in single crystal lateral size during melting, as shown in Fig. 4.16 (Kovacs *et al.*, 1977).

A recent report for another poly(ethylene oxide) fraction with a molecular weight of 4.25 kg/mol describes a detailed investigation of the morphological evolution of superheated crystals on a mica substrate (Zhu *et al.*, 2006). In this case, a superheated thrice-folded crystal (a fold number of three with four stems per chain molecule) was obtained with the entire periphery of the crystal surrounded by twice-folded chain growth. The thicker lamellar peripheral growth remained unchanged at the melting

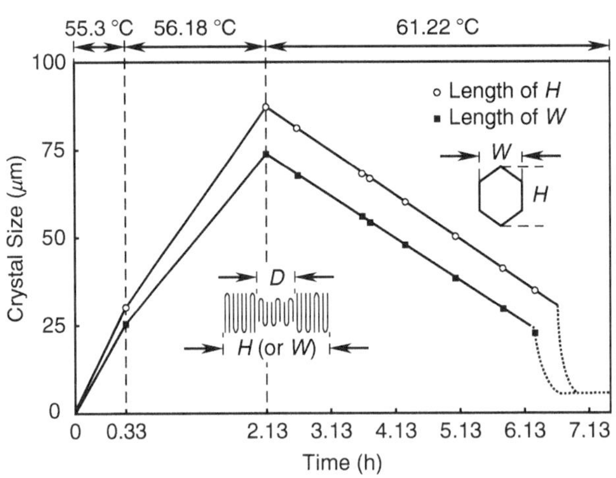

Figure 4.16 Change of lateral size in poly(ethylene oxide) single crystals during two-step crystallization and melting. The poly(ethylene oxide) fraction has a molecular weight of 6 kg/mol. The first step forms the twice-folded chain crystal at 55.3 °C, and then, the second step grows the once-folded chain crystal at 56.18 °C. The composite crystal is then heated to 61.22 °C. The twice-folded chain crystal is not affected until the once-folded crystal is melted [reprinted from Kovacs *et al.* (1977), with permission].

temperature of the unconfined thrice-folded chain crystal, and the thinner core only started thickening at the superheated conditions (Zhu *et al.*, 2006).

Lamellar thickness variation in different sectors of polyethylene has been suggested for a long time (Bassett *et al.*, 1959; Runt *et al.*, 1983). It has been recently reported that in solution-grown truncated lozenge-shaped polyethylene single crystals, the thickness of the {110} sector is over 1 nm thicker than the {200} sector after chain tilting in the different sectors has been taken into account. This thickness difference leads to about a 1 °C melting temperature difference between these two sectors as measured by real-time, temperature-controlled atomic force microscopy experiments (Hocquet *et al.*, 2003). However, in order to observe superheated crystals, we need to build up thicker crystals around the thinner ones. This procedure can be carried out by using the isochronous decoration method developed by Point *et al.* (1986) and Dosière *et al.* (1986). The thinner lamellae are confined within the thicker lamellae, and their melting temperature can, in principle, be experimentally determined. Unfortunately, this experiment has not been done.

In another case, sectorized rectangular-shaped single crystals of syndiotactic polypropylene were grown from the melt; the {100} and {010} sectors are shown in Fig. 4.17A (Bu *et al.*, 1996; Lotz *et al.*, 1988; Lovinger *et al.*, 1990, 1991, 1993; Zhou *et al.*, 2000, 2003). These two sectors have different thicknesses. The {010} sector is several nanometers thinner than the {100} sector. The thickness differences can be as large as 15–20% (Bu *et al.*, 1996; Zhou *et al.*, 2000). When the single crystal is at a temperature that is slightly higher than the melting temperature of the thinner {010} sector, the thinner sector starts to melt at a relatively slow rate. Simultaneously, isothermal thickening also takes place in this sector forming a "dam" (the boundaries drawn in Fig. 4.17B). Because the two-dimensional area of the crystal becomes fixed and no lattice expansion in the lateral direction is allowed, the remaining thin {010} sector cannot melt further (as shown in Fig. 4.17B). Even

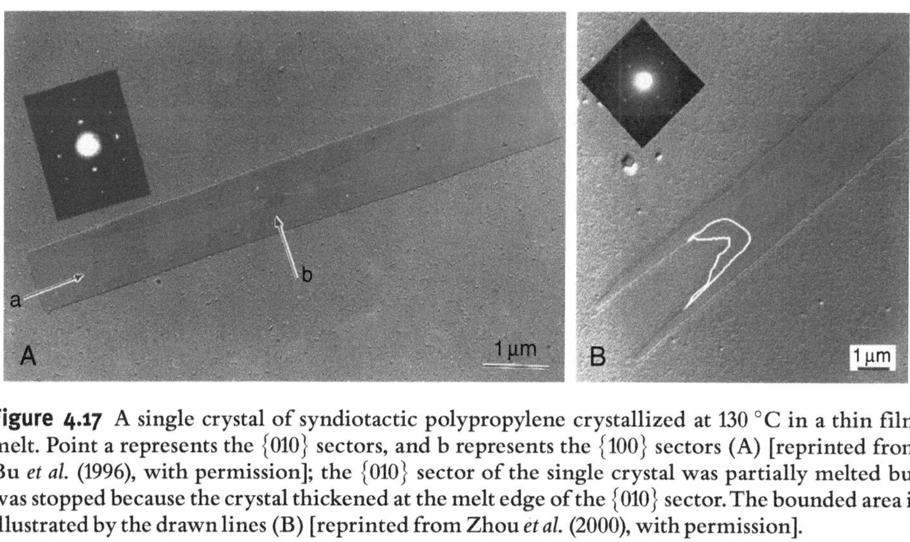

Figure 4.17 A single crystal of syndiotactic polypropylene crystallized at 130 °C in a thin film melt. Point a represents the {010} sectors, and b represents the {100} sectors (A) [reprinted from Bu *et al.* (1996), with permission]; the {010} sector of the single crystal was partially melted but was stopped because the crystal thickened at the melt edge of the {010} sector. The bounded area is illustrated by the drawn lines (B) [reprinted from Zhou *et al.* (2000), with permission].

after further heating of this crystal to several degrees above the melting temperature of the thinner {010} sector, the thinner part does not melt.

All of these examples clearly indicate that attaining equilibrium crystals is not necessary for demonstrating crystal superheating. Thin polymer single crystals can also be superheated while confined by thicker crystal edges. The melting temperatures of these single crystals are not, by any means, the thermodynamic equilibrium melting temperatures.

Recently, using calorimetric experiments, Toda *et al.* reported that the increase of the melting temperature with increasing heating rate is attributed to crystal superheating for several commodity polymers, such as polyethylene with short-chain branches, isotactic polypropylene, poly(ethylene terephthalate), and others (2002). More recently, nanocalorimetric experiments with ultrafast heating rates were conducted using a chip calorimeter to study superheating phenomena in metastable polymer lamellar crystals (Minakov *et al.*, 2007). The cooling and heating rates reached 100 k °C/sec. A power law between the ultimate superheating of the crystals and heating rate was observed in the heating rate range of 0.01 °C/sec to 10 k °C/sec. Upon further increase of the heating rate, the power law is no longer valid, and the superheating of the crystals reaches an upper limit. This superheating limit was observed to be critically associated with annealing conditions (Minakov *et al.*, 2007), or more precisely, with the metastability of the crystals. This ceiling may be the transition from heterogeneous melting to homogeneous melting for a given metastability of a polymer crystal and is in contrast to the case of an infinitely large equilibrium crystal where the superheating limit is constant.

Other approaches can also generate superheated crystals in polymers, but they are less controlled. From a thermodynamics point of view, as shown in Fig. 3.9 of Chapter 3, a strained isotropic liquid has its Gibbs free energy line shifted to higher temperatures, thus the melting temperature of the crystals increases. Some methods to achieve this melting temperature increase include applying a fast heating rate to a fiber sample under external tension to fix the fiber length or to heat the sample after chemical cross-linking. However, understanding crystal melting in highly oriented fibers is particularly difficult. This difficulty is not only due to the different ways that the samples are prepared (stretched versus unstretched, constrained versus unconstrained), but also due to the different processing histories. Some of the seminal work on ultrahigh molecular weight polyethylene fibers (Kwon *et al.*, 2000; Ratner *et al.*, 2004), poly(ethylene terephthalate) fibers (Miyagi and Wunderlich, 1972), polyamide 6 (Todoki and Kawaguchi, 1977a,b), and other semicrystalline polymers reflect the efforts to quantitatively understand crystal melting under superheating.

2.2. Irreversible polymer crystal melting

Although the thermodynamics of equilibrium crystal melting is straightforward, in almost every case, polymer crystal melting does not occur in equilibrium. This type of crystal melting is called irreversible melting to distinguish it from melting at thermodynamic equilibrium. Investigations concerning irreversible melting phenomena are much more difficult than those aimed at equilibrium crystal melting. Experimentally measured melting temperatures, heats of fusion, and other thermodynamic parameters can have very different values for semicrystalline polymers with identical chemical structures but exposed to different thermal and mechanical histories, provided that the polymer crystal melting processes are not at thermodynamic equilibrium.

Traditionally, studying an irreversible system using nonequilibrium thermodynamics requires significant effort. This problem can be approached by dividing the system into many subsystems, with each subsystem connected to the neighboring subsystems. However, how the system is divided into subsystems is a critical issue. The subsystems must be divided such that within each subsystem, equilibrium thermodynamics can be applied, or approximately applied, to describe its thermodynamic properties. To connect the subsystems, a set of boundary conditions is required. This process of developing subsystems and setting up boundary conditions is often very difficult to accomplish. However, if it can be done, the extensive thermodynamic properties of the whole system, such as volume and energy, can be determined by simply summing up these properties in each of the subsystems. This type of complicated analysis is not amenable to many systems.

We also know that metastable crystals exhibit irreversible melting. These metastable crystals can be isobarically annealed, reorganized, and/or recrystallized into more stable crystals during heating. The poorer the crystals, the more effective these annealing, reorganization, and recrystallization processes will be. When heating a system, one needs to be aware of metastability changes. Changes in metastability result from changes in the system's entropy, making it nearly impossible for the system to be analyzed thermodynamically, even when utilizing the complicated subsystem approach described. The questions then become: Can we find a relatively simple way to treat irreversible polymer crystal melting with equilibrium thermodynamics, and how do we design our experiments to make this kind of treatment possible? In order to answer these questions, we must find a structural or morphological parameter that can be correlated with crystal stability.

2.3. Determining crystal metastability

For general crystalline materials, a relationship between melting temperature (stability) and crystal size was derived almost 150 years ago; it is the Thomson–Gibbs equation:

$$T_{\mathrm{x}} = T_{\mathrm{x}}^{0}\left[1 - \left(\frac{\langle \gamma_{\mathrm{x}} \rangle V_{\mathrm{m}}}{\langle \ell \rangle \Delta h_{\mathrm{x}}}\right)\right] \tag{4.5}$$

where the subscript x represents the type of phase transition; T_{x} is the phase transition temperature for a phase of limited size and T_{x}^{0} is the phase transition temperature for a phase with infinite size. The average surface free energy is represented by $\langle \gamma_{\mathrm{x}} \rangle$, and Δh_{x} is the heat associated with this specific transition. The molar volume is V_{m}, and $\langle \ell \rangle$ is the average crystal dimension. This equation indicates that the phase transition temperature is indeed phase-size dependent.

Polymer crystals are lamellar in form, meaning that they have a large surface area but are relatively thin. It is known that lamellar crystals can have a range of thicknesses that are all metastable compared to equilibrium extended chain crystals. Furthermore, the thicker the lamella, the higher the melting temperature will be, so the crystal is more stable. The correlation between the melting temperature and crystal size for polymer crystals can then be simplified, based on Eq. (4.5), to the following form:

$$T_m = T_m^0\left[1 - \left(\frac{2\gamma_e}{\ell\Delta h_f}\right)\right] \tag{4.6}$$

where T_m represents the melting temperature of the crystal at a thickness of ℓ; T_m^0 is the equilibrium melting temperature of the corresponding crystal at infinite size; Δh_f is the equilibrium heat of fusion; and γ_e is crystal surface free energy. Therefore, if both the melting temperature of the crystal (T_m) and the crystal thickness (ℓ) can be experimentally determined, a correlation between the melting temperature and lamellar thickness can be established. Then, a plot of the melting temperature versus the reciprocal of lamellar thickness can be drawn. By extrapolating to a reciprocal lamellar thickness of zero, where the crystal is infinitely thick, the equilibrium melting temperature of the crystal can be obtained, as shown in Fig. 4.18. The slope for this melting temperature depression with respect to the reciprocal thickness is $-2T_m^0\gamma_e/\Delta h_f$.

Using small angle x-ray scattering for bulk samples or a microscopy technique for single crystals, the lamellar thickness can be obtained at room temperature in most cases or in real-time at different crystallization temperatures. Melting temperature measurements carried out using calorimetric methods require heating the crystals to high temperatures at a constant heating rate. It is inevitable that when the heating temperature exceeds the crystallization temperature, annealing, reorganization, and/or recrystallization will take place changing the metastability of the crystal. This is particularly true for thin crystals with low stability. The issue of

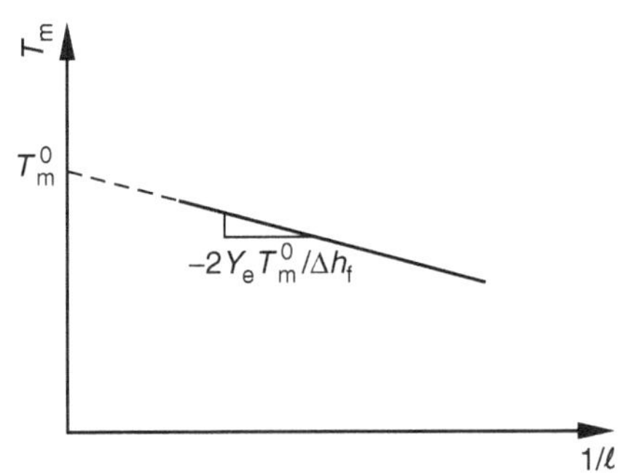

Figure 4.18 Schematic illustration of the extrapolation of the relationship between the melting temperature and reciprocal lamellar thickness. When the reciprocal lamellar thickness goes to zero (infinite thickness), the extrapolated melting temperature is the equilibrium melting temperature.

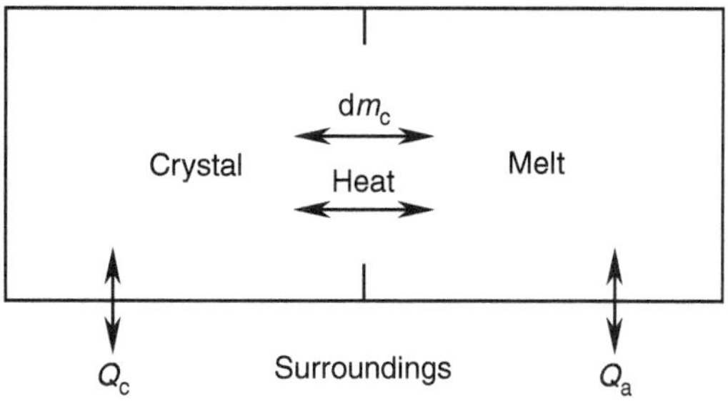

Figure 4.19 Schematic illustration of the melting and crystallization process of a lamellar crystal in a closed system with its surrounding [redrawn from Wunderlich (2005), with permission].

concern is the consequence of metastability changes when using the Thomson–Gibbs equation.

Let us first take a thermodynamic approach to analyze the effect of changing metastability (lamellar thickness) via annealing, reorganization, and/or recrystallization on the Thomson–Gibbs equation. The illustration in Fig. 4.19 follows the work published many years ago by Wunderlich (1980). Figure 4.19 is a schematic drawing of a crystalline solid–liquid transition during the heating or cooling of a closed system with only energy crossing the system boundary, such as in differential scanning calorimetry experiments. In this system, we assume the coexistence of two phases (the crystal and isotropic melt), and they are treated as an open system because both the mass and energy can be exchanged between them. Based on thermodynamics, both phases possess their own Gibbs free energies per unit volume, g_c and g_a, as well as their own masses, m_c and m_a. Assuming that this system is small enough to have one lamellar crystal with a thickness ℓ, the fold surface free energy is not negligible and is labeled γ_e. Therefore, the total Gibbs free energy for the lamellar crystal in this system is

$$G_c = m_c g_c + \frac{2m_c\gamma_e}{\rho\ell} \tag{4.7}$$

where ρ represents the crystal density. After completely melting the lamellar crystal in this system, the total Gibbs free energy for the isotropic melt is

$$G_a = m_a g_a \tag{4.8}$$

The driving force to melt the lamellar crystal is:

$$\Delta G = G_a - G_c = m_a g_a - m_c g_c - \frac{2m_c\gamma_e}{\rho\ell} \tag{4.9}$$

The heat exchange between the lamellar crystal and the surroundings is Q_c, and the heat exchange with the isotropic melt is Q_a. On lamellar crystal melting, the bulk Gibbs free energy change is $\Delta g_f = g_a - g_c$. Furthermore, overall mass is conserved,

$-\mathrm{d}m_c = \mathrm{d}m_a$. In crystal growth, the surroundings absorb the heat from the system, and the mass change is $\mathrm{d}m_c = -\mathrm{d}m_a$.

The entropy gain of the lamellar crystal generated by the heat exchange with the surroundings at constant temperature and pressure during melting at time $\mathrm{d}t$ can be expressed as:

$$\mathrm{d}S_e = \frac{\mathrm{d}Q_c + \mathrm{d}Q_a}{T} \tag{4.10}$$

This term can be measured by calorimetric experiments. On the other hand, because at a constant temperature all the enthalpy changes caused by crystal melting or crystallization are compensated for by heat exchange with the surroundings, the enthalpy production is zero. The entropy production within the system can thus be obtained by taking the first derivative of Eq. (4.9):

$$\mathrm{d}S_p = \frac{\Delta g_f \mathrm{d}m_c}{T} - \frac{2\gamma_e \mathrm{d}m_c}{T\rho\ell} + \frac{2m_c\gamma_e \mathrm{d}\ell}{T\rho\ell^2} \tag{4.11a}$$

Note that the quantity of the entropy produced, $\mathrm{d}S_p$, is not directly measurable. In Eq. (4.11a), the first two terms on the right side represent the lamellar crystal melting and the third term describes the change of lamellar thickness, therefore, representing the lamellar crystal reorganization. A further assumption is made that the temperature is not too far from the melting temperature. A still further assumption is that the driving force for the bulk crystal is $\Delta g_f = \Delta h_f \Delta T / T_m^0$. Equation (4.11a) can then be written as:

$$\mathrm{d}S_p = \frac{\Delta h_f \Delta T \mathrm{d}m_c}{T T_m^0} - \frac{2\gamma_e \mathrm{d}m_c}{T\rho\ell} + \frac{2m_c\gamma_e \mathrm{d}\ell}{T\rho\ell^2} \tag{4.11b}$$

where T_m^0 is the equilibrium melting temperature and ΔT is the supercooling. Based on the second law of thermodynamics, the entropy production must be equal to or larger than zero, $\mathrm{d}S_p \geq 0$.

Now let us conduct a detailed analysis of Eq. (4.11b). If the lamellar crystal is very thick, such as in the case of extended chain crystals of polyethylene, the second and third terms of equation (4.11b) approach zero. Because it deals with a crystal of effectively infinite size and negligible crystal surface area, this case is simply that of an equilibrium crystal either melting during heating or crystallizing during cooling.

$$\mathrm{d}S_p = \frac{\Delta g_f \mathrm{d}m_c}{T} = \frac{\Delta h_f \Delta T \mathrm{d}m_c}{T T_m^0} \geq 0 \tag{4.12}$$

If the entropy change in the system is zero, it represents equilibrium crystal melting and crystallization. If the crystal, due to superheating, melts above the equilibrium melting temperature, both ΔT and $\mathrm{d}m_c$ are smaller than zero leading to positive entropy production. This melting process is allowed by the second law of thermodynamics. With supercooling ($\Delta T = T_m^0 - T_x > 0$), all the parameters in Eq. (4.12) are positive; thus, the change in entropy is larger than zero. Therefore, crystallization is also permitted by the second law of thermodynamics.

The results are different when the crystal lamellar thickness is small. Let us first analyze the case when the crystal neither crystallizes nor melts, $dm_c = 0$. In Eq. (4.11b), the only term left is the third term on the right side of the equation, indicating that the crystal only changes by thickening. It is evident that the thickness increase during reorganization leads to a positive change, $d\ell > 0$, and therefore the entropy change is positive because all other parameters are larger than zero. This outcome indicates that the lamellar crystal can only increase its thickness, or at most, maintain its thickness based on thermodynamic stability, which fits what has been experimentally observed.

An alternative case is that of a thin lamellar crystal without reorganization. This case is most interesting to us because we are dealing with the metastability change of thin lamellar crystals. In this case, the third term on the right side of Eq. (4.11b) is zero, and thus:

$$dS_p = \left[\frac{\Delta g_f}{T} - \frac{2\gamma_e}{T\rho\ell}\right] dm_c = \left[\frac{\Delta h_f \Delta T}{T T_m^0} - \frac{2\gamma_e}{T\rho\ell}\right] dm_c \geq 0. \qquad (4.13)$$

The second term in the bracket on the right side of Eq. (4.13) is always positive. For crystallization, the first term in the bracket on the right side of this equation is also positive. However, because dm_c is larger than zero for crystallization, we need to have

$$\frac{\Delta h_f \Delta T}{T T_m^0} > \frac{2\gamma_e}{T\rho\ell} \qquad (4.14)$$

Here, crystallization can only occur if there is a sufficiently large supercooling to meet the requirement of Eq. (4.14). On the other hand, for thin lamellar crystals, melting below the equilibrium melting temperature only occurs when Eq. (4.14) is true. In this case, the entropy change is zero, and the metastable, thin lamellar crystal converts directly to the isotropic melt without a change in its metastability. This case further indicates that the melting of a thin lamellar crystal can be described using equilibrium thermodynamics when its metastability does not change during heating. As a result, it provides a useful analytical tool to understand thin lamellar crystal melting and avoids the complicated treatment of subdivided systems for irreversible melting.

Equation (4.14) leads to a more rewarding approach. At the melting temperature ($T = T_m$) at which the thin lamellar crystal melts without a change in metastability, the inequality sign of Eq. (4.14) becomes an equal sign. This equation can be rewritten as:

$$\Delta T = T_m^0 - T_m = \frac{2\gamma_e T_m^0}{\Delta h_f \rho\ell} \qquad (4.15)$$

which is equivalent to Eq. (4.6). This analytical process is the derivation of the Thomson–Gibbs equation.

It is evident that when we use the Thomson–Gibbs equation to obtain the equilibrium melting temperature, as shown in Fig. 4.18, several issues need to be carefully considered. Annealing, reorganization and recrystallization must be avoided during heating. These processes lead to an overestimation of the melting temperature of thin lamellar crystals. A relatively minor issue is the overestimation of the melting temperature of very thick lamellar crystals due to superheating. The issue then becomes how to maintain the metastability of a crystal during heating.

Another issue of equal importance is how to use the Thomson–Gibbs equation to describe the crystal size and stability of crystals with drastically different shapes (habits), such as the elongated (fibril) crystals, which are one-dimensionally thick (along the chain direction) but laterally small (in the directions perpendicular to the chain axis) or three-dimensionally small crystals? Equation (4.6) cannot be directly applied to these two cases. New versions of the Thomson–Gibbs equation must be derived for the elongated and three-dimensionally small crystals. For the elongated crystals, the equation can be modified as:

$$T_{\mathrm{m}} = T_{\mathrm{m}}^{0}\left[1 - \left(\frac{4\gamma}{a\Delta h_{\mathrm{f}}}\right)\right] \tag{4.16}$$

where γ represents the lateral surface free energy and a is the lateral size of the elongated crystal. The other variables are identical to those in Eq. (4.6). A plot of the melting temperature versus the reciprocal of the lateral size, a, is shown in Fig. 4.20. However, the slope of the melting temperature with respect to the lateral size is $4T_{\mathrm{m}}^{0}\gamma/\Delta h_{\mathrm{f}}$.

For three-dimensionally small crystals with equal lateral sizes of a and a thickness of ℓ, the Thomson–Gibbs equation can be modified to be:

$$T_{\mathrm{m}} = T_{\mathrm{m}}^{0}\left[1 - \frac{(2\gamma_{\mathrm{e}}/\ell + 4\gamma/a)}{\Delta h_{\mathrm{f}}}\right] \tag{4.17}$$

Now the plot must be three-dimensional because there are two extrapolations that need to be carried out. One is between the melting temperature and the reciprocal crystal thickness; the other is between the melting temperature and the reciprocal lateral size as shown in Fig. 4.21. The slopes of these two planes are $2T_{\mathrm{m}}^{0}\gamma_{\mathrm{e}}/\Delta h_{\mathrm{f}}$ and $4T_{\mathrm{m}}^{0}\gamma/\Delta h_{\mathrm{f}}$, respectively. In the three-dimensional plot, the plane represents these extrapolations. Maintaining the metastabilities of these small crystals is not easy.

Based on this analysis, it is clear that for the Thomson–Gibbs equation to work, the measured melting temperature of the metastable crystals must be representative of the original metastable state. In order to use this equation, no metastability changes are allowed.

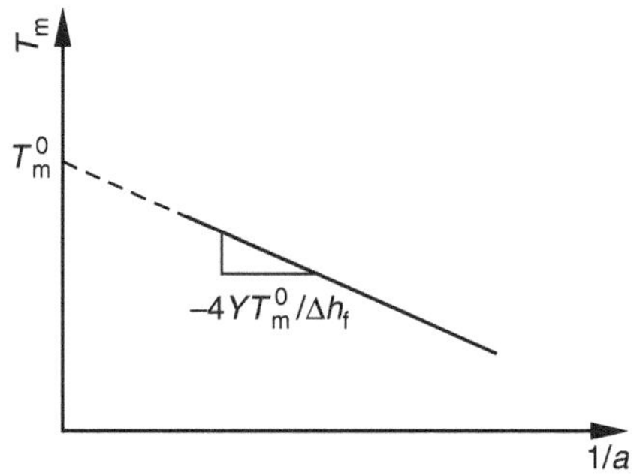

Figure 4.20 Schematic illustration of the extrapolation of the relationship between the melting temperature and reciprocal lateral size. When the reciprocal lateral size goes to zero, the extrapolated melting temperature is the equilibrium melting temperature.

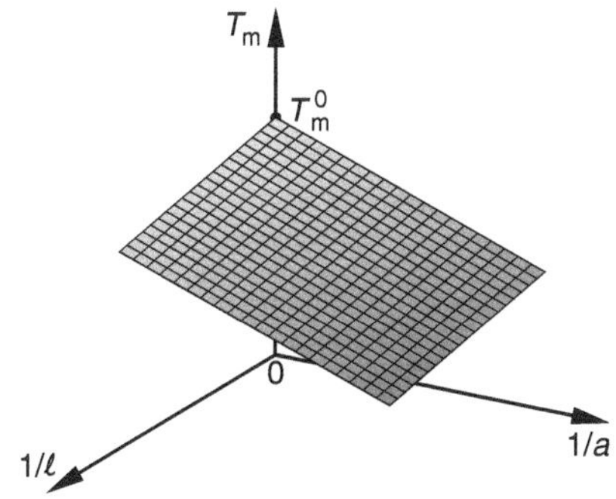

Figure 4.21 Schematic illustrations of the extrapolation of the relationship between the melting temperature, the reciprocal crystal size and reciprocal thickness. When the reciprocal crystal lateral size and thickness go to zero, the extrapolated melting temperature is the equilibrium melting temperature.

2.4. Ensuring constant metastability during heating

In order to ensure a constant metastability during isobaric heating, as shown in Fig. 3.9 of Chapter 3, the Gibbs free energy line must remain fixed from the initial temperature to the melting temperature. A drop in the Gibbs free energy line represents an increase in crystal stability, and this would make the Thomson–Gibbs analysis invalid. Also, a Gibbs free energy jump during heating is thermodynamically forbidden.

In the past half century, extensive experimental efforts have been exerted to prevent the change of crystal metastability during heating. Wunderlich, in his book *Macromolecular Physics, Volume III, Crystal Melting* (1980), summarized three different, but effective, approaches to maintain a constant metastability in small crystals during heating. These methods include very fast heating to prevent changes in crystal metastability, cross-linking to fix the crystal's bound amorphous fraction (thus stopping lamellar thickening), and chemical etching to eliminate the amorphous fraction, which permanently reduces the polymers to oligomers and makes them unable to reorganize. This section will only supplement these methods with some new observations and developments since the earlier summary given by Wunderlich (1980).

It is known that polymer crystal melting is time-dependent. There are two important experimental observations involving melting. First, polymer lamellar crystals are usually subject to annealing, reorganization, and recrystallization at heating rates normally used in calorimetric measurements. This indicates that even though polymer crystals are folded, long-chain lamellae, they have the ability to change their metastability as long as they have space and material to accommodate the change. Second, the less stable (the thinner) the crystals are, the stronger is the driving force to improve crystal stability.

Recent experiments on semicrystalline polymers with very fast heating rates have given new insights into how lamellar crystals reorganize themselves. In particular, instrumentation developments, such as ultrafast chip calorimetry, have allowed experiments with extremely fast heating rates of up to 100 k °C/sec, much faster than those attainable in traditional calorimetric measurements (Adamovsky and Schick, 2004; Adamovsky *et al.*, 2003; Efremov *et al.*, 2002, 2003; Minakov *et al.*, 2007). Schick's group reported that crystalline poly(ethylene terephthalate), an important engineering polymer which is widely used in containers, films, fibers, and tire cords, can rapidly reorganize even at very fast heating rates. When this polymer was crystallized at relatively low supercooling to grow thick lamellar crystals, the crystal metastability could be held constant at heating rates as low as 200 °C/min. It was determined that the melting temperature for the thicker lamellar crystals was around 240 °C, about 10 °C lower than the melting temperature measured at a heating rate of 10 °C/min. If the sample was crystallized at a high supercooling, such as at 130 °C, constant metastability could not be reached even at a heating rate of 2.7 k °C/sec. Figure 4.22 shows a set of

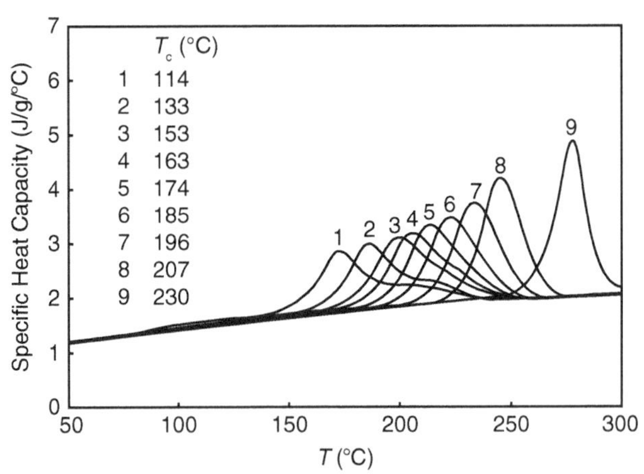

Figure 4.22 Set of heat capacity measurements at a heating rate of 2.7 k °C/sec for a poly(ethylene terephthalate) sample after isothermal crystallization at temperatures ranging from 114 to 230 °C [replotted from Minakov *et al.* (2004), with permission].

thermo-diagrams for poly(ethylene terephthalate) samples after they were completely crystallized at crystallization temperatures ranging between 113 and 230 °C and then heated to the isotropic melt at a rate of 2.7 k °C/sec. In all the diagrams, only one major melting endotherm could be observed. With increasing crystallization temperature, the melting peak continuously increases. It was concluded that the melting temperatures of isothermally crystallized poly(ethylene terephthalate) samples range from a few to tens of degrees above the isothermal crystallization temperatures (Minakov *et al.*, 2004). Therefore, the difference between the melting temperatures measured at extremely fast heating rates and those observed during normal calorimetric measurement at slower heating rates were caused by changes in metastability during crystal heating.

A similar observation was made for nylon 6, another important engineering polymer. After the nylon 6 samples were isothermally crystallized, the "real" initial crystal melting was observed at temperatures only slightly higher than the corresponding crystallization temperatures at a fast heating rate of 2 k °C/sec. This melting behavior was not found with normal calorimetric measurements. Furthermore, even a heating rate of 5 k °C/sec was not fast enough to completely avoid reorganization in nylon 6. This fast reorganization of nylon 6 to improve thermodynamic stability was judged to occur on the 0.01–0.1 sec time scale (Tol *et al.*, 2006). These observations seem to be common for many semicrystalline polymers, ranging from isotactic polypropylene (De Santis *et al.*, 2006) to isotactic polystyrene (Minakov *et al.*, 2006). It is now evident that the melting temperature of an isothermally crystallized metastable crystal is only slightly above (about 10 °C) its isothermal crystallization temperature (Minakov *et al.*, 2007).

It was not expected that polymer crystals could melt and recrystallize at such fast speeds. These observations indicate that both the crystal–crystal and/or crystal–melt–crystal transitions have a thermodynamic driving force to rapidly increase the crystal stability. Additionally, these transitions from preexisting ordered states have greatly decreased transition barriers. Considering the time scales associated with long-chain molecular dynamics, it can be speculated that these transitions may not involve large-scale molecular motion, but only local structural improvements. It is expected that these new experiments may revive interest in how semicrystalline polymers improve their stability during phase transitions.

At the other end of the spectrum, superheating of a crystal may also be observed when the heating rate is extremely fast and the crystal is relatively perfect. As described in Section 2.1 of this Chapter, crystals require heterogeneous (surface) "nucleation" to disorder near the melting temperature. Because in both cases we observe that crystal fusion takes place above the expected melting temperature, we need to carefully distinguish the superheating phenomena from instrument lag. Instrument thermal lag is caused by poor thermal conductivity of the samples, as well as the heat capacity of the sample, causing a delay in instrumental response. In particular, instrumental thermal lag occurs when the sample size is large (Wunderlich, 2005).

Using chemical cross-linking to immobilize the crystal and freeze the phase morphology has been a useful tool for many years in both single and multicomponent systems. Although there are many different uses for this technique, the principle is always identical. One of the phases in the system undergoes a chemical cross-linking reaction via irradiation with a high energy beam, such as γ-radiation or high energy electrons. This reaction makes one phase solid-like (or less mobile). To fix the metastability of a polymer crystal, the amorphous fraction cross-links, preventing the amorphous region from providing the material needed for the crystal to reorganize and stopping lamellar thickening, thus preventing the system from increasing crystallinity. On the other hand, chemical etching eliminates the amorphous folds at lamellar surfaces of semicrystalline polymers. In turn, this process cuts the molecular weight of the polymers to that of oligomers, whose molecular weight is equivalent to that needed for the stem lengths of the lamellar crystals. Therefore, the molecular weight of the oligomers is proportional to the lamellar thickness.

Most of the cross-linking and chemical etching studies on semicrystalline polymers were conducted about 30–50 years ago. The general observation was that with increasing irradiation and chemical etching time, the melting temperatures and heats of fusion decrease (see, e.g., Wunderlich, 1980). A few new reports using the cross-linking approach can be found from the 1980s. During this time period, the most systematic study focused on controlling the metastability of polyethylene lamellar crystals, as reported by the Bristol group (Keller and Ungar, 1983; Ungar, 1980; Ungar and Keller, 1980; Ungar *et al.*, 1980). These two methods assume that cross-linking and chemical etching reactions start in the amorphous region of the crystalline polymer samples. With increasing dose and time, the crystalline region begins to be affected. The major issue in using these two methods to lock in the metastability of the lamellar crystal is how to precisely determine the critical time and dose limit necessary to immobilize or remove the amorphous region without affecting the crystal.

A historical study on metastable states was reported for nylon 6 fibers before further drawing, after drawing, after annealing, and after cross-linking (Todoki and Kawaguchi, 1977a,b). Few studies have exceeded this investigation in terms of technical sophistication and in understanding the details of irreversible melting and the transitions between polymorphs. This work has been summarized by Wunderlich (1980, 2005). The most important finding in this study is that irradiation does effectively prevent metastability changes in these fibers. Crystals grown in fibers and fixed by irradiation exhibit melting temperatures up to 70 °C lower than the same samples without irradiation at the identical heating rate of 10 °C/min. This conclusion clearly indicates that annealing and reorganization of the crystals occurs in the fibers during heating.

2.5. Polymer crystal melting at elevated pressures

Although extensive studies on polymer crystal melting have been conducted at a constant pressure (usually atmospheric), fewer studies have been reported for this process at different pressures due, in large part, to the difficulties of carrying out experiments at elevated pressures. It is known that the effect of pressure on melting temperature should fit the Clausius–Clapeyron Eq. (3.1) of Chapter 3 with a positive slope when we plot melting temperature against pressure. However, there is a particularly interesting group of semicrystalline polymers in which their crystalline densities are lower than their amorphous glass densities. For example, poly(4-methyl-1-pentene) possesses an amorphous glass density of 0.830 g/cm^3 in comparison with its crystalline density of 0.813 g/cm^3 at room temperature (Frank *et al.*, 1959; Kusanagi *et al.*, 1978; Natta *et al.*, 1955). Another example is crystal phase III of syndiotactic poly(*p*-methylstyrene) with a density of 0.988 g/cm^3; while, the glass density of this polymer is 1.02 g/cm^3 (De Rosa *et al.*, 1995). In the δ-crystal modification of syndiotactic polystyrene, the crystal density is 0.977 g/cm^3; while, the glass density is 1.055 g/cm^3 (De Rosa *et al.*, 1997) at room temperature. The lower crystal densities are caused by specific stereohindrances and/or intra- and interchain interactions of relatively large side groups reducing the molecular packing density in the crystalline state.

It was reported for poly(4-methyl-1-pentene) that the crystal (which at room temperature has a tetragonal lattice) loses its crystalline order isothermally upon increasing pressure in two wide temperature regions. This amorphization and crystallization are reversible, which is indicated in Fig. 4.23. This figure is a proposed temperature–pressure equilibrium phase diagram based on both calorimetric and x-ray diffraction studies (Höhne *et al.*, 2000; Rastogi *et al.*, 1991). The first pathway indicated in this figure is isothermal and at room temperature, which is far away from the melting temperature of the polymer crystal. With increasing isothermal pressure, the crystal gradually reverts to a disordered phase. At each pressure, the transition between the crystal and disordered phase is reversible. The second pathway is isothermal at a temperature just above the equilibrium melting temperature of the crystals at atmospheric pressure but slightly lower than the melting temperature of the highest pressure reported. With increasing pressure, the polymer crystal transitions into the melt. In particular, based on the calorimetric data, there is a pressure that can be identified (P' in Fig. 4.23) below which the melting temperature of the polymer increases, while above that pressure, the melting temperature decreases with increasing pressure. The third pathway isothermally

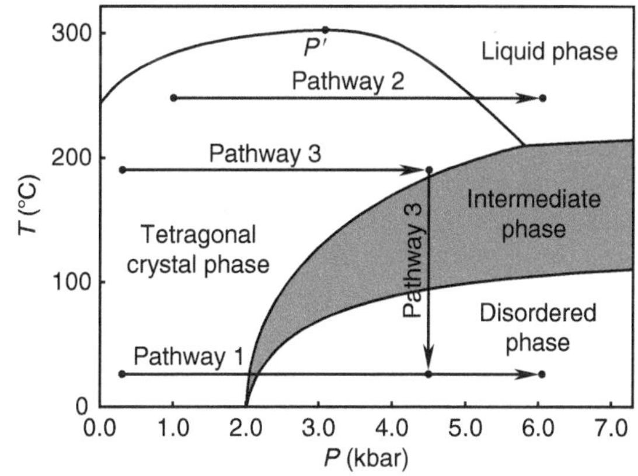

Figure 4.23 Proposed temperature–pressure equilibrium phase diagram of poly(4-methyl-1-pentene) [combined and replotted from Rastogi *et al.* (1991) and Höhne *et al.* (2000), with permission].

increases pressure at an elevated temperature below the transition temperature. The highest pressure reached is below the pressure where the intermediate phase transition takes place, as shown in Fig. 4.23. The temperature is then decreased toward room temperature. Surprisingly, the tetragonal crystal that passes through the intermediate phase loses its long-range order and transforms to the disordered phase (Rastogi *et al.*, 1993).

The disordering upon cooling observed in the third pathway of Fig. 4.23 is in contrast to the crystal melting behavior of most semicrystalline polymers. However, it can be explained by thermodynamics. First, when the crystal and amorphous densities are inverted, the melting temperature decreases with increasing pressure, such as in the cases of water and bismuth where the slope of the Clausius–Clapeyron equation is negative. This leads to isothermal amorphization with increasing pressure. Disordering upon cooling implies that the entropy of the crystal is higher than the entropy of the disordered amorphous phase (or the liquid). Consider poly(4-methyl-1-pentene) that has loosely packed molecules in the crystal lattice compared with its denser and configurationally-hindered amorphous melt. It is possible that a higher overall entropy exists in the geometrically more ordered phase. Specifically, there is an entropy change sign inversion at the transition. This entropy change inversion may also compete with a change in the interaction terms, an enthalpy change at the transition. Note that this enthalpy change might undergo a sign inversion as well. From the thermodynamic perspective, Tammann provided an explanation using "neutral lines" (Rastogi *et al.*, 1993; Tammann, 1903, see also Section 2.2 of Chapter 3). Although this polymer is a specific case, the experimentally observed crystal "melting" on cooling reflects general rules concerning phase transitions beyond polymers.

2.6. Polymer crystal melting kinetics

The final issue is the kinetics of polymer crystal melting. The kinetics here describe the heterogeneous (surface) "nucleation" process of melting. Compared to the extensive number of polymer crystal growth rate studies, this topic has been the focus of very few reports. At the melting temperature, polymer crystals melt very quickly making experimental observations difficult. The major reason is because polymer crystals are usually small making them metastable, so they do not melt at the equilibrium melting temperature.

As shown in Fig. 2.10 of Section 2.2 of Chapter 2, the polymer melting and crystallization kinetics are clearly discontinuous at the equilibrium temperature. To achieve the identical absolute rate of melting and crystallization, a much larger degree of supercooling is needed compared to superheating. However, if one carefully studies the melting kinetics exhibited by polymer crystals, different mechanisms may be revealed.

So far, the experimental data have shown linear melting rates at each particular superheating. However, the relationship between the linear melting rates and superheating is different for different polymers. In the case of low molecular weight poly (ethylene oxide) fractions, the melting of the extended and integral-folded chain single crystals exhibits an exponential relationship between the melting rate and superheating.

The conclusion is that the melting of these single crystals is "nucleation"-limited (Kovacs and Straupe, 1980; Kovacs *et al.*, 1975, 1977), as described in Section 2.2 of Chapter 2. On the other hand, the melting rate of polyethylene extended-chain crystals exhibits a different behavior. With a lamellar thickness on the order of a micrometer, this process is close to equilibrium melting. A linear relationship between the melting rate and superheating was observed (Wunderlich, 1979). This set of experimental data could be representative of another type of melting mechanism. Section 2.1 of Chapter 2 discusses that the rates of normal crystal growth, occurring on a rough surface, should be proportional to the supercooling in the low supercooling region [Eq. (2.16)]. The melting kinetics data just described may indicate a "rough-surface" melting process. Alternately, one may also need to consider a heat diffusion-limited melting process in the thick polyethylene extended-chain crystals.

The lack of experimental observations concerning isothermal crystal melting makes it difficult to generalize the mechanism of polymer crystal melting. This is further hampered by the melting behaviors we commonly observe in metastable polymer crystals. Another approach to understand polymer crystal melting is to employ nonisothermal experiments. The most frequent approach is to utilize metastable crystal superheating results obtained by calorimetric methods to illustrate melting kinetics. The simplest kinetic model for heterogeneous crystal melting follows the change of crystallinity with respect to time, $\mathrm{d}w(t)/\mathrm{d}t$, which depends upon the superheating ΔT_{sh} (Toda *et al.*, 2002):

$$\frac{\mathrm{d}w(t)}{\mathrm{d}t} = -A(\Delta T_{\mathrm{sh}})w(t) \tag{4.18}$$

where A is a rate coefficient,

$$A(\Delta T_{\mathrm{sh}}) = a(\Delta T_{\mathrm{sh}})^{x} \tag{4.19}$$

where a and x are constants. Here the melting rate depends nonlinearly on superheating. After integrating Eq. (4.18) at a constant heating rate, β, we have (Minakov *et al.*, 2007):

$$\ln\left(\frac{w}{w_0}\right) \sim \beta^{x} t^{x+1} \tag{4.20}$$

where w_0 represents the initial crystallinity of the polymer crystal before melting. The melting process thus occurs as soon as $\beta^{x}t^{x+1}$ is in the vicinity of unity. At $t = \Delta T_{\mathrm{sh}}/\beta$, we have:

$$\Delta T_{\mathrm{sh}} \sim \beta^{1/(x+1)} \tag{4.21}$$

This equation clearly indicates that a power law relationship exists between the superheating and heating rates, as has been experimentally observed and theoretically illustrated by Toda *et al.* (2002) and Minakov *et al.* (2007). There are differences in the value of $1/(x + 1)$ at various metastable states in several semicrystalline polymers. After differentiating the crystal superheating from the instrumental thermal lag, the power law of Eq. (4.21) indicates an activation process such as the "nucleation" process described in Section 2.2 of Chapter 2. Therefore, using the results from Eq. (4.21), we may estimate the melting kinetics, assuming a "nucleation" event is needed to initiate the melting process. One recent effort to understand the molecular aspects of polymer crystal melting kinetics focused on the consecutive segmental detachment of a chain from the crystalline surface. It was found that detaching a segment with a free chain end, compared to a segment having topological constraints, requires different activation energies and thus occurs at different temperatures. As a result, different crystal melting processes were suggested (Lippits *et al.*, 2006).

3. Metastable States in Phase-Separated Polymer Blends and Copolymers

3.1. Metastable states in phase-separated polymer blends

As discussed in earlier sections, polymers provide many examples of metastable states beyond the atomic length scale. The thermodynamic principles governing binary mixtures, such as liquid polymer blends, are very similar to small molecule and polymer solutions, as described in Section 3.3 of Chapter 2, except that volume fractions are used in polymer blends. We will focus on the phase morphological differences between polymer and small molecule systems. A typical upper critical solution temperature phase diagram is shown schematically in Fig. 4.24A. On sufficiently decreasing the temperature at the appropriate concentration, the upper critical point can be reached. In this section, no attention will be paid to the various fundamentally important features concerning behavior in the region of the critical point already discussed in Section 3.3 of Chapter 2, but rather we will focus on the situation at lower temperatures. Upon liquid–liquid phase separation, the system divides into two distinct phases with the volume fractions of the individual components within the phases (namely, the component concentrations) determined by the tie line end-points on the binodal curve. One phase contains more polymer A, the other more polymer B. By further lowering the temperature, the compositions of the two phases become increasingly different from each other. As shown in Fig. 4.24A, we can thermodynamically define the phase boundaries of the binodal and spinodal curves. The region between these two curves is the metastable region where phase separation must overcome a nucleation barrier to grow. Within the area bounded by the spinodal line, however, spontaneous density fluctuations dictate the barrier-free phase separation, as described in Section 3.4 of Chapter 2. Further readings are also available in many textbooks (e.g., Koningsveld *et al.*, 2001).

The equilibrium phase morphology of ultimate stability is the same for a polymer blend independent of the phase separation mechanism. The ultimately stable phase morphology is two layers, with the lower density, polymer B-rich phase on top of the higher density, polymer A-rich phase (like oil and water). An example is shown

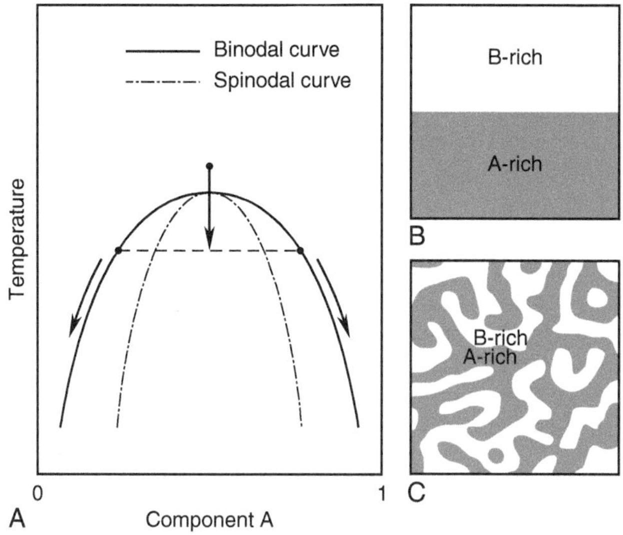

Figure 4.24 Schematic illustration of phase separation in an upper critical solution temperature polymer blend. At high temperatures, the two components are miscible, while at low temperatures, they phase separate within the binodal region (A). The equilibrium phase separation morphology is shown in (B), while the metastable phase separation morphology is shown in (C).

in Fig. 4.24B. However, the morphologies on the path toward this ultimate morphological stability are determined by the phase separation mechanism. It is commonly understood that the droplet morphology indicates nucleation-limited phase separation, while a bicontinuous morphology indicates spontaneous decomposition. There is an abundance of morphological variants within the two-phase separation mechanisms depending on the length of the ripening process at the time of observation. The volume fractions of the blend components, which denote the order parameter in a polymer blend, and the temperature significantly affect the observed phase morphology. The reason is that these parameters determine whether the system is in the nucleation or spinodal decomposition region, and within each given region, these parameters will crucially affect the size, shape, and dispersion of the phase domains in the transient stage of phase separation. The effect on phase domain dispersion is most easily seen for the nucleation process. For low volume fractions of polymer A, the A-rich phase will form droplets suspended within a continuous polymer B-rich matrix. There will be a matrix inversion for a high volume fraction of polymer A and a bicontinuous morphology for the intermediary volume fractions.

At this point, to a first approximation, we are dealing with two distinct processes: the establishment of the equilibrium partitioning of the two polymer liquids into the two phases and the evolution of the phase morphology. The volume fraction partition is thermodynamically determined by the phase diagram, as in Fig. 4.24A, while the transient phase morphology is determined by the mechanism of phase separation. Clearly, the two processes cannot be completely independent of each other, and this interdependence leads to issues that, to our knowledge, have not been fully understood and are seldom discussed. Nevertheless, it is usually true that the component partitioning is complete; while, the phase morphology is still moving toward equilibrium through ripening, as described in Section 3.4 of Chapter 2. In other cases, these two processes overlap. This concept introduces complications not only to the identification of the type of liquid–liquid phase separation in the initial stage, but also to the study of phase separation kinetics in the late stage of polymer blend phase separation.

Different from small molecule and dilute polymer/small molecule binary mixtures, polymer blends after separation do not achieve the state of ultimate stability, even after a prolonged coarsening process (Fig. 4.24C). The variety of phase morphologies observed indicates different degrees of metastability on this morphological length scale. These metastable states are technologically important to achieve the synergy critical to the improved material properties required for specific applications.

An example illustrating control over metastable liquid–liquid phase separation morphology is a two-step quench for a polymer blend system with one component, A, that is denser than the other component, B. The thermodynamic phase diagram of this system is shown in Fig. 4.25A. The two-step quench is carried out by initially quenching the blend and allowing it to equilibrate at the higher temperature T_1, then quenching and equilibrating the system at the lower temperature T_2. If we carefully carry out this two-step quench experiment and do not disturb the blend

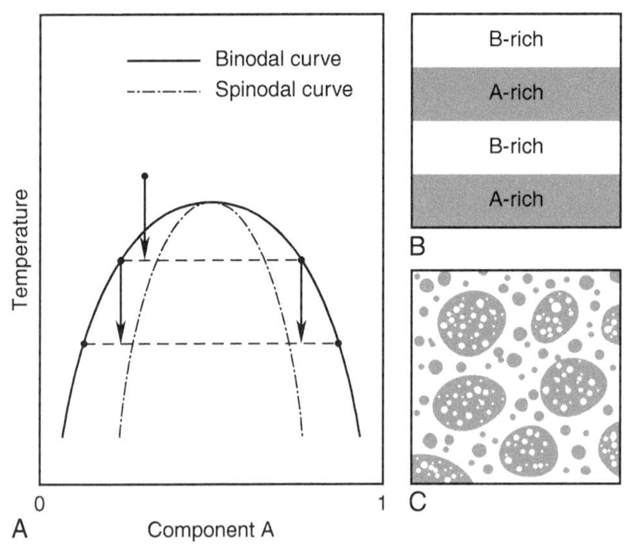

Figure 4.25 Phase diagram of a two-step quenched A–B polymer blend (A); the two-step quench phase morphology at the unstable four layer equilibrium morphology (B); and a liquid–liquid phase separation which does not reach the ultimate equilibrium and, thus, is in a morphological metastable state (C).

system, its ultimate equilibrium phase morphology should consist of four layers and three interfaces, as shown in Fig. 4.25B. This system equilibration occurs because at the higher temperature (T_1), the blend separates into two layers, one A-rich and the other B-rich, with the B-rich layer on the top. After the system is quenched to the lower temperature (T_2), each of these two layers undergoes a further phase separation to form two sublayers. Of these four liquid layers, two of them have the same A-rich volume fraction; the other two have the same B-rich volume fraction, which is determined by the tie line end-points on the binodal curve at temperature T_2. This morphological equilibrium is unstable. If we shake the system, these four layers merge to become a two-layer system identical to the case of a single-step quench to temperature T_2. Therefore, to indefinitely maintain this unstable four-layer morphology, one must be able to vitrify at least one of these two types of layers by quenching the system below its glass transition temperature.

Now, we consider the case when this A–B polymer blend does not reach the equilibrium phase morphology during liquid–liquid phase separation, as shown in Fig. 4.25B. In other words, the coarsening process of this phase separation is halted before reaching the final layer morphology. This process happens frequently when the polymer components are of a high molecular weight and/or when a polymerization reaction is carried out during and after the phase separation, such as in thermoset curing systems. For convenience, we will only consider the nucleation and growth mechanism in liquid–liquid phase separation.

After quenching the blend to the higher temperature of T_1, a droplet morphology forms with A-rich droplets embedded in a B-rich matrix. On the second quench to the lower temperature of T_2, further phase separation within the droplets and matrix will take place. The droplets with different average sizes appear in both phases, as shown in Fig. 4.25C, which leads to the droplet morphology on different length scales. The first quench leads to the droplets of larger size, while the second quench generates the droplets of smaller size. Furthermore, the smaller droplets are located within the larger droplets formed during the first quench and in the matrix. The droplets with multiple length scales are useful for improving polymer toughness and impact strength by changing the fracture mechanisms.

3.2. Kinetics of liquid–liquid phase separation in polymer blends

In principle, the kinetics of liquid–liquid phase separation in polymer blends should be identical to those of binary mixtures in small molecules, as described in Section 3.4 of Chapter 2. Liquid–liquid phase separation is usually monitored by real-time small angle scattering techniques over a wave-number vector scale corresponding to a real space scale that ranges from nanometers to micrometers. However, complications are introduced into polymer binary blends because of the very different dynamics of chain molecules compared to small, simple molecules. Moreover, within a one polymer binary blend, the two components can have very different self-diffusion coefficients.

Let us examine a simpler case of a polymer binary blend with dynamically symmetric, amorphous polymers, polybutadiene and polyisoprene, where the polybutadiene is deuterated for small angle neutron scattering experiments. The glass transition temperatures of both of the polymers are much lower than room temperature, and this blend possesses a lower critical solution temperature. Figure 4.26 schematically illustrates this phase diagram (Hashimoto, 2004). In this figure, the experiment starts with an equilibrium mixed phase. The system was then critically "quenched" by rapidly heating to 40 °C, which is inside the spinodal curve, so the mixed phase becomes thermodynamically unstable. The system undergoes a spinodal decomposition process. The final equilibrium of this blend, after the liquid–liquid phase separation, should be two phase-separated layers, one being deuterated polybutadiene-rich (at point 1 in Fig. 4.26) and the other being polyisoprene-rich (at point 2 in Fig. 4.26).

Using the scaling law described in Section 3.4 of Chapter 2, it is possible to plot a relationship between the reduced structure factor and the reduced scattering vector. The experimental data can be used to construct a universal master scattering curve for time t, as shown in Fig. 4.27. Therefore, in the late stage of the spinodal decomposition process, the morphological habit does not change, but the phase size increases. Compared with Fig. 2.15B of Chapter 2, it is evident that the time scale, as well as the quenching depth, in the small molecule liquid binary mixture was very different from that of this polymer blend by three orders of magnitude. Regardless of these differences, both systems can be well described by the general equations developed for binary mixtures in Section 3.4 of Chapter 2.

In polymer binary blends, several systems have exhibited early stage spinodal decomposition. Their phase separation characteristics are consistent with the description of Cahn's linearized theory based on the mean-field formulation. These behaviors include that $q_{max}(t)$ is independent of time over a long timescale or a reduced timescale, τ, when it is smaller than one. Figure 4.28 shows the

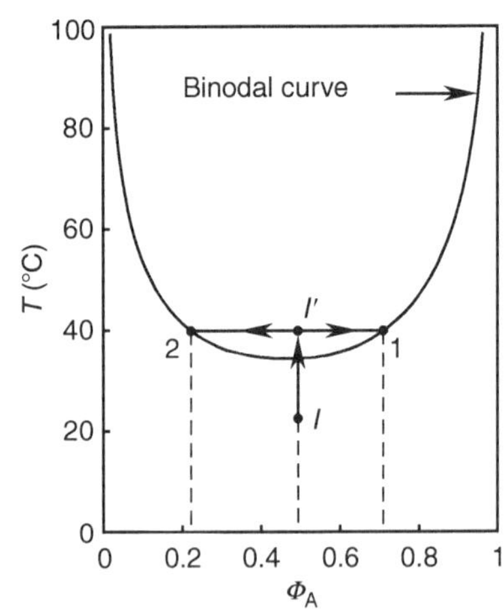

Figure 4.26 Schematic presentation of a lower critical solution temperature phase diagram for a blend of deuterated polybutadiene (component A) and protonated polyisoprene (component B). The binodal curve was calculated based on Flory-Huggins theory. [replotted from Hashimoto (2004), with permission].

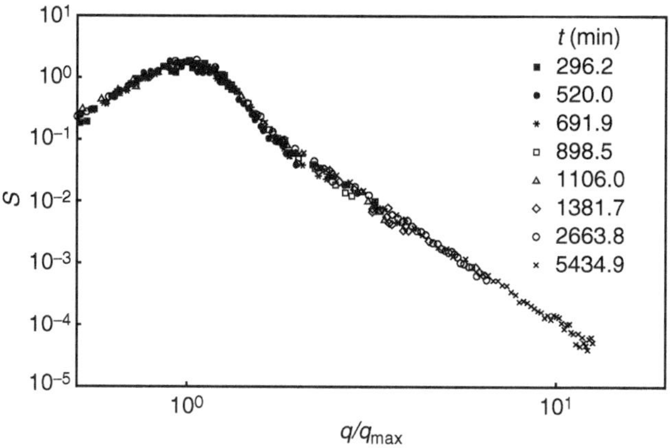

Figure 4.27 Relationship between the reduced structure factor, S, and reduced scattering vector, q/q_{max}, for a blend of deuterated polybutadiene and protonated polyisoprene with a volume composition of 47% and 53% at 40 °C, which is 3.9 °C higher than the critical temperature [replotted from Hashimoto (2004), with permission].

well-known example of early stage separation for a 50:50, by weight, blend of deuterated polycarbonate and poly(methyl methacrylate) monitored by small angle neutron scattering (Motowoka *et al.*, 1993). Because the volume fractions are 50:50, this is a critical quench. Figure 4.28A shows the very early stage of spinodal decomposition at times shorter than 45 min. The composition fluctuation dominates the scattering data, which can be determined from the intensity of the invariant scattering vector peak position in the structure factor. When the time is between 45 and 208 min, the system is in the early stage of spinodal decomposition. At times longer than 208 min, as shown in Fig. 4.28B, the intermediate stage of spinodal decomposition begins. This stage is represented by a decrease in the $q_{max}(t)$ value due to mode-coupling effects arising from nonlinearity in the time-evolution process. Longer than 270 min, the spinodal decomposition is in the late stage of coarsening the phase structure.

Dynamically asymmetric, polymer binary blends can be made using two polymers with different molecular weights. Their dynamic properties are strongly associated with stress–diffusion coupling and viscoelastic effects. It is, in theory, possible to investigate the nucleation-limited mechanism of phase separation in polymer blends, but experimentally, it is more difficult. The vast majority of experimental reports on the liquid–liquid phase separation of polymer blends have focused on understanding the spinodal decomposition mechanism. Based on the simplified illustration, the nucleation-limited process of liquid–liquid phase separation represents a relaxation from a metastable state in the area bounded between the spinodal and binodal curves, while the spinodal decomposition process is a relaxation process from an unstable state bounded by the spinodal curve, as described in Section 3.4 of Chapter 2. However, a recent experimental study to determine the critical length scales and the limits of metastability in phase-separating polymer blends has shown that the spinodal curve is not simply a sharp boundary line, but a zone due to

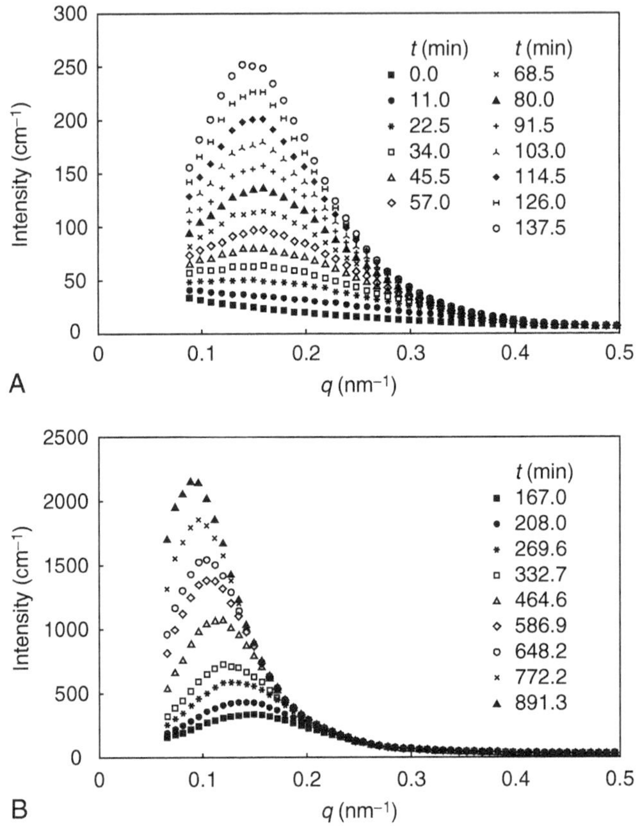

Figure 4.28 Real-time small angle neutron scattering results for a 50:50 weight blend of deuterated polycarbonate and poly(methyl methacrylate). The final temperature after quench was 70 °C. The very early and early stages of the spinodal decomposition are shown in (A) when the time is less than 137 min; while, the intermediate stage of the spinodal decomposition starts after 208 min. The later stage starts when the time is longer than 270 min (B) [replotted from Hashimoto (2004), with permission].

fluctuations (Lefebvre *et al.*, 2002). Some aspects of this experimental observation have been theoretically addressed by Wang (2002).

3.3. Metastable states in phase-separated block copolymers

The difference between block copolymers and polymer blends lies in the fact that although the chemically distinct blocks in a copolymer are not miscible, they are linked by chemical bonds. Thus, the macroscopic phase separation observed in polymer blends cannot be achieved in block copolymers. Flexible diblock copolymers have been extensively investigated. It is well-known that increasing Flory's interaction parameter, χ, (equivalent to lowering the temperature) and increasing the molecular weight (equivalent to losing some translational and conformational entropy) favor block segregation (Bates and Fredrickson, 1990). At high temperatures, the diblock copolymers are in a disordered state. When they are quenched into the strong segregation region, there are a variety of thermodynamic equilibrium phase morphologies observed, rather than metastable morphologies. The formation of these morphologies after quenching is a liquid–liquid phase separation on the nanometer scale.

It was found that the different morphologies resulting from block segregation are determined by the volume fractions of the blocks in the copolymer. When two blocks are nearly symmetric with approximately equal volume fractions, the lamellar phase morphology is formed. Increasing the volume fraction of one block causes the phase morphology to change to a hexagonal cylinder morphology. Further increasing the volume fraction of one block leads to a body-centered cubic sphere morphology. Figure 4.29 shows a polystyrene-*block*-polyisoprene system at different volume fractions, and the phase morphological changes (Khandpur *et al.*, 1995). The vertical axis is the product of Flory's interaction parameter and the degree of polymerization; it is inversely proportional to temperature. The horizontal axis

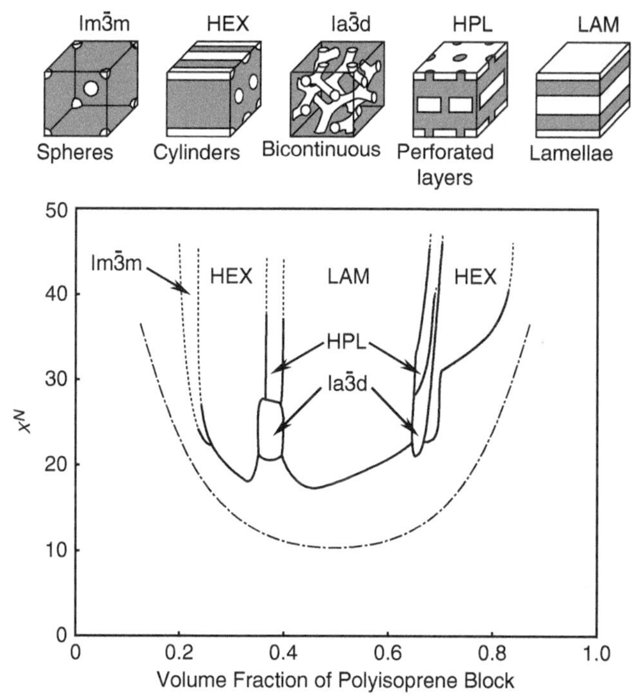

Figure 4.29 Phase diagram and structures for polystyrene-*block*-polyisoprene at different volume fractions. The dot–dash line is the prediction based on the mean-field theory. The phase structures shown on the top are in sequence from just above 20% up to 50% of polyisoprene block volume fractions. The phase structures form the symmetrically reversed polyisoprene block volume fraction region above 50%. Furthermore, the bicontinuous phase and perforated layer structures appear at nearly identical volume fractions but different temperatures [replotted from Khandpur *et al.* (1995), with permission].

is the volume fraction of polyisoprene. The dot-dashed line in this figure represents the mean-field theory prediction. The differences among the experimental data are caused by concentration fluctuations.

Why do those different morphologies occur at different block volume fractions? It was realized that the phase morphology change is critically associated with the interface of the phase-separated blocks. When a diblock copolymer is symmetric, the blocks separate to opposite sides of the interface. Because their volume fractions are identical, the chains on both sides of the interface feel the same pressure, thus leading to a flat interface and a lamellar morphology. Asymmetric diblock copolymers have blocks of unequal volume. Therefore, the blocks bracketing the interface do not occupy an equal volume. The interface starts to curve to balance the resulting pressure difference in order to achieve the minimum free energy. With increasing asymmetry between the two blocks, phase separation forms the hexagonal cylinder, then the body-centered cubic morphology. In this research area, this interfacial behavior is also sometimes called "intermaterial dividing surfaces" (see, e.g., Allen and Thomas, 1999).

From the molecular perspective, we can quantitatively explain the interface curvature of diblock copolymers using concepts from tethered chain molecules. When one kind of block is connected to an interface formed with another immiscible block, assuming the connecting covalent bond is located at the interface, one can define a tethering density (σ) that is inversely proportional to the unit area occupied by each block at the interface (S).

$$\sigma = \frac{1}{S} \tag{4.22}$$

The tethering density of blocks on either side of the interface is identical, regardless of the different phase morphologies, in diblock copolymers. In order to quantitatively explain the flat and curved interfaces of the phase morphologies in diblock copolymers, we define the reduced tethering density concept, Σ, as:

$$\Sigma = \sigma \pi R_{\mathrm{g}}^2 \tag{4.23}$$

where R_{g} is the radius of gyration of the tethered block at the interface (Kent, 2000). This is the R_{g} for a θ-solvent. As only one type of block exists on both sides of the interface, it is equivalent to the homopolymer in the melt. This reduced tethering density is independent of the molecular weight of the block in the melt (Chen *et al.*, 2004a,b; Zheng *et al.*, 2006). Therefore, the reduced tethering density quantifies the crowding, or the pressure, the blocks experience on either side of the interface. Physically, this concept represents how many chains are located within the projected area, A, of each block onto the interface ($A = \pi R_{\mathrm{g}}^2$).

It is evident that for asymmetric diblock copolymers, the reduced tethering densities of the two blocks are different because the volume occupied by the two blocks is not equal, and therefore, the radii of gyration for these two blocks are different. The resultant pressure differential on either side of the interface produces an unbalanced surface stress at the interface. The interface curves to relax the pressure on one side of the interface, allowing the system to settle into a free energy minimum. Recent experiments have shown that the onset of tethered chain crowding on a flat solid substrate is at the reduced tethering density of 3.7, and the onset of the highly stretched regime is at a reduced tethering density of 14.3 (Chen *et al.*, 2004a,b; Zheng *et al.*, 2006), as shown in Fig. 4.30. Therefore, in an asymmetric diblock copolymer, when the reduced tethering density on one side of the interface is larger than 3.7 and the density on another side is smaller than 3.7, the interface may start experiencing a pressure differential. Further increasing the reduced tethering density will generate a severe free energy penalty, causing a phase morphology change. The upper stability limit of a phase morphology could be in the vicinity of the 14.3 reduced tethering density. This understanding, in principle, also can be applied to explain the phase morphological changes in micelles, as described in Section 3.6 of Chapter 2 (Bhargava *et al.*, 2006, 2007).

There is a narrow volume fraction region in which a bicontinuous, double gyroid phase forms from two interpenetrating components with threefold nodes at each junction point in the high-temperature region. This phase symmetry has an $Ia\bar{3}d$ space group (Hajduk *et al.*, 1994; Schult *et al.*, 1994). Theoretical calculations have suggested that the double gyroid phase is an equilibrium phase structure in the weak-to-intermediate segregation limit (Matsen and Bates, 1997; Matsen and Schick, 1994).

In the low temperature region, a hexagonally perforated layer phase can be observed in diblock copolymer/homopolymer blends and diblock copolymers (Ahn and Zin, 2000; Almdal *et al.*, 1992; Disko *et al.*, 1993; Förster *et al.*, 1994; Hamley *et al.*, 1993; Spontak *et al.*, 1993; Thomas *et al.*, 1988; Vigild *et al.*, 1998; Zhu *et al.*, 2001b, 2002, 2003). This phase was first proposed to be a lamellar-catenoid

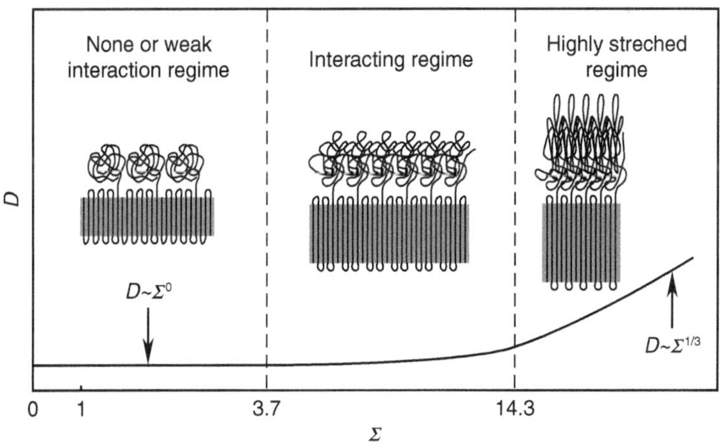

Figure 4.30 Schematic representation of the thickness of tethered chains (D) on a flat substrate in solution versus Σ. Note that the tethered chains start being squeezed by their neighbors at $\Sigma \sim 3.7$, namely, at the onset of overcrowding of the tethered chains. The onset of the highly stretched region is at $\Sigma \sim 14.3$.

structure in which both components were three-dimensionally connected (Thomas *et al.*, 1988). The stacking sequences of the hexagonal perforations in this phase were modeled as *ABCABC* (trigonal structure $R\bar{3}m$) and/or *ABAB* (hexagonal structure $P6_3/mmc$) layer arrangements. The most recent small angle x-ray diffraction and transmission electron microscopy experimental results for a polystyrene-*block*-poly (ethylene oxide) diblock copolymer show that this shear-induced hexagonally perforated layer phase contained a mixture of trigonal twins (~80%) and hexagonal (~20%) structures. Both structures had the identical orientation and in-plane unit cell parameters (*a* and α) but different out-of-plane (*c* axis) dimensions because the trigonal structure was constructed with three layers (*ABC*), while the hexagonal structure was constructed with two layers (*AB*). The formation mechanism was attributed to edge dislocations formed by "plastic deformation" under a large amplitude mechanical shear. The small angle x-ray diffraction from this hexagonally perforated layer phase is shown in Fig. 4.31 (Zhu *et al.*, 2001b, 2002, 2003).

Theoretical calculations using a broad range of symmetries, including face- and body-centered cubic structures, hexagonal close packing, trigonal structure, and others predicted the metastable nature of the hexagonally perforated layer phase (Laradji *et al.*, 1997; Olmsted and Milner, 1989; Qi and Wang, 1997). Experimental findings suggested that the hexagonally perforated layer phase is induced by mechanical shear. As such, it is a long-lived metastable phase or intermediate state (Ahn and Zin, 2000; Hajduk *et al.*, 1997; Vigild *et al.*, 1998; Zhu *et al.*, 2001b, 2003). In the

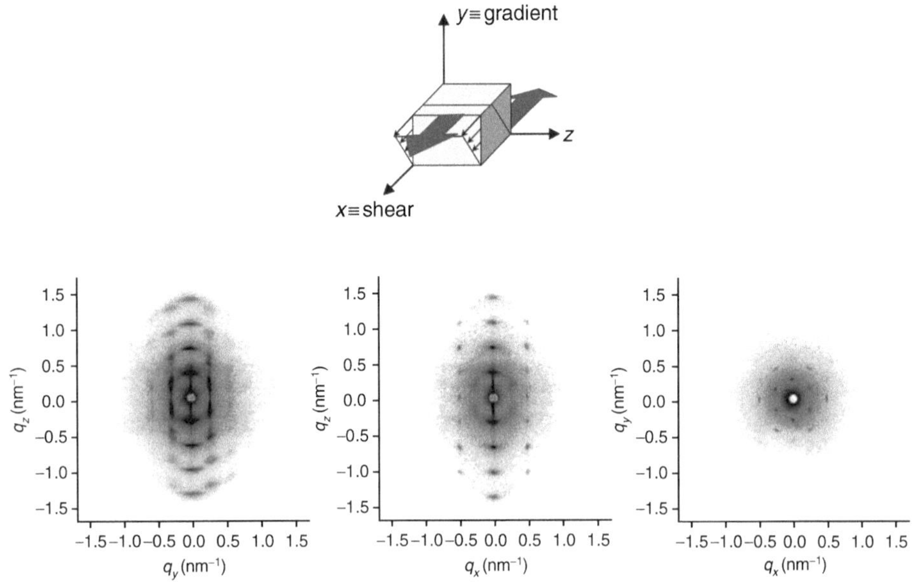

Figure 4.31 Set of small angle x-ray diffraction patterns of a shear-induced hexagonally perforated layer phase of polystyrene-*block*-poly(ethylene oxide) along three directions. These diffraction patterns result from two overlapped trigonal twins and hexagonal structures. The shearing geometry is also included [reprinted from Zhu *et al.* (2002), with permission].

polystyrene-*block*-poly(ethylene oxide) diblock copolymer, low-frequency rheological studies indicated that the hexagonally perforated layer phase transformed into a more stable double gyroid phase when the temperature exceeded 160 °C. The hexagonally perforated layer structure reappeared after the sample was again subjected to the large amplitude mechanical shear. Furthermore, the transformation from the hexagonally perforated layer phase to the double gyroid phase was also studied. It was found that the perforations in the hexagonal structure rearranged themselves first, followed by the rearrangement of the trigonal perforations (Zhu *et al.*, 2001b, 2003). Therefore, it can be concluded that the mechanical shearing facilitates the transformation toward the hexagonally perforated layer structure. As soon as the height of this barrier becomes lower than the barrier to form the double gyroid phase, the hexagonally perforated layer structure forms, even though the double gyroid phase is the thermodynamic equilibrium morphology. This structure serves as an illustration of Ostwald's stage rule for a phase morphological transition in diblock copolymers (Ostwald, 1897).

Some recent reports indicate the finding of new phase structures in diblock copolymers include the arrow-head morphology in a rod-coil diblock copolymers of polystyrene-*block*-poly(hexyl isocyanate) (Chen *et al.*, 1995, 1996), helical cylinders in polystyrene-*block*-poly(L-lactic acid) (Ho *et al.*, 2004), and a tetragonal perforated layer structure in a rod-coil polystyrene-*block*-poly([2,5-bis(4-methoxyphenyl)oxycarbonyl]-styrene) diblock copolymer (Tenneti *et al.*, 2005). The thermodynamic stabilities of these new phase structures are not known. The existence of these phase structures and the transformations between them in the weak-to-intermediate segregation limit, similar to the polymorphs in crystals, may provide interesting insights into understanding the role of metastable states in ordered block copolymers. Furthermore, a less systematic understanding of the equilibrium and metastable phase behaviors of block copolymers containing liquid crystals and other mesophases has been achieved. In particular, if one of the blocks in the copolymer forms not only an ordered phase but also polymorphs with monotropic phase behavior, one may have an opportunity to investigate metastable phase behavior confined within nanophase-separated environments based on the free energy pathways during the development of these phases.

From an even broader perspective, when there is a diblock copolymer with one crystallizable block, it may crystallize in dilute solution to form lamellar single crystals. These chain-folded single crystals are, by definition, metastable. Different from the single crystals of homopolymers grown in dilute solution, these block copolymer single crystals form sandwich-like structures: a single crystal formed by the crystallizable block at the center with amorphous blocks tethered at both crystal fold surfaces. We now need to consider the entropic effect of the tethered amorphous blocks on the overall free energy to obtain the minimum free energy of these metastable states (Chen *et al.*, 2004a,b; DiMarzio *et al.*, 1980; Zheng *et al.*, 2006). Furthermore, it is known that block copolymers in solution form micelles when block segregation takes place. The formation of these micelles is dominated by thermodynamics. Various phase morphologies formed by the micelles correspond to the minimum free energy states at fixed conditions. The mechanisms and kinetics of micelle morphological transitions with changing solvent concentration and/or temperature are not well understood. Their free energy landscapes during these transformations are also largely unknown. Efforts to understand aspects of these transformations are just beginning (Bhargava *et al.*, 2006, 2007).

3.4. Polymer crystallization in nano-confined environments using diblock copolymers as templates

As discussed in Section 2.2 of this Chapter, the smaller a crystal is, the farther away it is from the equilibrium stability; thus, it has a greater degree of metastability. One may ask: What is the stability limit of this crystal? This question then leads to a more general one: How small can a phase be and still be defined as a phase? A quantitative answer to this question does not presently exist. Let us assume that we can make a nanotube with the ability to change the diameter. If this nanotube can only accommodate a single polymer chain in an extended conformation, this polymer chain certainly does not form a phase. However, if one increases the diameter of this nanotube to allow two, three, four, or more polymer chains to be accommodated, one may logically ask how many molecules are required to form a physical phase. From the other perspective, we may ask, say, for an amorphous glassy polymer, what is the lower size limit for this phase to exhibit solid state properties?

These questions can be experimentally investigated and answered. The key issue is how to precisely control the nanophase-size and uniformly place polymers into the nanospace. During the last decade, researchers have tried many different ways to obtain a controlled phase-size. A few methods using porous inorganic materials, such as clay, ceramic, silicate gel, molecular sieves, and zeolite, are often used as templates to construct nano-confined environments (see, e.g., Giannelis *et al.*, 1999; Ogata *et al.*, 1997; Park and McKenna, 2000). These methods, in practice, are useful in controlling the phase size yet are not precise enough to meet the quantitative requirements for studying polymer phase transformations, structural changes, and phase stability in confined geometries.

As described in an earlier part of this section, in the strong and medium segregation limit regime, a variety of ordered phase morphologies, such as lamellae, double gyroid, hexagonal cylinders, and body-centered cubic spheres, can exist on the length scale of a few tens of nanometers. These phase-separated morphologies minimize the system's free energy by balancing the interfacial energy and chain-stretching while maintaining constant density (Bates and Fredrickson, 1990). Now, if we use phase-separated, crystalline–amorphous diblock copolymers to form these morphologies, they possess another structural level at the atomic length scale due to the crystallization of the crystallizable blocks within the phase-separated domains. These structural processes are associated with three thermal events: the self-organization of the diblock copolymer, vitrification of the amorphous block, and crystallization of the crystallizable block. Although these three processes possess different thermodynamic and kinetic origins, three temperature parameters are used to describe these events: the order–disorder transition temperature (T_{OD}), the glass transition of the amorphous blocks (T_{g}), and the crystallization temperature of the crystallizable blocks (T_{x}) (always lower than the melting temperature of the crystalline blocks) (Zhu *et al.*, 1999). These processes may compete or override one another in forming the final morphology at different temperatures and pressures.

In the case where the order–disorder transition temperature is the highest, followed by the glass transition temperature and the crystallization temperature ($T_{OD} > T_g > T_x$), the crystallization of the crystallizable blocks occurs within the ordered phase morphology, and the crystallization is confined by the vitrified amorphous block. Examples of this type of crystallization are found in polytetrahydrofuran-*block*-poly(methyl methacrylate) (Liu and Chu, 1999; Liu *et al.*, 1996) and polyethylene-containing diblock copolymers (Cohen *et al.*, 1990, 1994; Douzinas and Cohen, 1992; Hamely *et al.*, 1996a,b, 1998). Recently, a study of poly(ε-caprolactone)-*block*-polystyrene diblock copolymer was attempted in order to correlate the polystyrene block's vitrification with the poly(ε-caprolactone) block's crystallization within nanospheres (Loo *et al.*, 2000). Studies of crystallizable blocks as the matrix in the inverse hexagonal cylindrical phase morphology of block copolymers were also reported (Huang *et al.*, 2007; Loo *et al.*, 2000; Park *et al.*, 2000). Another approach was to cross-link the polybutadiene blocks of a poly(ε-caprolactone)-*block*-polybutadiene diblock copolymer (Chen *et al.*, 2001). The polybutadiene blocks become solid-like after cross-linking; thus, crystallization of the poly(ε-caprolactone) block can only take place in the confined phase morphology. As a result, the memory of the phase morphology is preserved.

Diblock copolymers have been used as templates to construct one-dimensional and two-dimensional nano-confined environments. A series of diblock copolymers with crystallizable poly(ethylene oxide) blocks and amorphous polystyrene blocks in the amorphous phase was used to study confined polymer phase transformations. To ensure a "solid" confinement environment, the glass transition of the polystyrene blocks needs to be higher than the melting temperature of the poly(ethylene oxide) block crystals. Furthermore, the order–disorder transition temperature of this series of copolymers is much higher than both the glass transition temperature of the polystyrene blocks and the melting temperature of the poly(ethylene oxide) block crystals. By adjusting the composition of these two components, one can obtain lamellar (one-dimensional) and hexagonal cylinder (two-dimensional) confinement when the amorphous polystyrene, glassy phase is the matrix. The size of the confinement can be controlled by changing the molecular weight of the block copolymers in the ordered morphology region.

In the case of the lamellar phase morphology in a poly(ethylene oxide)-*block*-polystyrene diblock copolymer with a 8.7 kg/mol poly(ethylene oxide) block and a 11 kg/mol polystyrene block, the poly(ethylene oxide) blocks crystallize in a one-dimensionally confined lamellar space with a thickness of 8.8 nm. This system was studied recently by small angle x-ray scattering and transmission electron diffraction (Zhu *et al.*, 2000). The sample was first mechanically sheared with large amplitude and at low frequency, then isothermally crystallized at different temperatures. The crystal orientation, represented by the *c* axis orientation of the poly(ethylene oxide) block crystals within these nanoscale-confined lamellae, was investigated by combining small and wide angle x-ray scattering experiments. The *c* axis orientation in the poly(ethylene oxide) block crystals has been observed to change only with crystallization temperature. When the sample was quickly quenched into liquid nitrogen, the *c* axis of the poly(ethylene oxide) block crystals was randomly oriented. The *c* axis of the crystals was preferentially oriented perpendicular to the lamellar normal direction (the homogeneous configuration) when the crystallization temperature was between −50 and −10 °C. Increasing the crystallization temperature between −50 and −10 °C resulted in gradually improved crystal orientation. The *c* axes of the poly(ethylene oxide) block crystals were preferentially tilted toward the lamellar normal direction when the crystallization temperature was between −5 and 30 °C. With increasing crystallization temperatures, the tilt-angle of the *c* axis with respect to the lamellar normal gradually decreased. Finally, the *c* axis of the poly(ethylene oxide) block crystals was oriented parallel to the lamellar normal, when the crystallization temperature equaled or exceeded 35 °C (the homeotropic configuration).

Figure 4.32 illustrates both the small and wide angle x-ray scattering data at different crystallization temperatures, indicating the crystal orientation change. The schematics included in this figure show the crystal orientation in real space. Detailed crystallographic analysis indicates that the *c* axis orientation at each crystallization temperature is uniform rather than a mixture of different crystal orientations (Zhu *et al.*, 2000). Furthermore, this one-dimensional confinement effect on crystal orientation is critically dependent on the confinement size (Huang *et al.*, 2004).

The hexagonal cylinder-forming poly(ethylene oxide)-*block*-polystyrene sample, with 8.8 kg/mol molecular weight poly(ethylene oxide) blocks and 24.5 kg/mol polystyrene blocks, was studied to determine the orientation of the crystals in a two-dimensional confined environment. In this system, the volume fraction of the poly (ethylene oxide) blocks was 0.26. Small angle x-ray scattering data showed that the poly(ethylene oxide) blocks formed cylinders with a diameter of 13.3 nm in a hexagonal lattice where $a = 25.1$ nm. It was found that the crystal orientation also changed with respect to crystallization temperature. At low crystallization temperatures (less than −30 °C), poly(ethylene oxide) crystals were randomly oriented within the confining cylinders. Starting at a crystallization temperature of −30 °C, the crystal orientation changed to tilted with respect to the cylinder axis. The tilt angle with respect to the cylinder axis continuously increased with increasing crystallization temperature, finally reaching 90° when the crystallization temperature reached 2 °C. Crystallographic analysis also indicates that the crystal *c* axis orientation at each crystallization temperature was uniform (Huang *et al.*, 2006). These observations were identical to a blend system of poly(ethylene oxide)-*block*-polystyrene mixed with homopolystyrene (Huang *et al.*, 2001). Figure 4.33 illustrates the poly(ethylene oxide) block crystal orientation changes with respect to isothermal crystallization temperature in this system. Again, the schematics included in this figure show the crystal orientation in real space at different crystallization temperatures.

For polymer crystallization in a three-dimensionally confined environment, discrete spheres of a block copolymer are needed. In the ideal case, the bulk matrix needs to be below its glass transition temperature to preserve the ridged confinement provided by the solid state. In a study by Loo *et al.* (2000), the diblock copolymer consisted of polyethylene blocks with styrene-ethylene-butene random terpolymer blocks. In this copolymer, the polyethylene blocks were short, with a 14.3% weight

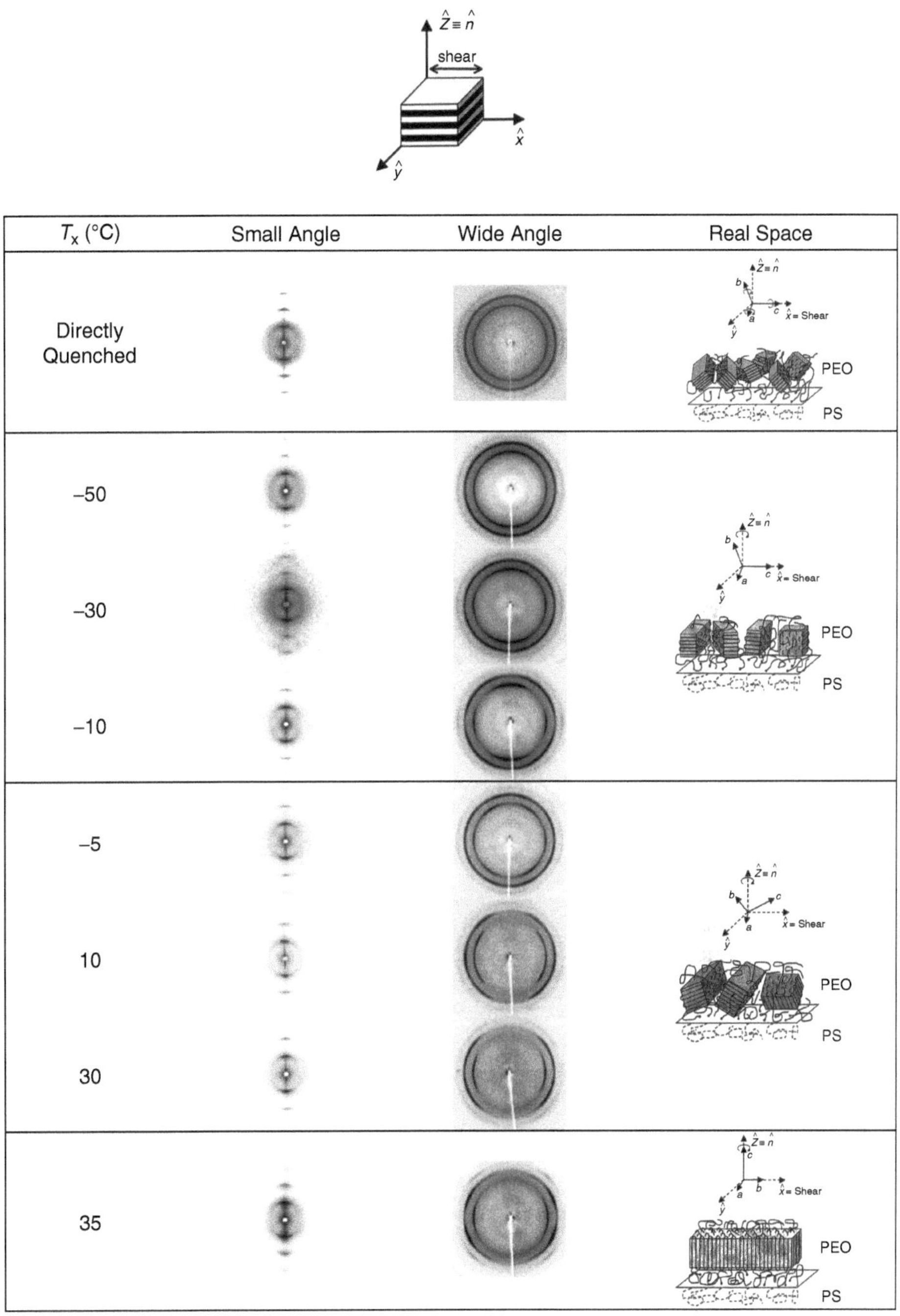

Figure 4.32 Small and wide angle x-ray scattering results for a lamella-forming poly(ethylene oxide)-*block*-polystyrene diblock copolymer after the samples were mechanically sheared and isothermally crystallized at different temperatures. All the two-dimensional patterns were taken along the *x* direction of the samples, and the mechanical geometry of the sample is also shown. The schematics represent the crystal orientation changes in real space at different crystallization temperatures, T_x [reprinted from Zhu *et al.* (2000), with permission].

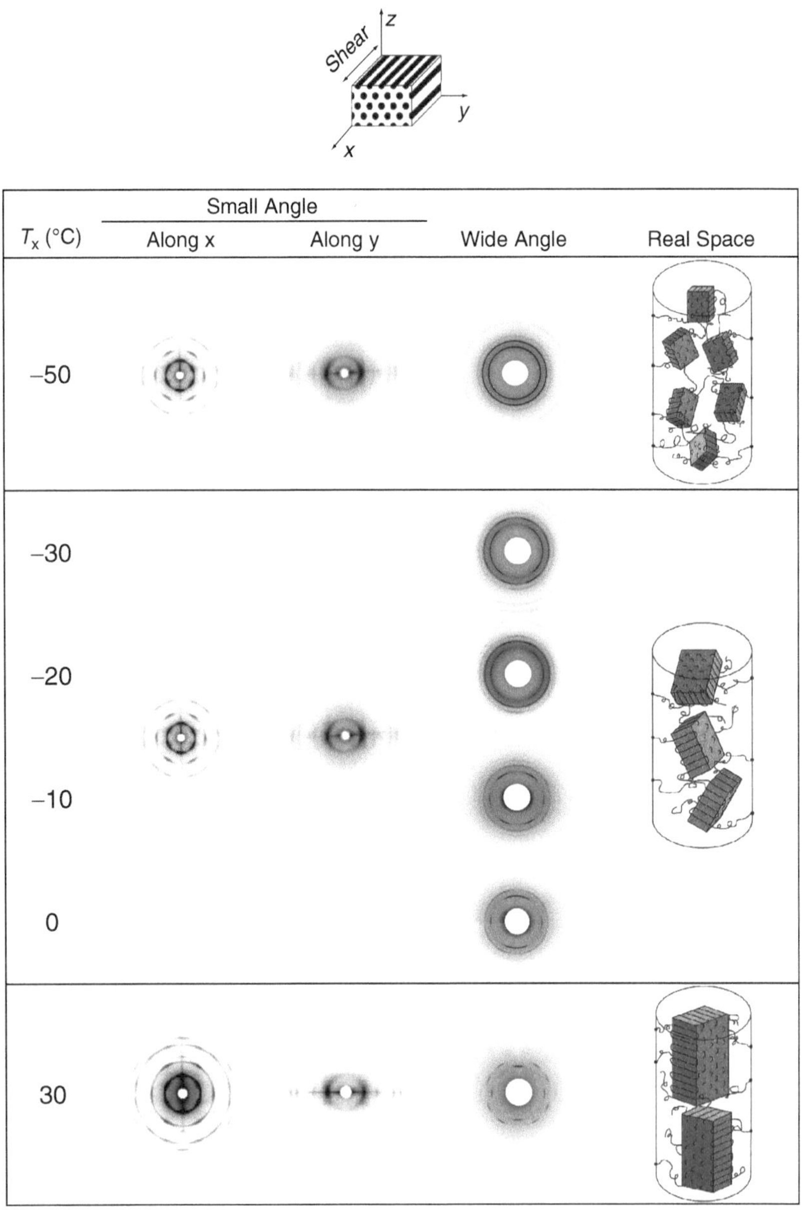

Figure 4.33 Two-dimensional small and wide angle x-ray scattering for a cylinder-forming poly (ethylene oxide)-*block*-polystyrene diblock copolymer blended with homopolystyrene after mechanical shearing and isothermal crystallization at different temperatures. Small angle x-ray scattering patterns were taken along the *x*- and *y*-directions of the samples. Wide angle x-ray diffraction patterns were taken along the *y*-direction. The schematics represent the crystal orientation in real space at different crystallization temperatures, T_x [reprinted from Huang *et al.* (2001), with permission].

fraction, and they formed spheres that packed into a cubic lattice below the order–disorder transition temperature. The amorphous terpolymer blocks had a glass transition temperature of 25 °C. Although the crystallization temperatures of the polyethylene blocks are higher than the glass transition temperature of the terpolymer block matrix, the cubic lattice phase morphology was preserved after crystallization due to the strong segregation between the blocks. No specific crystal orientation in the spheres was expected because in principle, each independent sphere of polyethylene blocks was required to have one primary nucleus to develop the crystal. Homogeneous nucleation during the isothermal crystallization within each sphere was found to follow first-order kinetics, differing from the kinetics normally exhibited in the quiescent crystallization of semicrystalline polymers (Loo *et al.*, 2000).

Another interesting example is the case of a poly(ethylene oxide)-*block*-polystyrene polymer that forms the hexagonally perforated layer phase, with the poly(ethylene oxide) blocks crystallizing in the confined phase structure. The orientation of the poly(ethylene oxide) crystals in this confined phase structure reflects both the crystal orientations in the lamellar and cylinder phases (Huang *et al.*, 2001, 2006; Zhu *et al.*, 2000). More importantly, however, in this case, metastable crystals are confined by a metastable hexagonally perforated layer phase structure and, therefore, form metastable states on two different length scales.

The key question is: What is the dominant factor that causes crystal orientation changes in these nano-confined environments? To answer this question, we need to approach the issue both experimentally and theoretically to understand how polymer nucleation and growth, as well as anisotropic thermal fluctuations in the supercooled melt, affect crystal orientation changes.

To experimentally determine whether primary nucleation or the crystal growth determines the crystal orientation in these nano-confined environments, specifically designed self-seeded crystallization experiments with two-dimensional wide and small angle x-ray scattering were carried out. The poly(ethylene oxide)-*block*-polystyrene diblock copolymer described previously with the lamellar phase morphology was used. The results obtained suggested that orientation of primary nuclei (seeds) do not affect the orientation of the poly(ethylene oxide) block crystals. It was found that the final crystal orientation in the nano-confined lamellae develops to maximize the crystallinity in the system. Studies of the correlation length (apparent crystallite sizes) along the {120} normal directions of the poly(ethylene oxide) block crystals at various crystal orientations indicated that the crystals formed in nano-confined lamellae undergo a change from one-dimensional to two-dimensional growth with increasing crystallization temperature (Zhu *et al.*, 2001a).

The theoretical approach needs to carefully consider the anisotropic thermal fluctuations in the supercooled melt. When the size of the isotropic melt is on the nanometer scale in one- or two-dimensions, it is smaller than the wavelength of the thermal fluctuations in the melt. Therefore, the solid confining geometry starts to influence the thermal fluctuations. As a result, the thermal fluctuations are anisotropic in these nano-confined environments. As has been pointed out in Section 1.5 of this Chapter, thermal (density) fluctuations are required to overcome the nucleation barrier in polymer crystallization (or even more generally, in any first-order phase transition). Therefore, the anisotropic thermal fluctuations may influence the orientation preference of primary nucleation. However, detailed studies along these lines have not been actively pursued. Some numerical calculations indicate that crystal orientation changes within nanolamellar and cylindrical confined environments occur at different crystallization temperatures (Hu and Frenkel, 2005; Wang *et al.*, 2006).

REFERENCES AND FURTHER READING

Adamovsky, S., and Schick, C. (2004). Ultra-fast isothermal calorimetry using thin film sensors. *Thermochim. Acta* **415**, 1–7.

Adamovsky, S. A., Minakov, A. A., and Schick, C. (2003). Scanning microcalorimetry at high cooling rate. *Thermochim. Acta* **403**, 55–63.

Ahn, J.-H., and Zin, W.-C. (2000). Structure of shear-induced perforated layer phase in styrene-isoprene diblock copolymer melts. *Macromolecules* **33**, 641–644.

Alamo, R. G., Blanco, J. A., Agarwal, P. K., and Randall, J. C. (2003). Crystallization rates of matched fractions of $MgCl_2$-supported Ziegler Natta and metallocene isotactic poly(propylene)s. 1. The role of chain microstructure. *Macromolecules* **36**, 1559–1571.

Alcazar, D., Thierry, A., Schultz, P., Kawaguchi, A., Cheng, S. Z. D., and Lotz, B. (2006). Determination of the extent of lateral spread and density of second nucleation density in polymer single crystal growth. *Macromolecules* **39**, 9120–9131.

Alemán, C., Lotz, B., and Puiggali, J. (2001). Crystal structure of the α-form of poly(L-lactide). *Macromolecules* **34**, 4795–4801.

Alfonso, G. C., and Russell, T. P. (1986). Kinetics of crystallization in semicrystalline/amorphous polymer mixtures. *Macromolecules* **19**, 1143–1152.

Allen, S. M., and Thomas, E. L. (1999). "The Structure of Materials." John Wiley & Sons, New York.

Almdal, K., Koppi, K. A., Bates, F. S., and Mortensen, K. (1992). Multiple ordered phases in a block copolymer melt. *Macromolecules* **25**, 1743–1751.

Angell, C. A. (1985). Strong and fragile liquids. *In* "Relaxation in Complex Systems." (K. L. Ngai, G. B. Wright, Eds.), Chapter 1, National Technical Information Series. US Department of Commerce, Springfield.

Arlie, J.-P., Spegt, P., and Skoulios, A. E. (1965). Variation discontinue du nombre de repliement des chaînes d'un polyoxyéthylène crystallize en masse. Comptes Rendus Hebdomadaires des Séances de l'Académie des. *Sciences* **260**, 5774–5777.

Arlie, J. P., Spegt, P. A., and Skoulios, A. E. (1966). Etude de la cristallisation des polymères. I. Structure lamellaire de polyoxyéthylhes de faible masse moléculaire. *Die Makromolekulare Chemie* **99**, 160–174.

Arlie, J. P., Spegt, P. A., and Skoulious, A. E. (1967). Etude de la cristallisation des polymères. II. Structure lamellaire et repliement des chaines du polyoxyéthylène. *Die Makromolekulare Chemie* **104**, 212–229.

Armistead, J. P., and Hoffman, J. D. (2002). Direct evidence of regimes I, II, and III in linear polyethylene fractions as revealed by spherulite growth rates. *Macromolecules* **35**, 3895–3913.

Armitstead, K., Goldbeck-Wood, G., and Keller, A. (1992). Polymer crystallization theories. *Adv. Polym. Sci.* **100**, 219–312.

Atkins, E. D. T., Isaac, D. H., and Keller, A. (1980). Conformation of polystyrene with special emphasis to the near all-trans extended-chain model relevant in polystyrene gels. *J. Polym. Sci., Polym. Phys. Ed.* **18**, 71–82.

Auriemma, F., de Ballesteros, O. R., De Rosa, C., and Corradini, P. (2000). Structural disorder in the α form of isotactic polypropylene. *Macromolecules* **33**, 8764–8774.

Bakai, A. S., and Fischer, E. W. (2004). Nature of long-range correlations of density fluctuations in glass-forming liquids. *J. Chem. Phys.* **120**, 5235–5252.

Bassett, D. C. (1981). "Principles of Polymer Morphology." Cambridge University Press, Cambridge.

Bassett, D. C., and Keller, A. (1962). The habits of polyethylene crystals. *Philos. Mag.* **7**, 1553–1584.

Bassett, D. C., Frank, F. C., and Keller, A. (1959). Evidence for distinct sectors in polymer single crystals. *Nature* **184**, 810–811.

Bassett, D. C., Olley, R. H., and Al Raheil, A. M. (1988). On isolated lamellae of melt-crystallized polyethylene. *Polymer* **29**, 1539–1543.

Bates, F., and Fredrickson, G. H. (1990). Block copolymer thermodynamics: Theory and experiment. *Annu. Rev. Phys. Chem.* **41**, 525–557.

Bhargava, P., Zheng, J. X., Li, P., Quirk, R. P., Harris, F. W., and Cheng, S. Z. D. (2006). Self-assembled polystyrene-block-poly(ethylene oxide) micelle morphologies in solutions. *Macromolecules* **39**, 4880–4888.

Bhargava, P., Tu, Y., Zheng, J. X., Xiong, H., Quirk, R. P., and Cheng, S. Z. D. (2007). Temperature-induced reversible micelle morphological changes of polystyrene-block - poly (ethylene oxide) micelles in solution. *J. Am. Chem. Soc.* **129**, 1113–1121.

Boda, E., Ungar, G., Brooke, G. M., Burnett, S., Mohammed, S., Proctor, D., and Whiting, M. C. (1997). Crystallization rate minima in a series of n-alkanes from $C_{194}H_{390}$ to $C_{294}H_{590}$. *Macromolecules* **30**, 4674–4678.

Brückner, S., and Meille, S. V. (1989). Non-parallel chains in crystalline γ-isotactic polypropylene. *Nature* **340**, 455–457.

Brückner, S., Allegra, G., and Corradini, P. (2002). Helix inversions in polypropylene and polystyrene. *Macromolecules* **35**, 3928–3936.

Bu, H. S., Pang, Y. W., Song, D. D., Yu, T. Y., Voll, T. M., Czornyj, G., and Wunderlich, B. (1991). Single-molecule single crystals. *J. Polym. Sci., Polym. Phys. Ed.* **29**, 139–152.

Bu, H. S., Gu, F., Bao, L., and Chen, M. (1998a). Influence of entanglements on crystallization of macromolecules. *Macromolecules* **31**, 7108–7110.

Bu, H. S., Cao, J., Zhang, Z. S., Zhang, Z., Festag, R., Joy, D. C., Kwon, Y. K., and Wunderlich, B. (1998b). Structure of single-molecule single crystals of isotactic polystyrene and their radiation resistance. *J. Polym. Sci., Polym. Phys. Ed.* **36**, 105–112.

Bu, Z., Yoon, Y., Ho, R.-M., Zhou, W., Jangchud, I., Eby, R. K., Cheng, S. Z. D., Hsieh, E. T., Johnson, T. W., Geerts, R. G., Palackal, S. J., Hawley, G. R., *et al.* (1996). Crystallization, melting, and morphology of syndiotactic polypropylene fractions. 3. Lamellar single crystals and chain folding. *Macromolecules* **29**, 6575–6581.

Bunn, C. W., Cobbold, A. J., and Palmer, R. P. (1958). The fine structure of polytetrafluoro-ethylene. *J. Polym. Sci.* **28**, 363–376.

Cartier, L., Okihara, T., and Lotz, B. (1997). Triangular polymer single crystals: Stereocomplexes, twins, and frustrated structures. *Macromolecules* **30**, 6313–6322.

Chen, H.-L., Hsiao, S.-C., Lin, T.-L., Yamauchi, K., Hasegawa, H., and Hashimoto, T. (2001). Microdomain-tailored crystallization kinetics of block copolymers. *Macromolecules* **34**, 671–674.

Chen, J. T., Thomas, E. L., Ober, C. K., and Hwang, S. S. (1995). Zigzag morphology of a poly (styrene-b-hexyl isocynate) rod-coil block copolymer. *Macromolecules* **28**, 1688–1697.

Chen, J. T., Thomas, E. L., Ober, C. K., and Mao, G.-P. (1996). Self-assembled smectic phases in rod-coil block copolymers. *Science* **273**, 343–346.

Chen, W. Y., Li, C. Y., Zheng, J. X., Huang, P., Zhu, L., Ge, Q., Quirk, R. P., Lotz, B., Deng, L., Wu, C., Thomas, E. L., and Cheng, S. Z. D. (2004a). Chemically shielded" poly(ethylene oxide) single crystal growth and construction of channel-wire arrays with chemical and geometric recognitions on a submicrometer scale. *Macromolecules* **37**, 5292–5299.

Chen, W. Y., Zheng, J. X., Cheng, S. Z. D., Li, C. Y., Huang, P., Zhu, L., Xiong, H., Ge, Q., Guo, Y., Quirk, R. P., Lotz, B., Deng, L., *et al.* (2004b). Onset of tethered chain overcrowding. *Phys. Rev. Lett.* **93**, 028301.

Cheng, S. Z. D. (2007). Materials science: Polymer crystals downsized. *Nature* **448**, 1006–1007.

Cheng, S. Z. D., and Chen, J. H. (1991). Nonintegral and integral folding crystal growth in low-molecular mass poly(ethylene oxide) fractions. III. Linear crystal growth rates and crystal morphology. *J. Polym. Sci., Polym. Phys. Ed.* **29**, 311–327.

Cheng, S. Z. D., and Li, C. Y. (2002). Structure and formation of polymer single crystal textures. *Mater. Sci. Forum* **408**, 25–37.

Cheng, S. Z. D., and Lotz, B. (2003). Nucleation control in polymer crystallization: Structural and morphological probes in different length- and time-scales for selection processes. *Philos. Transact. Math. Phys. Eng. Sci.* **361**, 517–537.

Cheng, S. Z. D., and Lotz, B. (2005). Enthalpic and entropic origins of nucleation barriers during polymer crystallization: The Hoffman–Lauritzen theory and beyond. *Polymer* **46**, 8662–8681.

Cheng, S. Z. D., and Wunderlich, B. (1986a). Molecular segregation and nucleation of poly(ethylene oxide) crystallized from the melt. I. Calorimetric study. *J. Polym. Sci., Polym. Phys. Ed.* **24**, 577–594.

Cheng, S. Z. D., and Wunderlich, B. (1986b). Molecular segregation and nucleation of poly(ethylene oxide) crystallized from the melt. II. Kinetic study. *J. Polym. Sci., Polym. Phys. Ed.* **24**, 595–617.

Cheng, S. Z. D., Bu, H. S., and Wunderlich, B. (1988). Molecular segregation and nucleation of poly (ethylene oxide) crystallized from the melt. III. Morphological study. *J. Polym. Sci., Polym. Phys. Ed.* **26**, 1947–1964.

Cheng, S. Z. D., Janimak, J. J., Zhang, A., and Hsien, E. T. (1991). Isotacticity effect on crystallization and melting in polypropylene fractions. I. Crystalline structures and thermodynamic property changes. *Polymer* **32**, 648–655.

Cheng, S. Z. D., Chen, J., and Heberer, D. P. (1992). Extended chain crystals-growth of low-molecular mass poly(ethylene oxide) and α,ω-methoxy poly(ethylene oxide) fractions near their melting temperatures. *Polymer* **33**, 1429–1436.

Cheng, S. Z. D., Li, C. Y., and Zhu, L. (2000). Commentary on polymer crystallization: Selection rules in different length scales of a nucleation process. *Eur. Phys. J. E Soft. Matter.* **3**, 195–197.

Chiu, F.-C., Wang, Q., Fu, Q., Honigfort, P., Cheng, S. Z. D., Hsiao, B. S., Yeh, F. J., Keating, M. Y., Hsieh, E. T., and Tso, C. C. (2000). Structural and morphological inhomogeneity of short-chain branched polyethylenes in multiple-step crystallization. *J. Macromol. Sci., Part B: Phys.* **39**, 317–331.

Cicerone, M. T., and Ediger, M. D. (1995). Relaxation of spatially heterogeneous dynamics in supercooled ortho-terphenyl. *J. Chem. Phys.* **103**, 5684–5692.

Cohen, R. E., Cheng, P. L., Douzinas, K., Kofinas, P., and Berney, C. V. (1990). Path-dependent morphologies of a diblock copolymer of polystyrene/hydrogenated polybutadiene. *Macromolecules* **23**, 324–327.

Cohen, R. E., Bellare, A., and Drzewinski, M. A. (1994). Spatial organization of polymer chains in a crystallizable diblock copolymer of polyethylene and polystyrene. *Macromolecules* **27**, 2321–2323.

Colet, M. Ch., Point, J. J., and Dosiere, M. (1986). Nucleation-controlled growth. Observations on (110) twinned polyethylene crystals. *J. Polym. Sci., Polym. Phys. Ed.* **24**, 1183–1206.

Dai, P. S., Cebe, P., Capel, M., Alamo, R. G., and Mandelkern, L. (2003). *In situ* wide- and small-angle X-ray scattering study of melting kinetics of isotactic poly(propylene). *Macromolecules* **36**, 4042–4050.

Davis, G. T., Weeks, J. J., Martin, G. M., and Eby, R. K. (1974). Cell dimensions of hydrocarbon crystals. Surface effects. *J. Appl. Phys.* **45**, 4175–4181.

Dawson, I. M. (1952). The study of crystal growth with the electron microscope. II. The observation of growth steps in the paraffin n-hectane. *Proc. R. Soc. Lond. A* **214**, 72–79.

De Rosa, C., Guerra, G., Napolitano, R., Petraccone, V., and Pirozzi, B. (1984). Conditions for the α_1–α_2 transition in isotactic polypropylene samples. *Eur. Polym. J.* **20**, 937–941.

De Rosa, C., Petraccone, V., Dal Poggetto, F., Guerra, G., Pirozzi, B., Di Lorenzo, M. L., and Corradini, P. (1995). Crystal-structure of form-III of syndiotactic poly(p-methylstyrene). *Macromolecules* **28**, 5507–5511.

De Rosa, C., Guerra, G., and Pirozzi, B. (1997). Crystal structure of the emptied clathrate form (δ_e form) of syndiotactic polystyrene. *Macromolecules* **30**, 4147–4152.

De Rosa, C., Auriemma, F., and Resconi, L. (2005). Influence of chain microstructure on the crystallization kinetics of metallocene-made isotactic polypropylene. *Macromolecules* **38**, 10080–10088.

De Santis, F., Adamovsky, S., Titomanlio, G., and Schick, C. (2006). Scanning nanocalorimetry at high cooling rate of isotactic polypropylene. *Macromolecules* **39**, 2562–2567.

Debye, P., and Bueche, A. M. (1949). Scattering by an inhomogeneous solid. *J. Appl. Phys.* **20**, 518–525.

DiMarzio, E. A., Guttman, C. M., and Hoffman, J. D. (1980). Calculation of lamellar thickness in a diblock copolymer, one of whose components is crystalline. *Macromolecules* **13**, 1194–1198.
Di Lorenzo, M. L. (2003). Spherulite growth rates in binary polymer blends. *Prog. Polym. Sci.* **28**, 663–689.
Disko, M. M., Liang, K. S., Behal, S. K., Roe, R. J., and Jeon, K. J. (1993). Catenoid-lamellar phase in blends of styrene-butadiene diblock copolymer and homopolymer. *Macromolecules* **26**, 2983–2986.
Dosière, M., Colet, M.-Ch., and Point, J. J. (1986). An isochronous decoration method for measuring linear growth rates in polymer crystals. *J. Polym. Sci., Polym. Phys. Ed.* **24**, 345–356.
Douzinas, K. C., and Cohen, R. E. (1992). Chain folding in EBEE semicrystalline diblock copolymers. *Macromolecules* **25**, 5030–5035.
Efremov, M. Y., Warren, J. T., Olson, E. A., Zhang, M., Kwan, A. T., and Allen, L. H. (2002). Thin-film differential scanning calorimetry: A new probe for assignment of the glass transition of ultrathin polymer films. *Macromolecules* **35**, 1481–1483.
Efremov, M. Y., Olson, E. A., Zhang, M., Zhang, Z., and Allen, L. H. (2003). Glass transition in ultrathin polymer films: Calorimetric study. *Phys. Rev. Lett.* **91**, 085703.
Ergoz, E., Fatou, J. G., and Mandelkern, L. (1972). Molecular weight dependence of the crystallization kinetics of linear polyethylene. I. Experimental results. *Macromolecules* **5**, 147–157.
Faraday Discussions of the Chemical Society (1979). "Organization of macromolecules in the condensed phase." The Royal Society of Chemistry, Norwich.
Fischer, E. W. (1957). Stufen- und spiralförmiges Kristallwachstum bei Hochpolymeren. *Zeitschrift für Naturforschung A: Astrophisik, Physik und Physikalische Chemie* **12**, 753–754.
Fischer, E. W. (1993). Light scattering and dielectric studies on glass forming liquids. *Physica A: Statistical Mechanics and its Applications* **201**, 183–206.
Fischer, E. W., Bakai, A., Patkowski, A., Steffen, W., and Reinhardt, L. (2002). Heterophase fluctuations in supercooled liquid and polymers. *J. Non-cryst. Solids* **307–310**, 584–601.
Flory, P. J. (1955). Theory of crystallization in copolymers. *Trans. Faraday Soc.* **51**, 848–857.
Flory, P. J. (1969). "Statistical Mechanics of Chain Molecules." Wiley-Interscience, New York.
Förster, S., Khandpur, A. K., Zhao, J., Bates, F. S., Hamley, I. W., Ryan, A. J., and Bras, W. (1994). Complex phase behavior of polyisoprene-polystyrene diblock copolymers near the order-disorder transition. *Macromolecules* **27**, 6922–6935.
Fourkas, J. T., Kivelson, D., Mohanty, U., and Nekson, K. A., eds. (1997). "Supercooled Liquids: Advances and Novel Applications." ACS, Washington DC.
Frank, F. C. (1949). The influence of dislocations on crystal growth. *Discuss. Faraday Soc.* **5**, 48–54.
Frank, F. C. (1974). Nucleation-controlled growth on a one dimensional growth of finite length. *J. Cryst Growth* **22**, 233–236.
Frank, F. C., and Tosi, M. (1961). The theory of polymer crystallization. *Proc. R. Soc. Lond. A* **263**, 323–339.
Frank, F. C., Keller, A., and O'Connor, A. (1959). Single crystals of an isotactic polyolefin: Morphology and chain packing in poly(4-methyl-pentene-1). *Philos. Mag.* **4**, 200–214.
Geil, P. H. (1963). "Polymer Reviews. Volume 5. Polymer Single Crystals." Wiley-Interscience, New York.
Geil, P. H. (2000). Some "overlooked problems" in polymer crystallization. *Polymer* **41**, 8983–9001.
Geil, P. H., Anderson, F. R., Wunderlich, B., and Arakawa, T. (1964). Morphology of polyethylene crystallized from the melt under pressure. *J. Polym. Sci., Part A* **2**, 3707–3720.
Geil, P. H., Yang, J., Williams, R. A., Petersen, K. L., Long, T.-C., and Xu, P. (2005). Effect of molecular weight and melt time and temperature on the morphology of poly(tetrafluorethylene). *Adv. Polym. Sci.* **180**, 89–159.
Giannelis, E. P., Krishnamoorti, R., and Manias, E. (1999). Polymer-silicate nanocomposites: Model systems for confined polymers and polymer brushes. *Adv. Polym. Sci.* **138**, 107–147.
Hajduk, D. A., Harper, P. E., Gruner, S. M., Honeker, C. C., Kim, G., Thomas, E. L., and Fetters, L. J. (1994). The gyroid: A new equilibrium morphology in weakly segregated diblock copolymers. *Macromolecules* **27**, 4063–4075.
Hajduk, D. A., Takenouchi, H., Hillmyer, M. A., Bates, F. S., Vigild, M. E., and Almdal, K. (1997). Stability of the perforated layer (PL) phase in diblock copolymer melts. *Macromolecules* **30**, 3788–3795.

Hamley, I. W., Koppi, K. A., Rosedale, J. H., Bates, F. S., Almdal, K., and Mortensen, K. (1993). Hexagonal mesophases between lamellae and cylinders in a diblock copolymer melt. *Macromolecules* **26**, 5959–5970.

Hamley, I. W., Fairclough, J. P. A., Ryan, A. J., Bates, F. S., and Towns-Andrews, E. (1996a). Crystallization of nanoscale-confined diblock copolymer chains. *Polymer* **37**, 4425–4429.

Hamley, I. W., Fairclough, J. P. A., Terrill, N. J., Ryan, A. J., Lipic, P. M., Bates, F. S., and Towns-Andrews, E. (1996b). Crystallization in oriented semicrystalline diblock copolymers. *Macromolecules* **29**, 8835–8843.

Hamley, I. W., Fairclough, J. P. A., Bates, F. S., and Ryan, A. J. (1998). Crystallization thermodynamics and kinetics in semicrystalline diblock copolymers. *Polymer* **39**, 1429–1437.

Hashimoto, T. (2004). Small-angle neutrin scattering studies of dynamics and hierarchical pattern formation in binary mixtures of polymers and small molecules. *J. Polym. Sci., Polym. Phys. Ed.* **42**, 3027–3062.

Hikosaka, M. (1987). Unified theory of nucleation of folded-chain crystals and extended-chain crystals of linear-chain polymers. *Polymer* **28**, 1257–1264.

Hikosaka, M. (1990). Unified theory of nucleation of folded-chain crystals (FCCS) and extended-chain crystals (ECCS) of linear-chain polymers. 2. Origin of FCC and ECC. *Polymer* **31**, 458–468.

Hikosaka, M., and Seto, T. (1973). The order of the molecular chains in isotactic polypropylene crystals. *Polym. J. (Jpn.)* **5**, 111–127.

Ho, R.-M., Chiang, Y.-W., Tsai, C.-C., Lin, C.-C., Chung, B.-T., and Huang, B.-H. (2004). Three-dimensionally packed nanohelical phase in chiral block copolymers. *J. Am. Chem. Soc.* **126**, 2704–2705.

Hobbs, J. K., Hill, M. J., and Barham, P. J. (2001). Crystallization and isothermal thickening of single crystals of $C_{246}H_{494}$ in dilute solution. *Polymer* **42**, 2167–2176.

Hocquet, S., Dosière, M., Thierry, A., Lotz, B., Koch, M. H. J., Dubreuil, N., and Ivanov, D. A. (2003). Morphology and melting of truncated single crystals of linear polyethylene. *Macromolecules* **36**, 8376–8384.

Hoffman, J. D. (1982). Role of reptation in the rate of crystallization of polyethylene fractions from the melt. *Polymer* **23**, 656–670.

Hoffman, J. D. (1983). Regime III crystallization in melt-crystallized polymers: The variable cluster model of chain folding. *Polymer* **24**, 3–26.

Hoffman, J. D. (1985a). The kinetic substrate length in nucleation-controlled crystallization in polyethylene fractions. *Polymer* **26**, 803–810.

Hoffman, J. D. (1985b). Theory of the substrate length in polymer crystallization: Surface roughening as an inhibitor for substrate completion. *Polymer* **26**, 1763–1778.

Hoffman, J. D. (1991). Transition from extended-chain to once-folded behavior in pure n-paraffins crystallized from the melt. *Polymer* **32**, 2828–2841.

Hoffman, J. D. (1992). The relationship of C_∞ to the lateral surface free energy σ: Estimation of C_∞ for the melt from rate of crystallization data. *Polymer* **33**, 2643–2644.

Hoffman, J. D., and Lauritzen, J. I., Jr. (1961). Crystallization of bulk polymers with chain folding: Theory of growth of lamellar spherulites. *J. Res. Natl. Bur. Stand. Sect. A* **65**, 297–336.

Hoffman, J. D., and Miller, R. L. (1989). Response to criticism of nucleation theory as applied to crystallization of lamellar polymers. *Macromolecules* **22**, 3502–3505.

Hoffman, J. D., and Miller, R. L. (1997). Kinetics of crystallization from the melt and chain folding in polyethylene fractions revisited: Theory and experiment. *Polymer* **38**, 3151–3212.

Hoffman, J. D., Lauritzen, J. I., Jr., Passaglia, E., Ross, G. S., Frolen, L. J., and Weeks, J. J. (1969). Kinetics of polymer crystallization from solution and the melt. *Kolloid-Zeitschrift & Zeitschrift für Polymere* **231**, 564–592.

Hoffman, J. D., Frolen, L. J., Gaylon, S. R., and Lauritzen, J. I., Jr. (1975). On the growth rate of spherulites and axialites from the melt in polyethylene fractions: Regime I and regime II crystallization. *J. Res. Natl. Bur. Stand. Sect. A* **79**, 671–699.

Hoffman, J. D., Davis, G. T., and Lauritzen, J. I., Jr. (1976). The rate of crystallization of linear polymers with chain folding. *In* "Treatise on Solid State Chemistry. Volume 3. Crystalline and Noncrystalline Solids" (N. B. Hannay, ed.), Chapter 7, Plenum Press, New York.

Hoffman, J. D., Guttman, C. M., and DiMarzio, E. A. (1979). On the problem of crystallization of polymers from the melt with chain folding. *Faraday Discuss. Chem. Soc.* **68**, 177–197.

Hoffman, J. D., Miller, R. L., Marand, H., and Roitman, D. B. (1992). Relationship between the lateral surface free energy σ and the chain structure of melt-crystallized polymers. *Macromolecules* **25**, 2221–2229.

Höhne, G. W. H., Rastogi, S., and Wunderlich, B. (2000). High pressure differential scanning caloremitry of poly(4-methyl-pentene-1). *Polymer* **41**, 8869–8878.

Hosier, I. L., Bassett, D. C., and Vaughan, A. S. (2000). Spherulitic growth and cellulation in dilute blends of monodisperse long n-alkanes. *Macromolecules* **33**, 8781–8790.

Hosier, I. L., Alamo, R. G., Esteso, P., Isasi, J. R., and Mandelkern, L. (2003). Formation of the α and γ polymorphs in random metallocene-propylene copolymers. Effect of concentration and type of comonomer. *Macromolecules* **36**, 5623–5636.

Hosier, I. L., Alamo, R. G., and Lin, J. S. (2004). Lamellar morphology of random metallocene propylene copolymers studied by atomic force microscopy. *Polymer* **45**, 3441–3455.

Hu, W. B., and Frenkel, D. (2005). Oriented primary crystal nucleation in lamellar diblock copolymer systems. *Faraday Discuss.* **128**, 253–260.

Huang, P., Zhu, L., Cheng, S. Z. D., Ge, Q., Qiurk, R. P., Thomas, E. L., Lotz, B., Hsiao, B. S., Liu, L., and Yeh, F. (2001). Crystal orientation changes in two-dimensionally confined nanocylinders in a poly(ethylene oxide)-b-polystyrene/polystyrene blend. *Macromolecules* **34**, 6649–6657.

Huang, P., Zhu, L., Guo, Y., Ge, Q., Jing, A. J., Chen, W. Y., Quirk, R. P., Cheng, S. Z. D., Thomas, E. L., Lotz, B., Hsiao, B. S., Avila-Orta, C. A., *et al.* (2004). Confinement size effect on crystal orientation changes of poly(ethylene oxide) blocks in poly(ethylene oxide)-b-polystyrene diblock copolymers. *Macromolecules* **37**, 3689–3698.

Huang, P., Guo, Y., Quirk, R. P., Ruan, J., Lotz, B., Thomas, E. L., Hsiao, B. S., Avila-Orta, C. A., Sics, I., and Cheng, S. Z. D. (2006). Comparison of poly(ethylene oxide) crystal orientations and crystallization behaviors in nano-confined cylinders constructed by a poly(ethylene oxide)-b-polystyrene diblock copolymer and a blend of poly(ethylene oxide)-b-polystyrene and polystyrene. *Polymer* **47**, 5457–5466.

Huang, P., Zheng, J. X., Leng, S., Van Horn, R., Hsiao, M.-S., Jeong, K.-U., Guo, Y., Quirk, R. P., Cheng, S. Z. D., Lotz, B., Thomas, E. L., and Hsiao, B. (2007). Poly(ethylene oxide) crystal orientation changes in an inversed cylindrical morphology constructed by a poly(ethylene oxide)-block-polystyrene diblock copolymer. *Macromolecules* **40**, 526–534.

Iguchi, M. (1973). Growth of needle-like crystals of polyoxymethyene during polymerization. *Br. Polym. J.* **5**, 195–198.

Iguchi, M., Kanetsuna, H., and Kawai, T. (1969). Growth of polyoxymethylene crystals in the course of polymerization of trioxane in solution. *Die Makromolekulare Chemie* **128**, 63–82.

Imai, M., Kaji, K., and Kanaya, T. (1993). Orientation fluctuations of poly(ethylene terephthalate) during the induction period of crystallization. *Phys. Rev. Lett.* **71**, 4162–4165.

Jaccodine, R. (1955). Observations of spiral growth steps in ethylene polymer. *Nature* **176**, 305–306.

Janimak, J. J., Cheng, S. Z. D., Giusti, P. A., and Hsieh, E. T. (1991). Isotacticity effect on crystallization and melting in poly(propylene) fractions. 2. Linear crystal growth rate and morphology study. *Macromolecules* **24**, 2253–2260.

Janimak, J. J., Cheng, S. Z. D., and Zhang, A. (1992). Isotacticity effect on crystallization and melting in polypropylene fractions. 3. Overall crystallization and melting behavior. *Polymer* **33**, 728–735.

Jing, A. J., Taikum, O., Li, C. Y., Harris, F. W., and Cheng, S. Z. D. (2002). Phase identifications and monotropic transition behaviors in a thermotropic main-chain liquid crystalline polyether. *Polymer* **43**, 3431–34340.

Keith, H. D., and Padden, F. J. Jr. (1964a). Spherulitic crystallization from the melt. I. Fractionation and impurity segregation and their influence on crystalline morphology. *J. Appl. Phys.* **35**, 1270–1285.

Keith, H. D., and Padden, F. J. Jr. (1964b). Spherulitic crystallization from the melt. II. Influence of fractionation and impurity segregation on the kinetics of crystallization. *J. Appl. Phys.* **35**, 1286–1296.

Keith, H. D., Padden, F. J. Jr. and Vadimsky, R. G. (1965). Intercrystalline links in bulk polyethylene. *Science* **150**, 1026–1027.

Keith, H. D., Padden, F. J. Jr. and Vadimsky, R. G. (1966a). Intercrystalline links in polyethylene crystallized from the melt. *J. Polym. Sci., Part A-2* **4**, 267–281.

Keith, H. D., Padden, F. J. Jr., and Vadimsky, R. G. (1966b). Further studies of intercrystalline links in polyethylene. *J. Appl. Phys.* **37**, 4027–4034.

Keith, H. D., Padden, F. J. Jr., and Vadimsky, R. G. (1971). Intercrystalline links: Critical evaluation. *J. Appl. Phys.* **42**, 4585–4592.

Keith, H. D., Padden, F. J. Jr., Lotz, B., and Wittmann, J. C. (1989). Asymmtries of habit in polyethylene crystals grown from the melt. *Macromolecules* **22**, 2230–2238.

Keller, A. (1957). A note on single crystals in polymers: Evidence of a folded-chain configuration. *Philos. Mag.* **2**, 1171–1175.

Keller, A. (1968). Polymer crystals. *Rep. Prog. Phys.* **31**, 623–704.

Keller, A. (1995). Aspects of polymer gels. *Faraday Discuss.* **101**, 1–49.

Keller, A., and Ungar, G. (1983). Radiation effects and crystallinity in polyethylene. *Radiat. Phys. Chem.* **22**, 155–181.

Kent, M. S. (2000). A quantitative study of tethered chains in various solution conditions using Langmuir diblock copolymer monolayers. *Macromol. Rapid Commun.* **21**, 243–270.

Khandpur, A. K., Forster, S., Bates, F. S., Hamlet, I. W., Ryan, A. J., Bras, W., Almdal, K., and Mortensen, K. (1995). Polyisoprene-polystyrene diblock copolymer phase diagram near the order-disorder transition. *Macromolecules* **28**, 8796–8806.

Khoury, F. (1963). Crystal habits and morphology of n-tetranonacontane (n-$C_{94}H_{190}$). *J. Appl. Phys.* **34**, 73–79.

Khoury, F., and Padden, F. J. Jr. (1960). Growth habits of twinned crystals of polyethylene. *J. Polym. Sci.* **47** 455–468.

Khoury, F., and Passaglia, E. (1976). The morphology of crystalline synthetic polymers. *In* "Treatise on Solid State Chemistry. Volume 3. Crystalline and Noncrystalline Solids" (N. B. Hannay, ed.), Chapter 6, Plenum Press, New York.

Kim, I., and Krimm, S. (1996). Raman longitudinal acoustic mode studies of a poly(ethylene oxide) fraction during isothermal crystallization from the melt. *Macromolecules* **29**, 7186–7192.

Koningsveld, R., Stockmayer, W. H., and Nies, E. (2001). "Polymer Phase Diagrams, A Textbook." University Press, Oxford.

Kovacs, A. J., and Gonthier, A. (1972). Crystallization and fusion of self-seeded polymers. II. Growth rate, morphology, and isothermal thickening of single crystals of low-molecular-weight poly (ethylene oxide) fractions. *Kolloid-Zeitschrift & Zeitschrift für Polymere* **250**, 530–551.

Kovacs, A. J., and Straupe, C. (1979). Isothermal growth, thickening, and melting of poly(ethylene oxide) single crystals in the bulk. Part 4. Dependence of pathological crystal habits on temperature and thermal history. *Faraday Discuss. Chem. Soc.* **68**, 225–238.

Kovacs, A. J., and Straupe, C. (1980). Isothermal growth, thickening and melting of polyethylene oxide single crystals in the bulk. III. Bilayer crystals and the effect of chain ends. *J. Cryst. Growth* **48**, 210–226.

Kovacs, A. J., Gonthier, A., and Straupe, C. (1975). Isothermal growth, thickening, and melting of poly(ethylene oxide) single crystals in the bulk. *J. Polym. Sci., Polym. Symp.* **50**, 283–325.

Kovacs, A. J., Straupe, C., and Gonthier, A. (1977). Isothermal growth, thickening, and melting of poly(ethylene oxide) single crystals in the bulk. II. *J. Polym. Sci., Polym. Symp.* **59**, 31–54.

Kusanagi, H., Takase, M., Chatani, Y., and Tadokoro, H. (1978). Crystal structure of isotactic poly (4-methyl-1-pentene). *J. Polym. Sci., Polym. Phys. Ed.* **16**, 131–142.

Kubo, S., and Wunderlich, B. (1971). Morphology of poly-p-xylylene crystallized from polymerization. *J. Appl. Phys.* **42**, 4558–4565.

Kubo, S., and Wunderlich, B. (1972). Crystallization during polymerization of poly-p-xylylene. *J. Polym. Sci., Part A-2* **10**, 1949–1966.

Kwon, Y. K., Boller, A., Pyda, M., and Wunderlich, B. (2000). Melting and heat capacity of gel-spun, ultra-high molar mass polyethylene fibers. *Polymer* **41**, 6237–6249.

Laradji, M., Shi, A. C., Desai, R. C., and Noolandi, J. (1997). Stability of ordered phases in weakly segregated diblock copolymer systems. *Phys. Rev. Lett.* **78**, 2577–2580.

Lauritzen, J. I. Jr., and Hoffman, J. D. (1960). Theory of formation of polymer crystals with chain-folded chains in dilute solution. *J. Res. Natl. Bur. Stand. Sect. A* **64**, 73–102.

Lauritzen, J. I. Jr., and Hoffman, J. D. (1973). Extension of theory of growth of chain-folded polymer crystals to large undercoolings. *J. Appl. Phys.* **44**, 4340–4352.

Lauritzen, J. I., Jr., and Passaglia, E. (1967). Kinetics of crystallization in multicomponent systems. II. Chain-folded polymer crystals. *J. Res. Natl. Bur. Stand. Sect. A* **71**, 261–275.

Lefebvre, A. A., Lee, J. H., Balsara, N. P., and Vaidyanathan, C. (2002). Determination of critical length scales and the limit of metastability in phase separating polymer blends. *J. Chem. Phys.* **117**, 9063–9073.

Leung, W. M., St. Manley, R. J., and Panaras, A. R. (1985). Isothermal growth of low molecular weight polyethylene single crystals from solution. 3. Kinetic studies. *Macromolecules* **18**, 760–771.

Lippits, D. R., Rastogi, S., and Höhne, G. W. H. (2006). Melting kinetics in polymers. *Phys. Rev. Lett.* **96**, 218–303.

Liu, J., Kim, D., Harris, F. W., and Cheng, S. Z. D. (1994a). Crystal structure, morphology, and phase transitions in aromatic polyimide oligomers. 2. Poly(1,4-phenylene-oxy-1,4-phenylene pyromellitimide). *Polymer* **35**, 4048–4056.

Liu, J., Kim, D., Harris, F. W., and Cheng, S. Z. D. (1994b). Crystal-structure, morphology, and phase-transitions in aromatic polyimide oligomers. 3. Poly(1,4-phenyleneoxy-1,3-phenylene pyromellitimide). *J. Polym. Sci., Polym. Phys. Ed.* **32**, 2705–2713.

Liu, J., Cheng, S. Z. D., Harris, F. W., Hsiao, B. S., and Gardner, K. H. (1994c). Crystal-structure, morphology, and phase-transitions in aromatic polyimide oligomers. I. Poly(4,4′- oxydiphenylene pyromellitimide). *Macromolecules* **27**, 989–996.

Liu, J., Cheng, S. Z. D., and Geil, P. H. (1996a). Morphology and crystal structure in single crystals of poly(p-phenylene terephthalamide) prepared by melt polymerization. *Polymer* **37**, 1413–1430.

Liu, J., Rybnikar, F., and Geil, P. H. (1996b). Morphology of solution- and melt-polymerized poly (p-oxybenzoate/2,6-naphthoate) copolymers: Single crystals, disclination domains, and superlattices. *J. Macromol. Sci., Part B: Phys.* **35**, 375–410.

Liu, J., Sidoti, G., Hommema, J. A., Geil, P. H., Kim, J. C., and Cakmak, M. (1998). Crystal structures and morphology of thin-film, melt-crystallized, and polymerized poly(ethylene naphthalate). *J. Macromol. Sci., Part B: Phys.* **37**, 567–586.

Liu, L., and Muthukumar, M. (1998). Langevin dynamics simulations of early-stage polymer nucleation and crystallization. *J. Chem. Phys.* **109**, 2536–2542.

Liu, L.-Z., and Chu, B. (1999). Crystalline structure and morphology of microphases in compatible mixtures of poly(tetrahydrofuran-methyl methacrylate) diblock copolymer and polytetrahydrofuran. *J. Polym. Sci., Polym. Phys. Ed.* **37**, 779–792.

Liu, L.-Z., Yeh, F., and Chu, B. (1996). Synchrotron SAXS study of crystallization and microphase separation in compatible mixtures of tetrahydrofuran-methyl methacrylate diblock copolymer and poly(tetrahydrofuran). *Macromolecules* **29**, 5336–5345.

Loo, Y. L., Register, R. A., and Ryan, A. J. (2000). Polymer crystallization in 25-nm spheres. *Phys. Rev. Lett.* **84**, 4120–4123.

Lotz, B. (2000). What can polymer crystal structure tell about polymer crystallization processes? *Eur. Phys. J. E Soft. Matter.* **3**, 185–194.

Lotz, B. (2005). Analysis and observation of polymer crystal structures at the individual stem level. *Adv. Polym. Sci.* **180**, 17–44.

Lotz, B., and Cheng, S. Z. D. (2005). A critical assessment of unbalanced surface stress as the mechanical origin of twisting and scrolling of polymer crystals. *Polymer* **46**, 577–610.

Lotz, B., Lovinger, A. J., and Cais, R. E. (1988). Crystal-structure and morphology of syndiotactic polypropylene single-crystals. *Macromolecules* **21**, 2375–2382.

Lotz, B., Graff, S., Straupe, C., and Wittmann, J. C. (1991). Single crystals of phase isotactic polypropylene: Combined diffraction and morphological support for a structure with nonparallel chains. *Polymer* **32**, 2902–2910.

Lovinger, A. J., Davis, D. D., and Padden, F. J. Jr. (1985). Kinetic analysis of the crystallization of poly (p-phenylene sulfide). *Polymer* **26** 1595–1604.

Lovinger, A. J., Lotz, B., and Davis, D. D. (1990). Interchain packing and unit-cell of syndiotactic polypropylene. *Polymer* **31**, 2253–2259.

Lovinger, A. J., Davis, D. D., and Lotz, B. (1991). Temperature-dependence of structure and morphology of syndiotactic polypropylene and eptitaxial relationship with isotactic polypropylene. *Macromolecules* **24**, 552–560.

Lovinger, A. J., Lotz, B., Davis, D. D., and Padden, F. J. Jr. (1993). Structure and defects in fully syndiotactic polypropylene. *Macromolecules* **26** 3494–3503.

Magill, J. H. (1964). Crystallization of poly(tetramethyl-p-silphenylene) siloxane [TMPS] polymers. *J. Appl. Phys.* **35**, 3249–3259.

Magill, J. H. (1967). Crystallization of poly(tetramethyl-p-silphenylene)-siloxane polymers. II. *J. Polym. Sci., Part A-2* **5**, 89–99.

Magill, J. H. (1969). Spherulitic crystallization studies of poly(tetramethyl-p-silphenylene) - siloxane (TMPS). III. *J. Polym. Sci., Part A-2* **7**, 1187–1195.

Mansfield, M. L. (1988). Solution of the growth equations of a sector of a polymer crystal including consideration of the changing size of the crystal. *Polymer* **29**, 1755–1760.

Mareau, V. H., and Prud'homme, R. E. (2003). Growth rates and morphologies of miscible PCL/PVC blend thin and thick films. *Macromolecules* **36**, 675–684.

Massa, M. V., and Dalnoki-Veress, K. (2004). Homogeneous crystallization of poly(ethylene oxide) confined to droplets: The dependence of the crystal nucleation rate on length scale and temperature. *Phys. Rev. Lett.* **92**, 255509.

Massa, M. V., Carvalho, J. L., and Dalnoki-Veress, K. (2003). Direct visualisation of homogeneous and heterogeneous crystallisation in an ensemble of confined domains of poly(ethylene oxide). *Eur. Phys. J. E Soft. Matter.* **12**, 111–117.

Mateva, R., Wegner, G., and Lieser, G. (1973). Growth of polyoxymethylene crystals during cationic polymerization of trioxane in nitrobenzene. *J. Polym. Sci., Polym. Lett. Ed.* **11**, 369–376.

Matsen, M. W., and Bates, F. S. (1997). Block copolymer microstructures in the intermediate-segregation regime. *J. Chem. Phys.* **106**, 2436–2448.

Matsen, M. W., and Schick, M. (1994). Stable and unstable phases of a linear multiblock copolymer melt. *Macromolecules* **27**, 7157–7163.

Matsuba, G., Kanaya, T., Saito, M., Kaji, K., and Nishida, K. (2000). Further evidence of spinodal decomposition during the induction period of polymer crystallization: Time-resolved small-angle X-ray scattering prior to crystallization of poly(ethylene naphthalate). *Phys. Rev. E* **62**, R1497–R1500.

Melillo, L., and Wunderlich, B. (1972). Extended chain crystals. VIII. Morphology of poly(tetrafluoroethylene). *Kolloid-Zeitschrift & Zeitschrift für Polymere* **250**, 417–425.

Miller, R. L. (1960). Existence of near-range order in isotactic polypropylenes. *Polymer* **1**, 135–143.

Miller, R. L., and Hoffman, J. D. (1991). Nucleation theory applied to polymer crystals with curved edges. *Polymer* **32**, 963–978.

Minakov, A. A., Mordvintsev, D. A., and Schick, C. (2004). Melting and reorganization of poly (ethylene terephthalate) on fast heating (1000 K/s). *Polymer* **45**, 3755–3763.

Minakov, A. A., Mordvintsev, D. A., Tol, R., and Schick, C. (2006). Melting and reorganization of the crystalline fraction and relaxation of the rigid amorphous fraction of isotactic polystyrene on fast heating (30,000 K/min). *Thermochim. Acta* **442**, 25–30.

Minakov, A. A., Wurm, A., and Schick, C. (2007). Superheating in linear polymers studied by ultrafast nanocalorimetry. *Eur. Phys. J. E Soft. Matter.* **23**, 43–53.

Miyagi, A., and Wunderlich, B. (1972). Superheating and reorganization on melting of poly(ethylene terephthalate). *J. Polym. Sci., Part A-2* **10**, 1401–1405.

Miyamoto, Y., Nakafuku, C., and Takemura, T. (1972). Crystallization of poly(chlorotrifluoroethylene). *Polym. J. (Jpn.)* **3**, 120–128.

Morgan, R. L., Barham, P. J., Hill, M. J., Keller, A., and Organ, S. J. (1998). The crystallization of the n-alkane $C_{294}H_{590}$ from solution: Inversion of crystallization rates, crystal thickening, and effects of supersaturation. *J. Macromol. Sci., Part B: Phys.* **37**, 319–338.

Motowoka, M., Jinnai, H., Hashimoto, T., Qiu, Y., and Han, C. C. (1993). Phase separation in deuterated polycarbonate/poly(methyl methacrylate) blend near glass transition temperature. *J. Chem. Phys.* **99**, 2095–2100.

Murry, R. L., Fourkas, J. T., Li, W.-X., and Keyes, T. (1999). Mechanisms of light scattering in supercooled liquids. *Phys. Rev. Lett.* **83**, 3550–3553.

Muthukumar, M. (2000). Commentary on theories of polymer crystallization. *Eur. Phys. J. E Soft. Matter.* **3**, 199–202.

Muthukumar, M. (2003). Molecular modelling of nucleation in polymers. *Philos. Transact. Math. Phys. Eng. Sci.* **361**, 539–556.

Napolitano, R., Pirozzi, B., and Varriale, V. (1990). Temperature dependence of the thermodynamic stability of the two crystalline α forms of isotactic polypropylene. *J. Polym. Sci., Polym. Phys. Ed.* **28**, 139–147.

NATO Advanced Science Institutes Series C (1993). "Mathematical and Physical Science. Crystallization of Polymers." (M. Dosière, ed.), Kluwer Academic, Dordrecht.

Natta, G., Corradini, P., and Bassi, I. W. (1955). The crystalline structure of several isotactic polymers of α-olefins. *Atti della Accademia nazionale dei Lincei. Rendiconti della Classe di scienze fisiche, matematiche e naturali* **19**, 404–411.

Natta, G., Peraldo, M., and Corradini, P. (1959). Modificazione mesomorpha smettica del polipropilene isotattico. *Atti della Accademia nazionale dei Lincei. Rendiconti della Classe di scienze fisiche, matematiche e naturali* **26**, 14–17.

Ogata, N., Kawakage, S., and Ogihara, T. (1997). Structure and thermal/mechanical properties of poly (ethylene oxide)-clay mineral blends. *Polymer* **38**, 5115–5118.

Olmsted, P. D., and Milner, S. T. (1989). Strong segregation theory of bicontinuous phases in block copolymers. *Macromolecules* **31**, 4011–4022.

Olmsted, P. D., Poon, W. C. K., McLeish, T. C. B., Terrill, N. J., and Ryan, A. J. (1998). Spinodal-assisted crystallization in polymer melts. *Phys. Rev. Lett.* **81**, 373–376.

Organ, S. J., Ungar, G., and Keller, A. (1989). Rate minimum in solution crystallization of long paraffins? *Macromolecules* **22**, 1995–2000.

Organ, S. J., Barham, P. J., Hill, M. J., Keller, A., and Morgan, R. L. (1997). A study of the crystallization of the n-alkane $C_{246}H_{494}$ from solution: Further manifestations of the inversion of crystallization rates with temperature. *J. Polym. Sci., Polym. Phys. Ed.* **35**, 1775–1791.

Ostwald, W. (1897). Studien über die Bildung und Umwandlung fester Körper. *Zeitschrift für Physikalische Chemie, Stöchiometrie und Verwandschaftslehre* **22**, 289–300.

Pardey, R., Zhang, A. Q., Gabori, P. A., Harris, F. W., Cheng, S. Z. D., Adduci, J., Facinnelli, J. V., and Lenz, R. W. (1992). Monotropic liquid crystal behavior in two poly(ester imides) with even and odd flexible spacers. *Macromolecules* **25**, 5060–5068.

Pardey, R., Shen, D. X., Gabori, P. A., Harris, F. W., Cheng, S. Z. D., Adduci, J., Facinnelli, J. V., and Lenz, R. W. (1993). Ordered structures in a series of liquid crystalline poly(ester imiders). *Macromolecules* **26**, 3687–3697.

Pardey, R., Wu, S. S., Chen, J. H., Harris, F. W., Cheng, S. Z. D., Keller, A., Adduci, J., Facinnelli, J. V., and Lenz, R. W. (1994). Liquid crystal transition and crystallization kinetics in poly(ester imide)s. *Macromolecules* **27**, 5794–5802.

Park, C., De Rosa, C., Fetters, L. J., and Thomas, E. L. (2000). Influence of an oriented glassy cylindrical microdomain structure on the morphology of crystallizing lamellae in a semicrystalline block terpolymer. *Macromolecules* **33**, 7931–7938.

Park, J.-Y., and McKenna, G. B. (2000). Size and confinement effects on the glass transition behavior of polystyrene/o-terphenyl polymer solutions. *Phys. Rev. B* **61**, 6667–6676.

Passaglia, E., and Khoury, F. (1984). Crystal growth kinetics and the lateral habits of polyethylene crystals. *Polymer* **25**, 631–644.

Pérez, E., Bello, A., and Fatuo, J. G. (1984). Effect of molecular weight and temperature on the isothermal crystallization of poly(oxetane). *Colloid. Polym. Sci.* **262**, 605–610.

Petraccone, V., De Rosa, C., Guerra, G., and Tuzi, A. (1984). On the double peak shape of melting endotherms of isothermally crystallized isotactic polypropylene samples. *Die Makromolekulare Chemie, Rapid Communications* **5**, 631–634.

Petraccone, V., Guerra, G., De Rosa, C., and Tuzi, A. (1985). Extrapolation to the equilibrium melting temperature for isotactic polypropylene. *Macromolecules* **18**, 813–814.

Phillips, P. J. (1990). Polymer crystals. *Rep. Prog. Phys.* **53**, 549–604.

Phillips, P. J. (2003). Polymer morphology and crystallization. *Mater. Sci. Technol.* **19**, 1153–1160.

Point, J. J. (1979a). Reconsideration of kinetic theories of polymer crystal growth with chain folding. *Faraday Discuss. Chem. Soc.* **68**, 167–176.

Point, J. J. (1979b). A new theoretical approach to the secondary nucleation at high supercooling. *Macromolecules* **12**, 770–775.

Point, J. J., Colet, M. C., and Dosière, M. (1986). Experimental criterion for the crystallization regime in polymer crystals grown from dilute solution: Possible limitation due to fractionation. *J. Polym. Sci., Polym. Phys. Ed.* **24**, 357–388.

Psarski, M., Piorkowska, E., and Galeski, A. (2000). Crystallization of polyethylene from melt with lowered chain entanglements. *Macromolecules* **33**, 916–932.

Putra, E. G. R., and Ungar, G. (2003). In situ solution crystallization study of n-$C_{246}H_{494}$: Self-poisoning and morphology of polymethylene crystals. *Macromolecules* **36**, 5214–5225.

Qi, S., and Wang, Z. G. (1997). On the nature of the perforated layer phase in undiluted diblock copolymers. *Macromolecules* **30**, 4491–4497.

Rastogi, S., and Ungar, G. (1992). Hexagonal columnar phase in 1,4-trans-polybutadiene: Morphology, chain extension, and isothermal phase reversal. *Macromolecules* **25**, 1445–1452.

Rastogi, S., Newman, M., and Keller, A. (1991). Pressure-induced amorphization and disordering on cooling in a crystalline polymer. *Nature* **353**, 55–57.

Rastogi, S., Newman, M., and Keller, A. (1993). Unusual pressure-induced phase behavior in crystalline poly-4-methly-pentene-1. *J. Polym. Sci., Polym. Phys. Ed.* **31**, 125–139.

Ratner, S., Weinberg, A., Wachtel, E., Mona Moret, P., and Marom, G. (2004). Phase transitions in UHMWPE fiber compacts studied by in situ synchrotron microbeam WAXS. *Macromol. Rapid Commun.* **25**, 1150–1154.

Reiter, G., and Sommer, J.-U. (1998). Crystallization of adsorbed polymer monolayers. *Phys. Rev. Lett.* **80**, 3771–3774.

Reiter, G., and Sommer, J.-U. (2000). Polymer crystallization in quasi-two dimensions. I. Experimental resultss. *J. Chem. Phys.* **112**, 4376–4383.

Runt, J., Harrison, I. R., Varnell, W. D., and Wang, J. I. (1983). An examination of the longitudinal acoustic mode of polyethylene crystals. *J. Macromol. Sci., Part B: Phys.* **22**, 197–212.

Sadler, D. M. (1986). When is a nucleation theory not a nucleation theory? *Polym. Commun.* **27**, 140–145.

Sadler, D. M. (1987a). New explanation for chain folding in polymers. *Nature* **326**, 174–177.

Sadler, D. M. (1987b). On the growth of two-dimensional crystals 1. Fluctuations and the relation of step free energies to morphology. *J. Chem. Phys.* **87**, 1771–1784.

Sadler, D. M. (1987c). On the growth of two-dimensional crystals 2. Assessment of kinetic theories of crystallization of polymers. *Polymer* **28**, 1440–1454.

Sadler, D. M. (1987d). Preferred fold lengths in polymer crystals: Predications of minima in growth rates. *Polym. Commun.* **28**, 242–246.

Sadler, D. M., and Gilmer, G. H. (1984). A model for chain folding in polymer crystals: Rough growth faces are consistent with the observed growth rates. *Polymer* **25**, 1446–1452.

Sadler, D. M., and Gilmer, G. H. (1986). Rate theory model of polymer crystallization. *Phys. Rev. Lett.* **56**, 2708–2711.

Sadler, D. M., and Gilmer, G. H. (1988). Selection of lamellar thickness in polymer crystal growth: A rate-theory model. *Phys. Rev. B* **38**, 5684–5693.

Sadler, D. M., Barber, M., Lark, G., and Hill, M. J. (1986). Twin morphology: 2. Measurements of the enhancement in growth due to re-entrant corners. *Polymer* **27**, 25–33.

Sanchez, I. C. (1977). Problems and theories of polymer crystallization. *J. Polym. Sci., Polym. Symp.* **59**, 109–120.

Sanchez, I. C., and DiMarzio, E. A. (1972). Dilute solution theory of polymer crystal growth: Fractionation effects. *J. Res. Natl. Bur. Stand. Sect. A* **76**, 213–223.

Sanchez, I. C., and Eby, R. K. (1973). Crystallization of random copolymers. *J. Res. Natl. Bur. Stand. Sect. A* **77**, 353–358.

Sanchez, I. C., and Eby, R. K. (1975). Thermodynamics and crystallization of random copolymers. *Macromolecules* **8**, 638–641.

Sasaki, S., and Asakura, T. (2003). Helix distortion and crystal structure of the α-form of poly(L-lactide). *Macromolecules* **36**, 8385–8390.

Schmidt-Rohr, K., and Spiess, H. W. (1991). Nature of nonexponential loss of correlation above the glass transition investigated by multidimensional NMR. *Phys. Rev. Lett.* **66**, 3020–3023.

Schulz, M. F., Bates, F. S., Almdal, K., and Mortensen, K. (1994). Epitaxial relationship for hexagonal-to-cubic phase transition in a block copolymer mixture. *Phys. Rev. Lett.* **73**, 86–89.

Spontak, R. J., Smith, S. D., and Ashraf, A. (1993). Dependence of the OBDD morphology in diblock copolymer molecular weight in coplolymer/homopolymer blends. *Macromolecules* **26**, 956–962.
Stranski, I. (1949). Forms of equilibrium of crystals. *Discuss. Faraday Soc* **5**, 13–21.
Strobl, G. (2000). From the melt via mesomorphic and granular crystalline layers to lamellar crystallites: A major route followed in polymer crystallization? *Eur. Phys. J. E Soft. Matter.* **3**, 165–183.
Sutton, S. J., Vaughan, A. S., and Bassett, D. C. (1996). On the morphology and crystallization kinetics of monodisperse polyethylene oligomers crystallized from the melt. *Polymer* **37**, 5735–5738.
Tammann, G. (1903). Kristallisieren und Schmelzen. Verlag Johann Ambrosius Barth, Leipzig.
Tenneti, K. K., Chen, X., Li, C. Y., Tu, Y., Wan, X., Zhou, Q.-F., Sics, I., and Hsiao, B. S. (2005). Perforated layer structures in liquid crystalline rod-coil block copolymers. *J. Am. Chem. Soc.* **127**, 15481–15490.
Thomas, E. L., Anderson, D. M., Henkee, C. S., and Hoffman, D. (1988). Periodic area-minimizing surfaces in block copolymers. *Nature* **334**, 598–601.
Till, P. H. Jr. (1957). The growth of single crystals of linear polyethylene. *J. Polym. Sci.* **24** 301–306.
Toda, A. (1991). Rounded lateral habits of polyethylene single crystals. *Polymer* **32**, 771–780.
Toda, A. (1992). Growth of polyethylene single crystals from the melt: Change in lateral habit and regime I–II transition. *Colloid. Polym. Sci.* **270**, 667–681.
Toda, A. (1993). Growth mode and curved lateral habits of polyethylene single crystals. *Faraday Discuss.* **95**, 129–143.
Toda, A., and Keller, A. (1993). Growth of polyethylene single crystals from the melt: Morphology. *Colloid. Polym. Sci.* **271**, 328–342.
Toda, A., and Kiho, H. (1989). Crystal growth of polyethylene from dilute solution: Growth kinetic of {110} twins and diffusion-limited growth of single crystals. *J. Polym. Sci., Polym. Phys. Ed.* **27**, 53–70.
Toda, A., Hikosaka, M., and Yamada, K. (2002). Superheating of the melting kinetics in polymer crystals: A possible nucleation mechanism. *Polymer* **43**, 1667–1679.
Todoki, M., and Kawaguchi, T. (1977a). Origin of double melting peaks in drawn nylon 6 yarns. *J. Polym. Sci., Polym. Phys. Ed.* **15**, 1067–1075.
Todoki, M., and Kawaguchi, T. (1977b). Melting of constrained drawn nylon 6 yarns. *J. Polym. Sci., Polym. Phys. Ed.* **15**, 1507–1520.
Tol, R. T., Minakov, A. A., Adamovsky, S. A., Mathot, V. B. F., and Schick, C. (2006). Metastability of polymer crystallites formed at low temperature studied by ultra fast calorimetry: Polyamide 6 confined in sub-micrometer droplets vs. bulk PA6. *Polymer* **47**, 2172–2178.
Tracht, U., Wilhelm, M., Heuer, A., Feng, H., Schmidt-Rohr, K., and Spiess, H. W. (1998). Length scale of dynamic heterogeneities at the glass transition determined by multidimensional nuclear magnetic resonance. *Phys. Rev. Lett.* **81**, 2727–2730.
Turnbull, D., and Fisher, J. C. (1949). Rate of nucleation in condensed systems. *J. Chem. Phys.* **17**, 71–73.
Turner-Jones, A., Aizlewood, J. M., and Beckett, D. R. (1964). Crystalline forms of isotactic polypropylene. *Die Makromolekulare Chemie* **75**, 134–158.
Umemoto, S., and Okui, N. (2005). Power law and scaling for molecular weight dependence of crystal growth rate in polymeric materials. *Polymer* **46**, 8790–8795.
Ungar, G. (1980). Effect of radiation on the crystals of polyethylene and paraffins. 2. Phase separation in γ-irradiated paraffins. *Polymer* **21**, 1278–1283.
Ungar, G., and Keller, A. (1980). Effect of radiation on the crystals of polyethylene and paraffins. 1. Formation of the hexagonal lattice and the destruction of crystallinity in polyethylene. *Polymer* **21**, 1273–1277.
Ungar, G., and Keller, A. (1986). Time-resolved synchrotron X-ray study of chain-folded crystallization of long paraffins. *Polymer* **27**, 1835–1844.
Ungar, G., and Keller, A. (1987). Inversion of the temperature dependence of crystallization rates due to onset of chain folding. *Polymer* **28**, 1899–1907.
Ungar, G., and Zeng, X. (2001). Learning polymer crystallization with the aid of linear, branched and cyclic model compounds. *Chem. Rev.* **101**, 4157–4188.
Ungar, G., Grubb, D. T., and Keller, A. (1980). Effect of radiation on the crystals of polyethylene and paraffins. 3. Irradiation in the electron microscope. *Polymer* **21**, 1284–1291.

Ungar, G., Stejny, J., Keller, A., Bidd, I., and Whiting, M. C. (1985). The crystallization of ultralong normal paraffins: The onset of chain folding. *Science* **229**, 386–389.
Ungar, G., Mandal, P., Higgs, P. G., de Silva, D. S. M., Boda, E., and Chen, C. M. (2000). Dilution wave and negative-order crystallization kinetics of chain molecules. *Phys. Rev. Lett.* **85**, 4397–4400.
Ungar, G., Putra, E. G. R., de Silva, D. S. M., Shcherbina, M. A., and Waddon, A. J. (2005). The effect of self-poisoning on crystal morphology and growth rates. *Adv. Polym. Sci.* **180**, 45–87.
Vasanthakumari, R., and Pennings, A. J. (1983). Crystallization kinetics of poly(L-lactic acid). *Polymer* **24**, 175–178.
Vigild, M. E., Almdal, K., Mortensen, K., Hamley, I. W., Fairclough, J. P. A., and Ryan, A. J. (1998). Transformations to and from the gyroid phase in a diblock copolymer. *Macromolecules* **31**, 5702–5716.
Vonnegut, B. (1948). Variation with temperature of the nucleation rate of supercooled liquid tin and water drops. *J. Colloid Sci.* **3**, 563–569.
Wagner, J., and Phillips, P. J. (2001). The mechanism of crystallization of linear polyethylene, and its copolymers with octene, over a wide range of supercoolings. *Polymer* **42**, 8999–9013.
Wang, M., Hu, W., Ma, Y., and Ma, Y.-Q. (2006). Confined crystallization of cylindrical diblock copolymers studied by dynamic Monte Carlo simulations. *J. Chem. Phys.* **124**, 244901.
Wang, Z.-G. (2002). Concentration fluctuation in binary polymer blends: χ parameter, spinodal and Ginzburg criterion. *J. Chem. Phys.* **117**, 481–500.
Weber, C. H. M., Chiche, A., Krausch, G., Rosenfeldt, S., Ballauff, M., Harnau, L., Göttker-Schnetmann, I., Tong, Q., and Mecking, S. (2007). Single lamella nanoparticles of polyethylene. *Nano Lett.* **7**, 2024–2029.
Wegner, G. (1969). Topochemical reactions of monomers with conjugated triple bonds. I. Polymerization of derivatives of 2,4-hexadiyne-1,6-diols in the crystalline state. *Zeitschrift für Naturforschung. Teil B. Anorganische Chemie, Organische Chemie, Biochemie, Biophysik, Biologie* **24**, 824–832.
Wegner, G. (1972). Topochemical polymerization of monomers with conjugated triple bonds. *Die Makromolekulare Chemie* **154**, 35–48.
Wegner, G. (1979). Introductory lecture: Solid-state polymerization. *Faraday Discuss. Chem. Soc.* **68**, 494–508.
Wittmann, L. C., and Kovacs, A. J. (1970). Vielartige zwillinge in polyäthylen einkristallen. *Berichte der Bunsen-Gesellschaft* **74**, 901–904.
Wunderlich, B. (1968a). Crystallization during polymerization. *Angewandte Chemie, International Edition in English* **7**, 912–919.
Wunderlich, B. (1968b). Crystallization during polymerization. *Fortschritte der Hochpolymeren-Forschung* **5**, 568–619.
Wunderlich, B. (1973). "Macromolecular Physics. Volume I. Crystal Structure, Morphology, Defects." Academy Press, New York.
Wunderlich, B. (1976). "Macromolecular Physics. Volume II. Crystal Nucleation, Growth, Annealing." Academy Press, New York.
Wunderlich, B. (1979). Molecular nucleation and segregation. *Faraday Discuss. Chem. Soc.* **68**, 239–243.
Wunderlich, B. (1980). "Macromolecular Physics. Volume III. Crystal Melting." Academic Press, New York.
Wunderlich, B. (2005). "Thermal Analysis of Polymeric Materials." Springer, Berlin.
Wunderlich, B., and Arakawa, T. (1964). Polyethylene crystallized from the melt under elevated pressure. *J. Polym. Sci., Part A* **2**, 6397–6706.
Yamazaki, S., Hikosaka, M., Toda, A., Wataoka, I., and Gu, F. (2002). Role of entanglement in nucleation and 'melt relaxation' of polyethylene. *Polymer* **43**, 6585–6593.
Yamazaki, S., Gu, F., Watanabe, K., Okada, K., Toda, A., and Hikosaka, M. (2006). Two-step formation of entanglement from disentangled polymer melt detected by using nucleation rate. *Polymer* **47**, 6422–6428.
Zannetti, R., Celotti, G., Fichera, A., and Francesconi, R. (1969). The structural effects of annealing time and temperature on the paracrystal-crystal transition in isotactic polypropylene. *Die Makromolekulare Chemie* **128**, 137–142.

Zheng, J. X., Xiong, H., Chen, W. Y., Lee, K., Van Horn, R. M., Quirk, R. P., Lotz, B., Thomas, E. L., Shi, A.-C., and Cheng, S. Z. D. (2006). Onsets of tethered chain overcrowding and highly stretched brush regime via crystalline-amorphous diblock copolymers. *Macromolecules* **39**, 641–650.

Zhou, W., Cheng, S. Z. D., Putthanarat, S., Eby, R. K., Reneker, D. H., Lotz, B., Magonov, S., Hsieh, E. T., Geerts, R. G., Palackal, S. J., Hawley, G. R., and Welch, M. B. (2000). Crystallization, melting and morphology of syndiotactic polypropylene fractions. 4. *In situ* lamellar single crystal growth and melting in different sectors. *Macromolecules* **33**, 6861–6868.

Zhou, W. W., Weng, X., Jin, S., Rastogi, S., Lovinger, A. J., Lotz, B., and Cheng, S. Z. D. (2003). Chain orientation and defects in lamellar single crystals of syndiotactic polypropylene fractions. *Macromolecules* **36**, 9485–9491.

Zhu, D.-S., Liu, Y.-X., Shi, A.-C., and Chen, E.-Q. (2006). Morphology evolution in superheated crystal monolayer of low molecular weight poly(ethylene oxide) on mica surface. *Polymer* **47**, 5239–5242.

Zhu, D.-S., Liu, Y.-X., Chen, E.-Q., Li, M., Chen, C., Sun, Y.-H., Shi, A.-C., Van Horn, R. M., and Cheng, S. Z. D. (2007). Crystal growth mechanism changes in pseudo-dewetted poly(ethylene oxide) thin layers. *Macromolecules* **40**, 1570–1578.

Zhu, L., Chen, Y., Zhang, A., Calhoun, B. H., Chun, M., Quirk, R. P., Cheng, S. Z. D., Hsiao, B. S., Yeh, F., and Hashimoto, T. (1999). Phase structures and morphologies determined by competitions among self-organization, crystallization, and vitrification in a disordered poly(ethylene oxide)-b-polystyrene diblock copolymer. *Phys. Rev. B* **60**, 10022–10031.

Zhu, L., Cheng, S. Z. D., Calhoun, B. H., Ge, Q., Quirk, R. P., Thomas, E. L., Hsiao, B. S., Yeh, F., and Lotz, B. (2000). Crystallization temperature-dependent crystal orientations within nanoscale confined lamellae of a self-assembled crystalline-amorphous diblock copolymer. *J. Am. Chem. Soc.* **122**, 5957–5967.

Zhu, L., Calhoun, B. H., Ge, Q., Quirk, R. P., Cheng, S. Z. D., Thomas, E. L., Hsiao, B. S., Yeh, F., Liu, L., and Lotz, B. (2001a). Initial-stage growth controlled crystal orientations in nanoconfined lamellae of a self-assembled crystalline-amorphous diblock copolymer. *Macromolecules* **34**, 1244–1251.

Zhu, L., Huang, P., Cheng, S. Z. D., Ge, Q., Quirk, R. P., Thomas, E. L., Lotz, B., Wittmann, J.-C., Hsiao, B. S., Yeh, F., and Liu, L. (2001b). Dislocation-controlled perforated layer phase in a PEO-b-PS diblock copolymer. *Phys. Rev. Lett.* **86**, 6030–6033.

Zhu, L., Huang, P., Chen, W. Y., Ge, Q., Quirk, R. P., Cheng, S. Z. D., and Thomas, E. L. (2002). Lotzed crystalline morphology in confined hexagonally perforated layers of a self-assembled PS-b-PEO diblock copolymer. *Macromolecules* **35**, 3553–3562.

Zhu, L., Huang, P., Chen, W. Y., Weng, X., Cheng, S. Z. D., Ge, Q., Quirk, R. P., Senador, T., Shaw, M. T., Thomas, E. L., Lotz, B., Hsiao, B. S., *et al.* (2003). Plastic deformation" mechanism and phase transformation in a shear-induced metastable hexagonally perforated layer phase of a polystyrene-b-poly(ethylene oxide) diblock copolymer. *Macromolecules* **36**, 3180–3188.

CHAPTER FIVE

Metastable States Observed Due to Phase Transformation Kinetics in Polymers

Contents

This chapter deals with metastable states that appear due to competing kinetic pathways caused by the existence of multiple free energy pathways to reach the equilibrium phase. The determining factor for experimentally observing these metastable states is their growth rate. The chapter begins by pointing out several interesting experimental observations: (1) very different crystal growth rates along distinctly different crystallographic planes resulting from different nucleation barriers; (2) the observation of initial transient non-integral-folded chain crystals before the formation of integral-folded chain crystals in uniform and nearly uniform oligomer crystallization due to their fast growth rates; and (3) the nucleation and growth rates being affected by the global chain conformation. Major attention is paid to the appearance of crystal polymorphs due solely to kinetic reasons at different temperatures, pressures, and under external fields. This principle can be clearly illustrated in the case of monotropic phase behavior where the transition temperature of a metastable phase is lower than the crystal melting temperature. Surface-induced metastable states have also been frequently reported from modifications of the surface physics and/or chemistry, as well as the hard or soft epitaxial growth of these phases. Hard epitaxy requires lattice matching between the metastable crystals and the substrate, while soft epitaxy requires only orientational order. On the larger length scale of crystal morphology, some curved, scrolled, and twisted-helix single crystals or lamellae in banded spherulites, are metastable states caused by unbalanced surface stresses on the opposite chain-folded surfaces. The crystal chain packing in these types of crystals deviates from the usual parallel scheme even though they possess the identical crystal structure as the corresponding flat crystal.

1. Appearance of Metastable States Based on Their Crystal Nucleation Barrier

1.1. Crystal growth rates along different growth planes

We will now focus on the growth of semicrystalline polymer single crystals. In the 1950s, polymer single crystals with crystallographically defined, facetted habits were observed in dilute solution using optical microscopy and transmission electron microscopy (Fischer, 1957; Jaccodine, 1955; Keller, 1957; Till, 1957). Lamellar crystals were also observed as the building blocks to form spherulites in the melt (Fischer, 1957; Kobayashi, 1962). These early observations implied that the crystal growth is a nucleation-limited process. Figure 5.1A shows a transmission electron microscopy bright field image of a polyethylene single crystals grown in dilute solution. Electron diffraction (Fig. 5.1B) provided direct experimental evidence of how the long chain molecules crystallize into the lamellar form. The concept of chain folding was proposed at that time by Keller (1957); while, the origin of this concept can be traced back to Storks (1938). This concept is widely accepted today despite some controversies concerning how the chains fold. At the same time, the observation of polymeric spherulites in the melt (Keith and Padden, 1959a,b, Keller, 1959; Point, 1955; Price, 1959) and dendrites in solution (Fischer and Lorenz, 1963; Khoury, 1966; Wunderlich and Sullivan, 1962; Wunderlich *et al.*, 1964) were also reported. These are analogous to crystal aggregates in small molecules. These

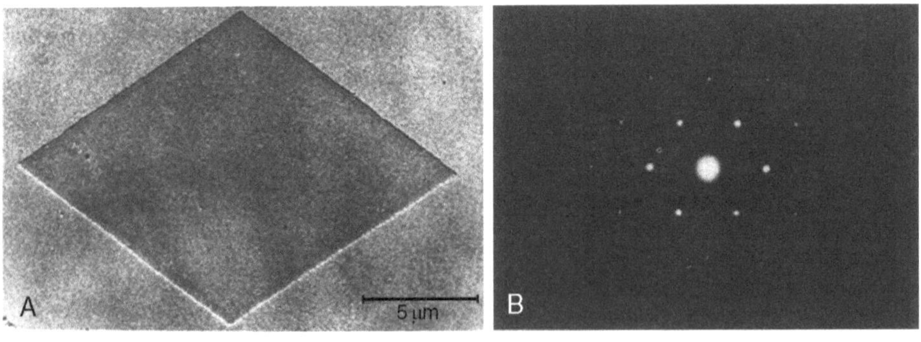

Figure 5.1 Single crystal of polyethylene grown in dilute solution (A) and its electron diffraction (B) (courtesy of S. J. Organ and A. Keller).

experimental observations also reveal that lamellar single crystals are the basic building blocks of these polymer crystal aggregates.

With the large number of lamellar single crystals observed and their crystal unit cell dimensions and symmetry groups determined, it was found that polymer single crystals possess a variety of habits ranging from elongated ribbon-like (kinetically anisotropic) to square- or hexagonal-shaped (kinetically isotropic) in both the melt and solution (Bassett, 1981; Cheng and Li, 2002; Geil, 1963; Keller, 1968; Khoury and Passaglia, 1976; Wunderlich, 1973). The formation of anisotropic single crystal morphologies implies that the growth rate along one specific crystallographic direction is faster than the others. Therefore, the nucleation barrier for polymers to crystallize on specific crystallographic planes is different. Isotropic single crystal habits are generated by identical growth rates along all the growth fronts because they belong to the same set of {hkl} planes and possess the same nucleation barrier. Generally speaking, a unit cell with lower rotational symmetry around the *c* axis generates an anisotropic single crystal habit, while a unit cell with higher rotational symmetry generates an isotropic single crystal habit (Cheng and Li, 2002; Wunderlich, 1973). Examples include single crystals of polyethylene in the orthorhombic phase grown in solution (Khoury, 1979; Khoury and Bolz, 1980, 1985; Organ and Keller, 1985a,b,c) and in the melt (shown in Fig. 5.2A; Bassett *et al.*, 1988; Keith *et al.*, 1989; Toda, 1992, 1993; Toda and Keller, 1993); isotactic polypropylene in the monoclinic α-phase grown from solution (shown in Fig. 5.2B; Lotz *et al.*, 1991; Sauer *et al.*, 1965; Wittmann and Lotz, 1985); syndiotactic polypropylene in the high-temperature orthorhombic phase grown in the melt (shown in Fig. 4.17A; Bu *et al.*, 1996; Lotz *et al.*, 1988; Lovinger *et al.*, 1990, 1991, 1993; Zhou *et al.*, 2000, 2003); poly(vinylidene fluoride) in the orthorhombic α-phase in the melt (Briber and Khoury, 1993; Toda *et al.*, 2001); the orthorhombic phase in poly(ε-caprolactone) in the melt (Beekmans and Vancso, 2000; Mareau

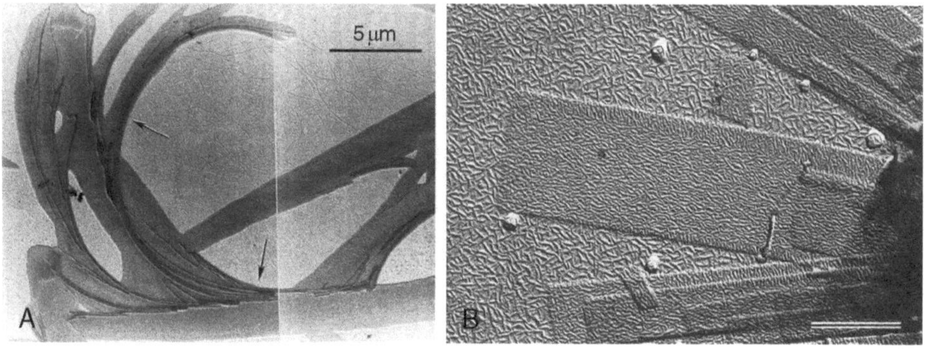

Figure 5.2 Elongated single crystals of polyethylene with an orthorhombic lattice crystallized from a thin film melt (A) [reprinted from Keith *et al.* (1989), with permission] and single crystals of isotactic polypropylene with a monoclinic lattice (α-phase) crystallized from solution on which polyethylene was decorated on the single crystal basal surface to illustrate the chain folding direction (see below in Section 4.1 of this Chapter). The scale bar represents 1 μm (B) [reprinted from Wittmann and Lotz (1985), with permission].

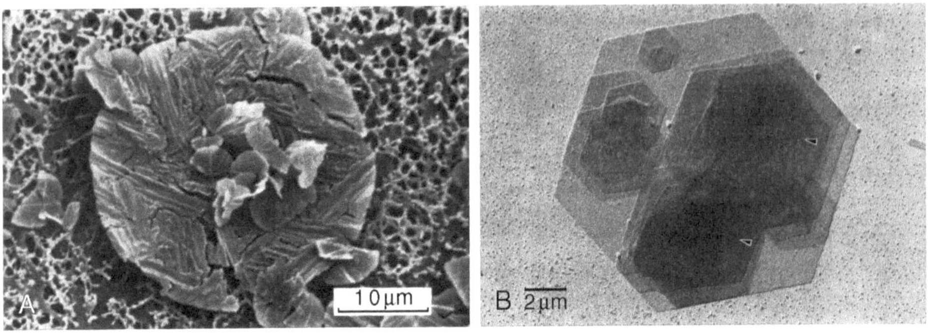

Figure 5.3 Isotropic single crystal habit of polyethylene with a hexagonal lattice crystallized at elevated pressure and temperature (A) [reprinted from DiCorleto and Bassett (1990), with permission] and a single crystal of isotactic polystyrene with a trigonal lattice crystallized from the thin film melt (B) [reprinted from Keith and Padden (1987), with permission].

and Prud'homme, 2005); and others. All examples possess unit cells having 2_1 rotational symmetry about the *c* axis; therefore, they exhibit elongated single-crystal habits. On the other hand, polyethylene single crystals in the high-pressure hexagonal phase (Bassett and Turner, 1972; DiCorleto and Bassett, 1990), as shown in Fig. 5.3A, poly(oxymethylene) in the trigonal phase grown in solution (Geil *et al.*, 1959; Khoury and Barnes, 1974a), isotactic polystyrene in the trigonal phase grown in the melt, as shown in Fig. 5.3B (Keith, 1964; Keith and Padden, 1987), poly (4-methyl-1-pentene) in the tetragonal phase in the melt and solution (Bassett, 1964; Bassett *et al.*, 1964; Frank *et al.*, 1959; Khoury and Barnes, 1972; Patel and Bassett, 1994), and others possess higher unit cell symmetry about their *c* axis (3_1, 4_1, or 6_1) and exhibit regular polygonal lamellar habits.

The habit of a polymer single crystal is determined by the slowest growing crystallographic plane because the faster growing planes are exhausted first during crystal growth. We are specifically interested in the formation of these anisotropic single crystal habits. Figure 5.4 shows a set of real-time atomic force microscopic images of syndiotactic polypropylene single crystals grown at 120 °C in a thin film melt. These single crystals possess an elongated ribbon-like habit. Based on electron diffraction experiments, the long growth fronts of this single crystal are the (100) planes; the short growth fronts are the (010) planes as shown in Fig. 4.17A (Bu *et al.*, 1996). Evidently, the growth rate along the (010) normal is much faster than that of the (100) normal. Quantitative data at a crystallization temperature of 120 °C shows that the linear growth rate along the long axis of the single crystal (the (010) normal) was 0.18 μm/min, while the rate along the short axis of the single crystal (the (100) normal) was 0.018 μm/min (Zhou *et al.*, 2000). This difference implies that along these two growth fronts, the nucleation barrier for growth in the isotropic thin film melt of the (010) planes is much lower than that of the (100) planes. Furthermore, it should also be pointed out, as described in Section 2.1 of Chapter 4, that the lamellar thicknesses in these two sectors are different, although the two sectors join to form a single crystal (Bu *et al.*, 1996; Zhou *et al.*, 2000). This observation is not unique to syndiotactic polypropylene. In polyethylene single crystals, different thicknesses were

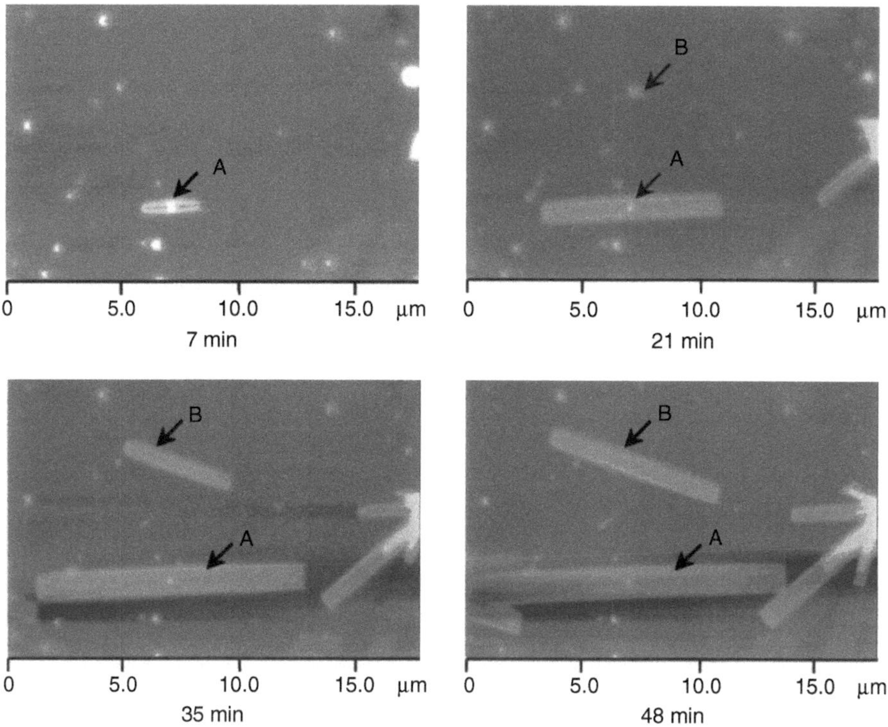

Figure 5.4 Set of real-time atomic force microscopic images of syndiotactic polypropylene grown at 120 °C from the thin-film melt at different times. The linear growth rates can thus be monitored [reprinted from Zhou *et al.* (2000), with permission].

also observed in different sectors. However, this lamellar thickness difference may not be the determining factor for the difference in growth rates. As shown in Fig. 5.4, the {010} thinner sectors of syndiotactic polypropylene grow faster, while in the case of polyethylene, the thicker {110} sectors grow faster in the melt (Toda, 1992, 1993). This difference implies that the growth rates are determined by the nucleation barrier that contains enthalpic and entropic parts, as described in Section 1.5 of Chapter 4. Kinetics alone determines the appearance of the metastable crystals, rather than their stabilities.

Unfortunately, for most semicrystalline polymers, only spherulite radial growth rates are available. They correspond to the inherent fastest growth direction of the constituent radiating lamellae but offer no insights into growth rates along other crystallographic dimensions. Anisotropic crystal growth rates can only be determined from single crystal growth rate studies.

An interesting issue on this topic, in connection with lateral lamellar crystal growth rates, is that in some cases the growth rates along two crystallographic planes may invert at different crystallization temperatures. One case is polyethylene single crystals grown either in the solution (Khoury, 1979; Khoury and Bolz, 1980, 1985; Organ and Keller, 1985a,b,c) or in the isotropic melt (Toda, 1992, 1993; Toda and Keller, 1993). Figure 5.5 shows the first observation of a set of polyethylene single

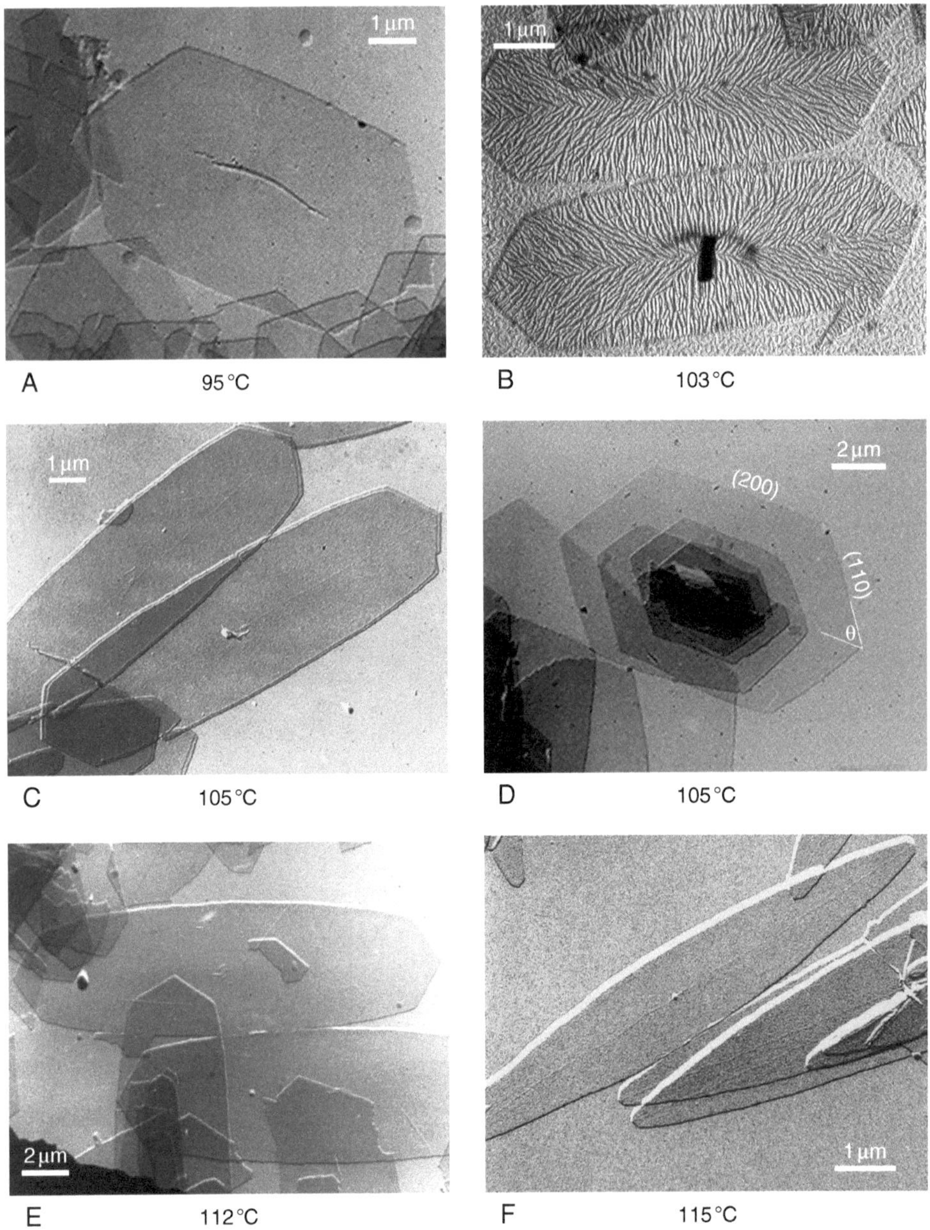

Figure 5.5 Set of polyethylene single crystals grown from dilute solutions at different crystallization temperatures in two solvents. The polyethylene fraction has a number average molecular weight of 11.4 kg/mol and a polydispersity of 1.19. Figures (A) to (C) are the single crystal habits obtained in heptyl acetate (a better solvent) and (D) to (F) are obtained in dodecanol (a poorer solvent). The crystal habit changes can be characterized by the aspect ratio between the dimensions along the *a* and *b* axes. In Figure B, polyethylene decoration has been performed on the single crystal surface to illustrate the chain folding direction (see in Section 4.1 of this Chapter) (Courtesy of F. Khoury).

crystals grown in dilute solution at different crystallization temperatures and in different solvents by Khoury and Bolz (1980, 1985). It is known that the polyethylene single crystal habit is bounded by only the (110) crystallographic planes when crystallized at low crystallization temperatures in xylene (Fig. 5.1A). In heptyl acetate with increasing temperature, the (200) planes start to be observed, creating a new boundary in the crystal habit, as shown in Fig. 5.5A, B, and C. When this occurs, the single crystal habit changes from lozenge to truncated lozenge. When polyethylene is in dodecanol, a poorer solvent, the aspect ratio, which is defined to be between the dimensions along the *a* and *b* axes, increases with increasing crystallization temperature. The truncated lozenge shape now becomes an almost lenticular habit (Fig. 5.5D, E, and F). In this figure, the angle θ is also defined between the (110) plane and the *b* axis (the long axis of the growth). The growth rate along the (110) plane normal can thus be deduced by experimentally measuring the size increase along the *b* axis with time. For polyethylene single crystals grown in the melt with increasing crystallization temperature, as shown in Fig. 5.6 (Toda, 1992, 1993; Toda and Keller, 1993), the single crystal habit changes from truncated lozenge to lenticular-shaped with a continuous increase of the aspect ratio. For the single crystals grown in the melt, the lozenge-shaped single crystals cannot grow even at very low crystallization temperatures. Note that the θ-angle changes with the crystal habits.

These single crystal habit changes of polyethylene originate from growth rate changes with crystallization temperature along the (200) and (110) normals (Passaglia and Khoury, 1984). At low crystallization temperatures, the growth rate along the (200) normal is faster than the growth rate along the (110) normal; thus, only the (110) planes create the single crystal boundary, causing them to become lozenge-shaped. With increasing temperature, the growth rate along the (200) normal slows down to become approximately equal to the growth rate along the (110) normal. Therefore, the single crystals are bounded by both the (110) and (200) planes, and as a result, they exhibit a truncated lozenge shape (Fig. 5.5). With a further increase in temperature, the growth rate along the (200) normal slows down faster than the growth rate along the (110) normal. Note that the growth rate along the *b* axis in the single crystal is represented by the ratio between the growth rate along the (110) normal and sinθ, where θ is defined as the angle between the (110) plane and the *b* axis, as shown in Fig. 5.5 (Passaglia and Khoury, 1984). The system gradually reaches the point where the growth rate along the (100) normal and along the *b* axis is much faster than it is along the (200) normal. The single crystal shape becomes lenticular (Fig. 5.6). Nevertheless, the growth tips are still the (110) planes, while the rest of the single crystal is bound by the curved surfaces that consist of a number of serrated, microscopic (200) planes.

Based on these experimental observations, we infer that the nucleation barriers for crystal growth on these two planes possess different supercooling dependencies. Detailed growth rate measurements of a polyethylene fraction with an average molecular weight of 11 kg/mol and a polydispersity of 1.16 in *n*-octane at a concentration of 10^{-4} wt% indicate that the single crystal habit changes in dilute solution as a result of growth rate inversion between the (110) and (200) planes, as shown in Fig. 5.7 (Toda *et al.*, 1986). The rate inversion temperature was observed

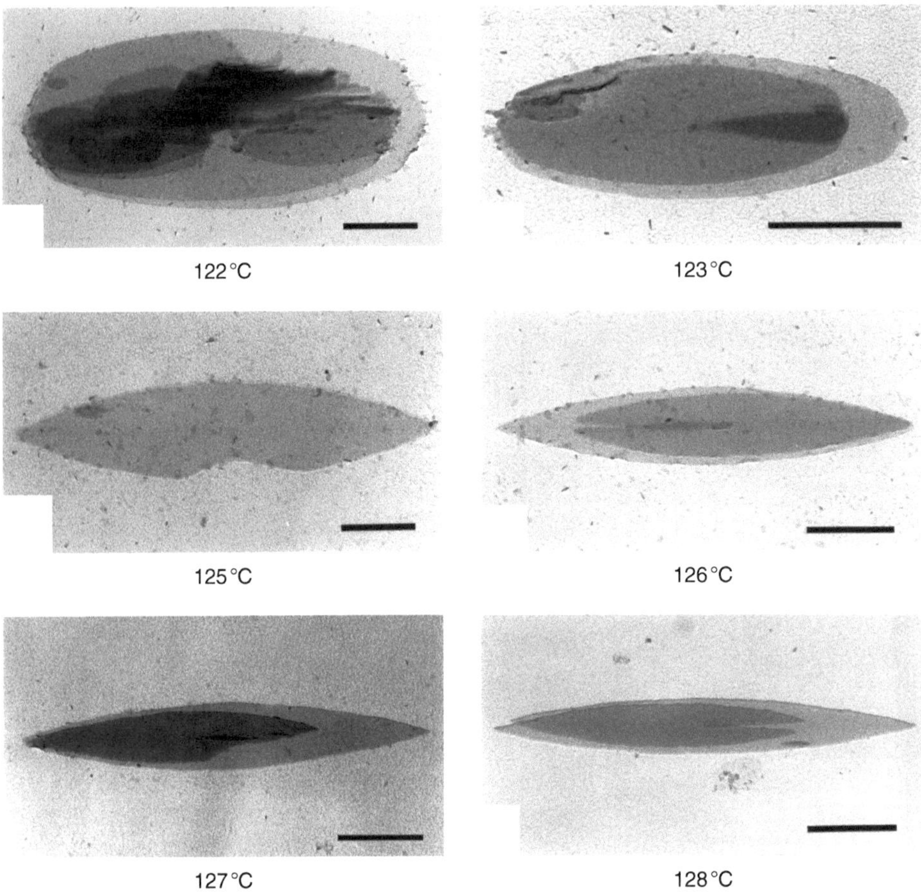

Figure 5.6 Set of polyethylene single crystals grown from the melt at different crystallization temperatures (from 122 to 128 °C). The polyethylene fraction had an average number molecular weight of 11 kg/mol and a polydispersity of 1.16. The crystal habit changes from a truncated lozenge to a lenticular shape with increasing temperature. The scale bar represents 1 μm [reprinted from Toda (1993), with permission].

to be about 92 °C for this fraction. Note that the inversion temperature is also concentration dependent. This rate inversion phenomenon is also expected in bulk melt-grown single crystals, although the growth rate inversion in the melt cannot be directly observed in experiments because the single crystal growth in the melt is limited to the high crystallization temperature region (Fig. 5.6). This growth rate inversion is expected to occur at or slightly below 119 °C (Toda, 1993). Again, this rate inversion implies that the height of the nucleation barrier along these two growth directions also inverts.

The growth rate difference along different crystallographic planes is also observed in single crystals of low molecular weight poly(ethylene oxide) fractions in the low supercooling region where the once-folded and extended chain crystals form. The transition is less visually dramatic than for single crystals of polyethylene in both the

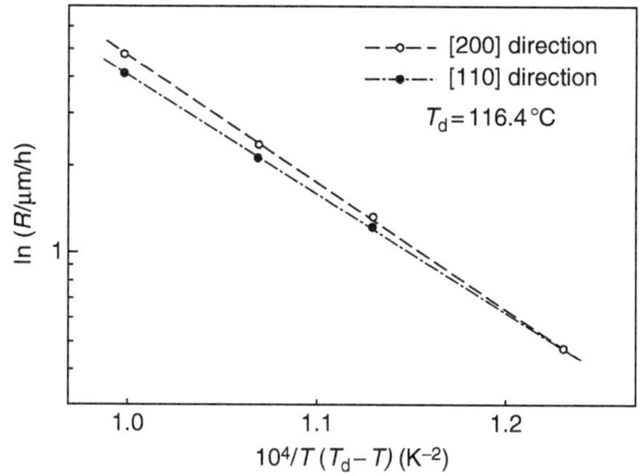

Figure 5.7 Polyethylene linear single crystal growth rates along both of the (110) and (200) normals at different crystallization temperatures for a polyethylene fraction with a number average molecular weight of 11 kg/mol and a polydispersity of 1.16 dissolved in *n*-octane with a concentration of 10^{-4} wt% [replotted from Toda *et al.* (1986), with permission].

bulk melt and solution. Single crystals of these poly(ethylene oxide) fractions are bounded by four (120) planes and two (100) crystallographic planes (Shcherbina and Ungar, 2007). Figure 5.8 shows a set of single crystal images grown at different crystallization temperatures for a poly(ethylene oxide) fraction with a number average molecular weight of 3 kg/mol in the melt. These morphologies were observed under polarized optical microscopy after the single crystals were self-decorated (Cheng and Chen, 1991). We can thus use an aspect ratio parameter to characterize the growth anisotropy, which is the ratio of the dimensions along the axis (vertical direction) and the $a^*(a\sin\beta)$ axis (horizontal direction) of the single crystals. Figure 5.9 illustrates the change of this aspect ratio with respect to crystallization temperature. It is evident that the aspect ratio increases first from unity for isotropic growth at 47 °C to 1.28 at 50 °C, indicating that along the *b* axis, the single crystal growth rate is faster than it is along the a^* axis. However, at a crystallization temperature of 51 °C, the aspect ratio suddenly drops back to unity and then gradually decreases until it reaches 0.8 above 55 °C (Cheng and Chen, 1991). This decrease indicates an inversion of the growth rates along the (120) and (100) normals; therefore, the nucleation barrier is crystallographic plane specific. The nucleation barriers of the (120) and (100) planes possess their own supercooling dependencies. Assuming that the metastabilities of the two different types of single crystal sectors grown along these two directions are identical, the growth rates are again determined by the nucleation barriers rather than by their stabilities.

The most intriguing example of anisotropic growth is a blend of poly(L-lactic acid) and poly(D-lactic acid) with equal molecular weights and concentrations, as described in Section 1.5 of Chapter 4. This blend leads to the formation of a stereocomplex crystal with a trigonal unit cell with $R3c$ (or $R\bar{3}c$) symmetry, and it is a frustrated cell (Cartier *et al.*, 1996). This stereocomplex crystal possesses a melting temperature about 50 °C higher than those of the corresponding enantiomeric

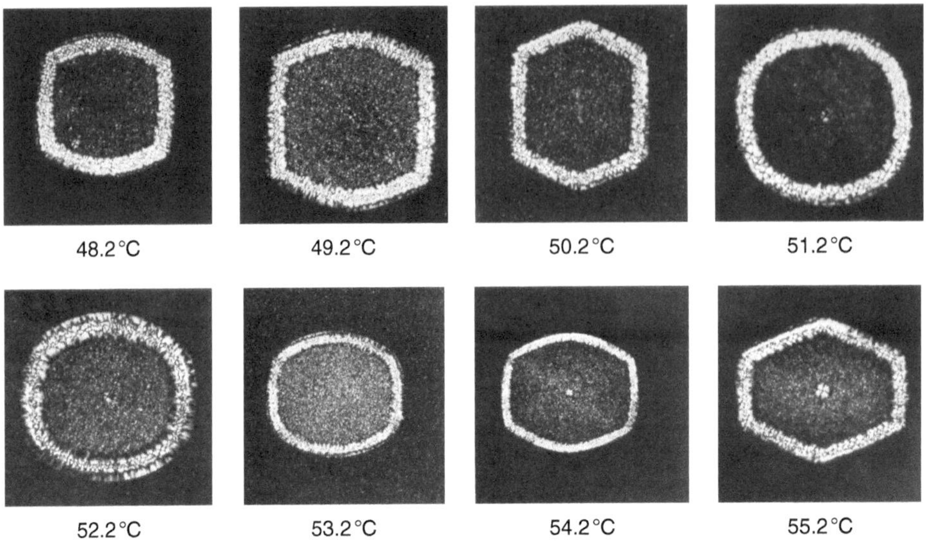

Figure 5.8 Set of poly(ethylene oxide) single crystals observed in polarized optical microscopy. The sample has a number average molecular weight of 3 kg/mol and polydispersity of 1.02 and was crystallized at different temperatures in the melt. The single crystal size is around 100 μm [reprinted from Cheng and Chen (1991), with permission].

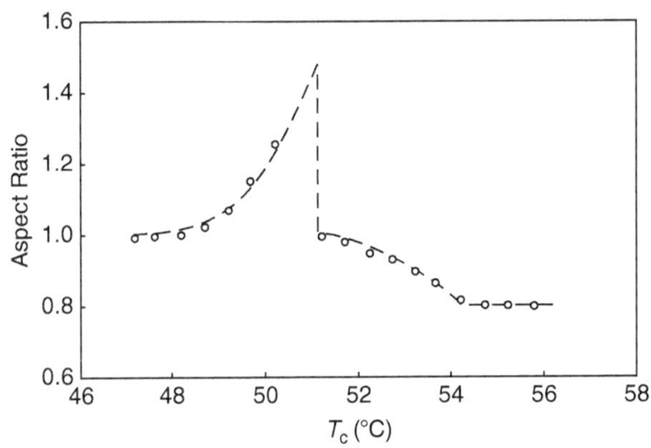

Figure 5.9 Aspect ratio changes with crystallization temperature of the single crystals shown in Fig. 5.8 [replotted from Cheng and Chen (1991), with permission].

poly(L-lactic acid) or poly(D-lactic acid) crystals. The most striking aspect of these stereocomplex crystals is their triangular shape (Brizzolara *et al.*, 1996; Okihara *et al.*, 1988, 1991), as shown in Fig. 5.10. It was recognized that these triangular single crystals must be related to the cocrystallization of two different molecular species, poly(L-lactic acid) with a 3_2 left-handed conformation and poly(D-lactic acid) with a 3_1 right-handed conformation, in the stereocomplex crystals. The

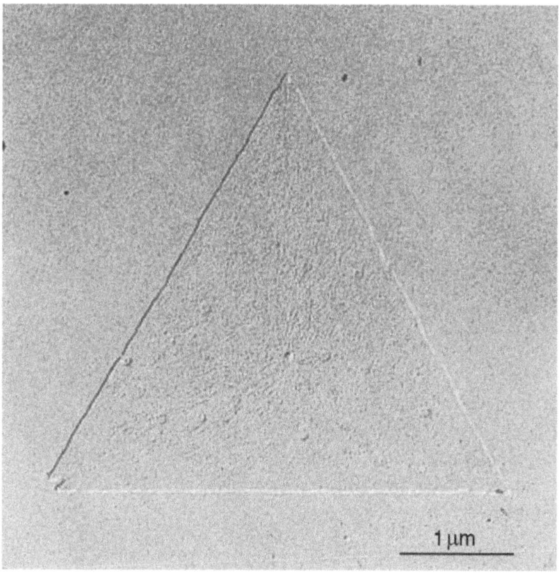

Figure 5.10 A triangular single crystal of the poly(L-lactic acid) and poly(D-lactic acid) stereocomplex with equal number average molecular weights of 7 kg/mol. Pure poly(L-lactic acid) was first solution-cast on one glass side and an equal molar blend was cast on another glass slide. After solvent evaporation, the two glass slides were then pressed together and heated to 250 °C. Then, they were cooled to 200 °C to crystallize for 3 h [reprinted from Cartier *et al.* (1997), with permission].

question is: How do the competing growth processes of these two different molecular species form this type of triangular single crystal?

In order to study the formation mechanism of the triangular single crystals, Cartier *et al.* (1997) specifically designed an experiment to create an environment in which the crystal grows with equal molecular weight chains but an imbalance in concentrations of poly(L-lactic acid) and poly(D-lactic acid). The triangular single crystals, as shown in Fig. 5.10, were observed. In the stereocomplex single crystal, the two types of chain helices arrange themselves into the crystal, as shown in Fig. 5.11. The growth planes are the {110} set of planes. On the front (110) surface, the right-handed poly(D-lactic acid) chain deposits in a notch, while the left-handed poly(L-lactic acid) chain deposits on a flat growth surface. On the opposite side of the (110) plane, the deposition position is reversed, such that the left-handed poly(L-lactic acid) chain deposits in a notch, while the right-handed poly(D-lactic acid) chain deposits on a flat growth surface. Therefore, from the molecular perspective, there are four deposition rates for triangular single crystal growth (Cartier *et al.*, 1997). This experimental setup generates local imbalances in composition on opposite sides of the {110} planes to the point that the composition imbalances determine the four rates of deposition. As soon as the relative growth rates start to deviate from one side to another and exceed a certain ratio, the triangular single crystals form uniformly. This deviation occurs because the slowest deposition rates on the opposite {110} growth surfaces determine the single crystal morphology. This example

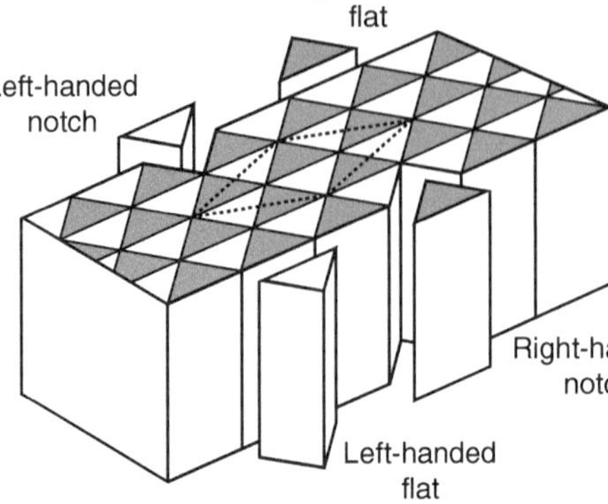

Figure 5.11 Illustration of two types of chain helices, right-handed (shadowed) and left-handed (unshadowed) arranged into the stereocomplex single crystal with $R3c$ (or $R\bar{3}c$) symmetry. This generates the polar nature of the crystal with respect to the handedness of the helical stems [redrawn from Cartier *et al.* (1997), with permission].

illustrates that a specific single crystal morphology forms because of competing growth rates caused by local composition imbalances that significantly affect the nucleation barriers of each of these four deposition rates (the "poisoning" effect described in Section 1.5 of Chapter 4). The only difference from the previous examples is that in this case, the nucleation barrier heights are controlled by the experimental growth environment instead of their characteristic supercooling dependences.

1.2. Initial transient state in polymer crystallization

In what follows, we focus on another set of examples in which metastable states appear due to competing phase transformation rates. Among them, the most striking experimental observations are that the crystallization kinetics of some model systems are affected by different chain conformations at and near the crystal growth front. These polymer crystallization model systems are strictly uniform oligomers with different chain lengths. Such systems make it possible, in principle, to connect the crystallization of short chains with the crystallization of high polymers. The central questions are: At what length do chains start folding, and how does this transition take place? This investigation, which is the obvious starting point for polymer crystallization studies, was, in practice, restricted 30 years ago by the unavailability of suitable materials. The molecules available in uniform lengths were not long enough to fold, and those which did show folding were far too long for this purpose and were highly polydisperse. Eventually, suitably monodisperse *n*-alkanes with long enough chains were synthesized (Bidd and Whiting, 1985; Bidd *et al.*, 1987; Lee and Wegner, 1985). Low molecular weight poly(ethylene oxide) fractions with narrow molecular weight distributions were also suitable.

As briefly described in Sections 1.5 and 2.1 of Chapter 4, *n*-alkanes and poly (ethylene oxide)s can crystallize into integral-folded chain crystals with a lamellar thickness almost identical to the lamellar thickness of the equivalent oligomers and that is close to an integral fraction of the chain length (*L*). An extended chain crystal has length (*L*), while the once-folded crystal has an integral fractional thickness of (*L*/2), a twice-folded crystal has a thickness of (*L*/3), and a thrice-folded crystal has a thickness of (*L*/4). The *n*-alkanes crystallize into integral-folded chain crystals, as shown experimentally from small angle x-ray scattering and in the longitudinal acoustic mode of Raman spectroscopy (Ungar *et al.*, 1985). For poly(ethylene oxide), the integral-folded chain crystals were first reported by Arlie *et al.* (1965, 1966, 1967). Later, Kovacs and his coworkers made a series of important findings with a set of low molecular weight poly(ethylene oxide)s with narrow molecular weight distributions based not only on small angle x-ray scattering experiments but also on single crystal habit changes and their formation kinetics (Kovacs and Gonthier, 1972; Kovacs and Straupe, 1979, 1980; Kovacs *et al.*, 1975, 1977).

The initial appearance of these integral-folded chain crystals in low molecular weight poly(ethylene oxide) fractions led to speculation as to why they form. One idea was that because poly(ethylene oxide) chains possess two hydroxyl end groups, they can form hydrogen bonds with neighboring chain ends, generating integral-folded chain crystals. However, poly(ethylene oxide) integral-folded chain crystals were also grown with different end groups (Cheng *et al.*, 1993). Work in *n*-alkanes also clearly pointed out that searching for the integral-folded crystal formation mechanism should focus on chain conformation rather than the chemistry of end groups. Thus, the question is whether the chain molecules form an integral-folded chain conformation during crystallization, or are they "blind" to the presence of other chains until their neighboring segments have crystallized. If the former case is true, the integral-folded chain crystal would form directly from the isotropic melt, whereas the latter case would indicate that the integral-folded chain crystals can only form after crystallization starts via rearrangement and reorganization of the chain conformation in the crystals. If so, non-integral-folded chain crystals must be an initial transient state.

The advancement of synchrotron real-time small angle x-ray scattering techniques led to new findings concerning *n*-alkane crystallization in the isotropic melt. When the crystallization occurs in the moderate to high supercooling region, below the crystallization temperatures where the extended chain crystals form, it was observed that the initial lamellar thickness does not fit any of the integral fractions of the chain length, thus indicating that integral-folded chain crystals do not initially form. Instead, the chain molecules crystallize into non-integral-folded chain crystals. These non-integral-folded chain crystals subsequently transform into integral-folded chain crystals via isothermal thickening or thinning (Ungar and Keller, 1986, 1987). Since then, a number of observations have been reported that found the non-integral-folded chain crystals to be an initial transient state in the crystallization of *n*-alkanes (Tracz and Ungar, 2005; Ungar and Zeng, 2001; Ungar *et al.*, 1998; Zeng and Ungar, 1998; Zeng *et al.*, 2005). An example of real-time, synchrotron radiation small angle x-ray scattering experiments monitoring lamellar thickness changes in the isothermal crystallization of an end-deuterated $C_{12}D_{25}C_{192}H_{384}C_{12}HD_{24}$ sample at 115 °C is shown in Fig. 5.12. In this figure, the strong diffraction peak rapidly forms with a lamellar periodicity of 18 nm. This periodicity is the thickness of the

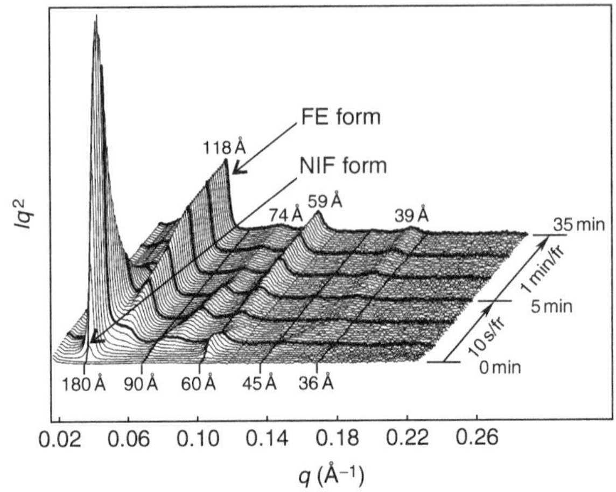

Figure 5.12 *In situ* small angle x-ray scattering results collected during the isothermal crystallization of an end-deuterated $C_{12}D_{25}C_{192}H_{384}C_{12}HD_{24}$ sample at 115 °C. The scattering vector q is defined as $4\pi\sin\theta/\lambda$, where λ is the wavelength of the x-ray (Courtesy of G. Ungar).

crystal at the initial stage of crystallization, and it does not correspond to an integral fraction of the chain length. The intensity of this diffraction peak decreases with increasing isothermal crystallization time and finally disappears, indicating that the non-integral-folded chain crystal rearranges to form crystals with a thickness of 11.8 nm. This crystal is a mixture of once-folded chain and extended chain crystals formed via a thinning process. The mixed crystal has also been confirmed by small angle neutron scattering experiments (Zeng *et al.*, 2005). Recently, detailed small angle x-ray scattering analysis shows that non-integral-folded chain crystals of *n*-alkanes consist of mixtures of folded and non-folded chain molecules.

The existence of non-integral-folded chain crystals in low molecular weight poly(ethylene oxide)s has also been shown by real-time small angle x-ray scattering experiments via synchrotron radiation. Over a wide range of supercooling, non-integral-folded chain crystals initially form, then transform into integral-folded chain crystals (Cheng and Chen, 1991; Cheng *et al.*, 1991b,c,d, 1992a,b, 1993). Figure 5.13 shows the formation of poly(ethylene oxide) non-integral-folded chain crystals with a molecular weight of 3 kg/mol crystallized at 43 °C. It was observed that a non-integral-folded chain crystal appears first having a lamellar thickness of 13.6 nm, which is between the thickness of the extended chain crystal (19.3 nm) and the once-folded chain crystal (10.0 nm). Within a crystallization time of 3.5 min, overall crystallization is accomplished. However, the transformation from non-integral-folded chain crystals to the integral-folded chain crystals continues, as shown in Fig. 5.13. Isothermal thinning toward the once-folded chain crystal is evident in this figure. Only at later times does the extended chain crystal start to develop, indicating an isothermal thickening process. Although the non-integral-folded chain crystal is thermodynamically less stable, it grows faster kinetically.

Similar changes in lamellar thickness are found for the same poly(ethylene oxide) fraction crystallized at different temperatures. The thickness of the initially formed

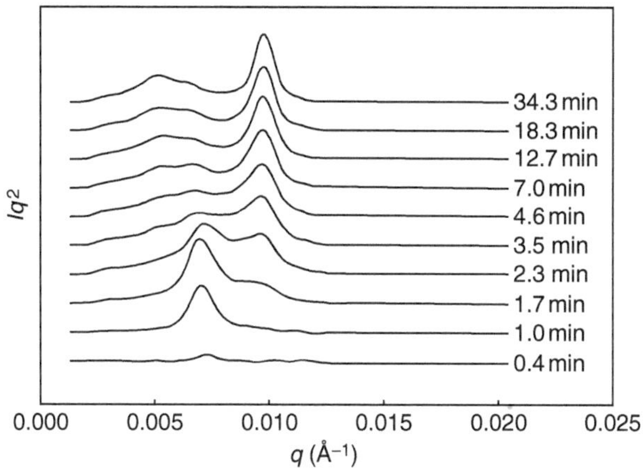

Figure 5.13 Set of Lorentz-corrected real-time small angle x-ray scattering data for a low molecular weight poly(ethylene oxide) fraction with a molecular weight of 3 kg/mol crystallized at 43 °C [replotted from Cheng *et al.* (1993), with permission].

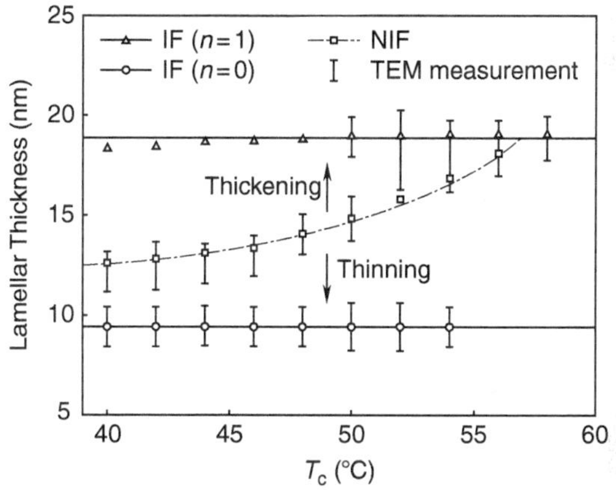

Figure 5.14 Set of lamellar thicknesses of the non-integral-folded chain and integral-folded chain crystals at different crystallization temperatures for a low molecular weight poly(ethylene oxide) with a number average molecular weight of 3 kg/mol. The thicknesses were detected by both small angle x-ray scattering and transmission electron microscopy experiments [replotted from Cheng *et al.* (1992a), with permission].

non-integral-folded chain crystals was found to be crystallization temperature dependent, as shown in Fig. 5.14. Results were obtained from both real-time small angle scattering and transmission electron microscopy experiments (Cheng *et al.*, 1992a). Again, both isothermal thinning toward the once-folded chain and crystal thickening toward the extended chain crystal are observed at each crystallization temperature. The existence of non-integral-folded chain crystals, as well as their isothermal transformation, have also been independently shown by longitudinal acoustic mode Raman spectroscopy (Kim and Krimm, 1996; Song and Krimm, 1989, 1990a,b).

We know that the lamellar thickness is critically associated with the crystal's thermodynamic stability, as described in Section 2.3 of Chapter 4. As a result, the thickening process is thermodynamically justified because the crystals anneal to a more stable form. However, the thinning process is an issue which needs to be discussed further. When we think of thermodynamic and morphological criteria for these two processes, one expects that the Gibbs free energies of these crystals follow:

$$G(\mathrm{NIF}) > G(\mathrm{IF},\ n = i + 1) > G(\mathrm{IF},\ n = i), \tag{5.1}$$

and that their lamellar thicknesses are

$$\ell(\mathrm{IF},\ n = i) > \ell(\mathrm{NIF}) > \ell(\mathrm{IF},\ n = i + 1) \tag{5.2}$$

where NIF and IF represent non-integral- and integral-folded chain crystals. At a constant temperature and pressure, both the thickening and thinning processes can take place. In the case of $i = 0$ and 1, the implication is that they are extended chain, IF($n = 0$), and once-folded chain, IF($n = 1$), crystals, respectively. On the other hand, when the thermodynamic and morphological criteria become:

$$G(\mathrm{IF},\ n = i + 1) > G(\mathrm{NIF}) > G(\mathrm{IF},\ n = i) \tag{5.3}$$

and

$$\ell(\mathrm{IF},\ n = i) > \ell(\mathrm{NIF}) > \ell(\mathrm{IF},\ n = i + 1), \tag{5.4}$$

we return to the common case of polymer lamellar crystals in which the thinning process is forbidden. The explanation of the higher Gibbs free energy for the non-integral-folded chain crystals in Eq. (5.1) is due to the inclusion of chain-end defects within the crystals, in addition to dangling chain ends causing rough folded surfaces with entropic effects on the crystals. Both factors destabilize the crystals and increase the Gibbs free energy of the system. Consequently, the non-integral-folded chain crystals are the least stable crystal among these three states, even though they are thicker than the once-folded crystal (Cheng *et al.*, 1992a). As a result, it can be understood that the free energy barrier to form the non-integral folded chain crystals must be the lowest among these three crystals; thus, the chain molecules are trapped in this metastable state after they overcome this barrier. Two free energy pathways exist for the non-integral-folded chain crystals to relax to a lower free energy state. One path is to the once-folded crystal, and the other path is to the extended chain crystal. The extended chain crystal is the ultimately stable state of these crystals. Furthermore, non-integral-folded chain crystals can also be found in the cases in which the fold number exceeds one (Cheng *et al.*, 1992b). Therefore, the observation of step-wise lamellar thickening with decreasing supercooling in these oligomers and low molecular weight fractions results from thickening and/or thinning processes, which are determined by both a thermodynamic driving force for the transformation from the non-integral- to integral-folded chain crystals and sufficient chain mobility along the crystallographic c axis.

With increasing molecular weight, the lamellar thickness of these crystals does not change substantially. The fold numbers, thus, increase with increasing chain length, and the difference between the thicknesses of two neighboring integral-folded chain crystals becomes vanishingly small. The thickness of the non-integral-folded chain crystal sits in between these two thicknesses. The driving force that dominates the transformation of the non-integral- to integral-folded chain crystals thus decreases. With increasing fold numbers, the chain mobility along the *c* axis also becomes increasingly small. Because both factors for the thinning and thickening processes are sufficiently reduced at high molecular weights, non-integral-folded chain crystals become permanent. This process serves at least as a qualitative explanation of why the lamellar thickness of polymer crystals continuously increases with decreasing supercooling (Cheng *et al.*, 1992a).

1.3. Nucleation and growth rates affected by chain conformation

When the crystallization temperature is near the equilibrium melting temperature, *n*-alkanes and low molecular weight poly(ethylene oxide)s crystallize into once-folded or extended chain crystals. The non-integral-folded chain conformations quickly adjust to become integral-folded chain conformations at or very near the surface of the growth front. Between these two integral-folded chain crystals, the once-folded chain crystals are metastable with respect to the extended chain crystals. As such, the melting temperature of the once-folded chain crystals is lower than that of the extended chain crystals. The relationship between these growth rates is very interesting because of the concept of competing growth rates between metastable and stable phases introduced in Fig. 3.8 of Section 3 of Chapter 3.

The first example addressing this concept involves a poly(ethylene oxide) fraction with a number average molecular weight of 6 kg/mol. Figure 5.15 shows

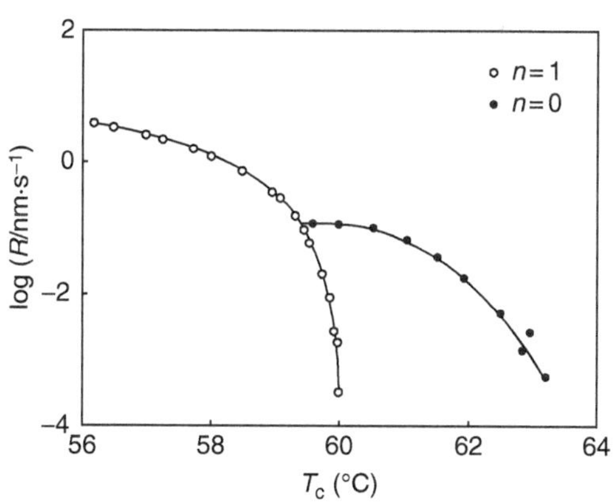

Figure 5.15 Linear growth rates of the once-folded ($n = 1$) and extended ($n = 0$) chain crystals of a poly(ethylene oxide) fraction with a number average molecular weight of 6 kg/mol. The melting temperature of the once-folded chain crystals is 60.3 °C and for the extended chain crystals is 63.7 °C. The crossover temperature at which both of the growth rates are identical is at 59.4 °C [replotted from Kovacs *et al.* (1977), with permission].

crystal growth rate data for this fraction in the low supercooling region (Kovacs *et al.*, 1977). The melting temperature of the once-folded chain crystals is 60.3 °C, and it is 63.7 °C for extended chain crystals. As shown in this figure, there are two growth rate branches. The lower temperature is for the once-folded chain crystal and the high temperature is for the extended chain crystal. At 59.4 °C, both of the growth rates are identical, and above this temperature the extended chain crystals grow faster (Kovacs *et al.*, 1975). This temperature is the crossover temperature in Fig. 3.8. This crossover is not because the crystals have different forms, but rather because they possess different lamellar thicknesses with two distinct conformations.

In a narrow supercooling region where the crystallization temperature is slightly higher than the crossover temperature, between 59.4 and 60.3 °C, both the once-folded and extended chain crystals can grow, but the growth rates of the once-folded chain crystal are slower than those of the extended chain crystal. This data indicates that the surface nucleation barrier of the once-folded chain crystal is lower than that of the extended chain crystal below the crossover temperature (59.4 °C). However, this barrier drastically increases to exceed the surface nucleation barrier of the extended chain crystal above the crossover temperature. Although these two branches seem to be independent, no growth rate minimum at or near the crossover temperature can be observed, as briefly described in Section 1.5 of Chapter 4.

A weak minimum in the single crystal growth rates of a 3 kg/mol number average molecular weight poly(ethylene oxide) fraction with methoxy end groups in the melt was reported in the vicinity of the crossover supercooling where the growth changes from the once-folded crystal to the extended chain crystal, as shown in Fig. 5.16 (Cheng and Chen, 1991). The reason a rate minimum is not observed in poly(ethylene oxide) fractions with two hydroxyl end groups may be because both end groups can form hydrogen bonds in both the solid and liquid states (Cheng *et al.*, 1991a). Therefore, in this case, the folded and extended conformations are difficult

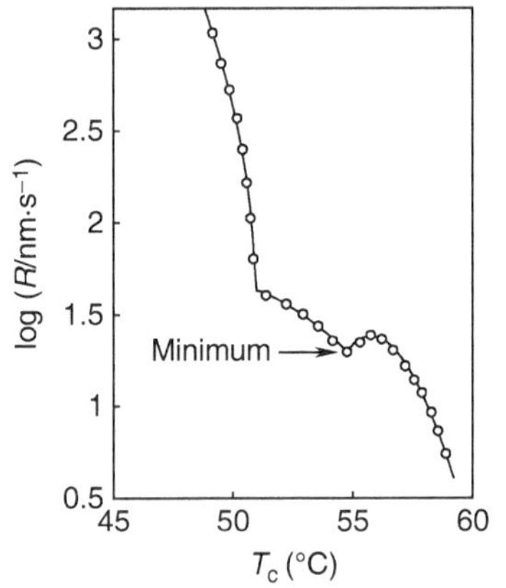

Figure 5.16 Linear growth rates of the once-folded ($n = 1$) and extended ($n = 0$) chain crystals of a poly(ethylene oxide) fraction having methoxy end groups with a number average molecular weight of 3 kg/mol and a polydispersity of 1.02 [replotted from Cheng and Chen (1991), with permission].

to differentiate. This case does not apply in the poly(ethylene oxide) fraction having methoxy end groups (Cheng and Chen, 1991).

The first reported and most evident rate minimum phenomenon was observed in the case of *n*-alkanes. The earliest reports showed that the nucleation and growth rates exhibit minima for monodisperse $C_{198}H_{398}$ and $C_{246}H_{494}$ in the supercooled melt and solution (Organ *et al.*, 1989, 1997; Ungar and Keller, 1986, 1987) when crystallized in the supercooling region of once-folded and extended chain crystals. The growth rates for each fold number are depicted in Fig. 5.17. As described in Section 1.5 of Chapter 4, the rate minimum is due to the "self-poisoning" effect. When an *n*-alkane chain with a once-folded conformation deposits onto an extended chain crystal growth front, this location becomes "poisoned," thus hampering further extended chain crystal growth. The correction of this "mistake" by either extending the folded conformation or removing this chain from this "poisoned" spot is necessary for further growth. Therefore, both the nucleation and growth rates decrease. Since the initial report, more sophisticated direct growth rate measurements have been made to prove the existence of this rate minimum phenomenon (Putra and Ungar, 2003; Ungar *et al.*, 2005). Moreover, investigations into the crystallization of *n*-alkanes and their blends have provided additional evidence for this "self-poisoning" effect (Boda *et al.*, 1997; Hobbs *et al.*, 2001; Hosier *et al.*, 2000; Morgan *et al.*, 1998; Putra and Ungar, 2003; Sutton *et al.*, 1996; Ungar and Zeng, 2001; Ungar *et al.*, 2000, 2005).

These experimental observations imply that crystals with identical structures but different conformations (here, once-folded versus extended chains) can also construct different metastable states due to variations in crystal thickness for samples of a particular molecular weight. The fastest growth rate is not for the ultimate

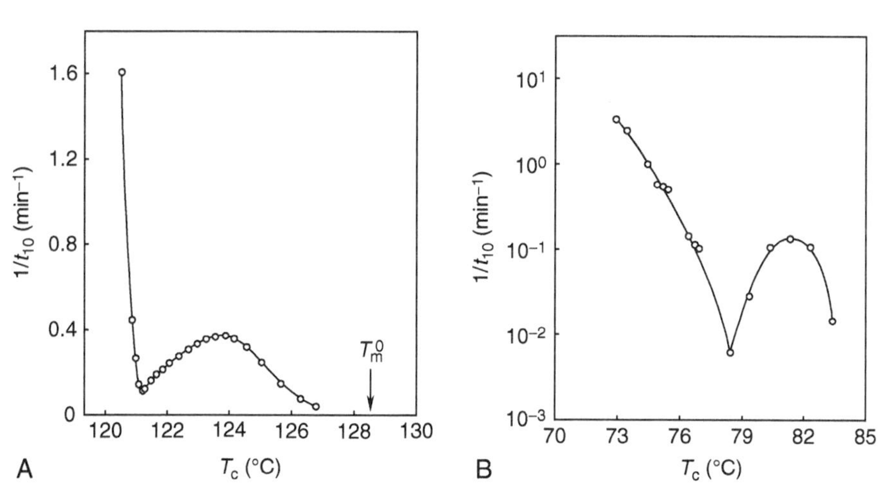

Figure 5.17 Crystallization rates of the once-folded ($n = 1$) and extended ($n = 0$) chain crystals of $C_{246}H_{494}$ crystallized from the melt (A) and $C_{198}H_{398}$ crystallized from solution (B). The rate minima can be clearly observed at the transition between the once-folded ($n = 1$) and extended ($n = 0$) chain crystals [replotted from Ungar and Keller (1987) and Organ *et al.* (1989), with permission].

thermodynamic equilibrium state, but is dependent on the barrier heights of the kinetic pathway. The observed rate minimum phenomenon indicates that defective chain conformations deposited onto the growth front reduce the crystal growth rates. This case is another example of competing kinetics leading to crystal formation with different chain conformations.

2. Polymorphs and Competing Formation Kinetics

2.1. Phase stability changes in polymorphs at atmospheric pressure

We now investigate a slightly more complicated case where polymorphs exist and possess competing formation kinetics. At a particular pressure and temperature, there is only one stable thermodynamic equilibrium phase, while the rest are metastable, as described in Section 3 of Chapter 3. These different ordered polymorphs have been experimentally observed to coexist. The appearance of more than one ordered state is due to kinetically competing processes resulting from the existence of more than one barrier. Because the ability to overcome a barrier relies on thermal (density) fluctuations, the barrier which possesses the lowest height is always overcome first. We emphasize again, if the phase formed after overcoming the lowest barrier happens not to be the ultimate equilibrium state, it is a metastable state. How fast the equilibrium state will be reached from this metastable state depends on the barrier between these two states, as illustrated in Section 1 of Chapter 3. The appearance of the metastable state is typically dominated by kinetics.

First let us recall cases of crystal polymorphs in polymers. Many semicrystalline polymers exhibit polymorphic behavior. Examples include the monoclinic α-phase, the trigonal frustrated β-phase, and the orthorhombic γ-phase of isotactic polypropylene (Brückner and Meille, 1989; Dorset *et al.*, 1998; Lotz and Wittmann, 1986; Lotz *et al.*, 1986, 1991, 1996; Natta and Corradini, 1959; Stocker *et al.*, 1998; Turner-Jones *et al.*, 1964); the trigonal phase I, the trigonal phase II, and the orthorhombic phase III of isotactic poly-1-butene (Cojazzi *et al.*, 1976; Dorset *et al.*, 1994; Natta *et al.*, 1960a; Petraccone *et al.*, 1976); the trigonal α-phase and the orthorhombic β-phase and their limited order modifications in the melt, in addition to solvent induced γ- and δ-phases of syndiotactic polystyrene (Cartier *et al.*, 1998; De Rosa, 1996; De Rosa *et al.*, 1992; Guerra *et al.*, 1990, 1991; Pradere and Thomas, 1990; Tosaka *et al.*, 1999); the orthorhombic α-phase, the orthorhombic polar β-phase, the monoclinic γ-phase, and the orthorhombic δ-phase of poly (vinylidene fluoride) with different chain conformations and/or packing schemes in the crystal unit cells (Davis *et al.*, 1978; Hasegawa *et al.*, 1972a,b; Lovinger, 1981a,b, 1982; Weinhold *et al.*, 1979, 1980) and the high- and low-temperature orthorhombic phases with different unit cell dimensions in poly(ether ketone ketone) (Avakian *et al.*, 1990; Blundell and Newton, 1991; Gardner *et al.*, 1992, 1994; Ho *et al.*, 1994a,b, 1995a,b). Polymorphs are also frequently observed in some polyoxides, such as poly(oxymethylene) and poly(trimethylene oxide), many aliphatic polyamides, such as poly-L-alanine and nylons, and others. All these crystal structures and symmetries have been determined utilizing x-ray and electron diffraction methods (see, e.g., Dorset, 1995; Geil, 1963; Tadokoro, 1979; Wunderlich, 1973).

An interesting area is solvent-induced polymorphs in polymer crystals. One type of solvent-induced crystal forms via polymer–solvent complexation, such as γ- and δ-phases in syndiotactic polystyrene. Another type does not form a complex, such as in the case of poly(4-methyl-1-pentene). This polymer possesses a tetragonal unit cell with a 7_2 helical chain conformation in crystals grown from both the melt and the solution (Bassett, 1964; Bassett *et al.*, 1964; Frank *et al.*, 1959; Khoury and Barnes, 1972; Patel and Bassett, 1994) and is usually called modification I. However, it also has three other crystal modifications, depending on crystallization conditions and procedures. Modification II was obtained in dilute xylene (Tanda *et al.*, 1966) as well as in tetramethyltin and tetraethyltin at high supercoolings (Charlet and Delmas, 1984), and it has a tetragonal unit cell with a 4_1 helical chain conformation. Modification III possesses the same unit cell lattice and helical chain conformation but a slightly different unit cell size. It has been grown from solution with alkane and cycloalkane solvents. Modification IV was obtained by annealing the original tetragonal crystals at 200 °C at an elevated pressure of 4.5 kbar to form a hexagonal unit cell with a 4_1 helical chain conformation (Hasegawa *et al.*, 1970). Furthermore, another modification of poly(4-methyl-1-pentene) occurs in a cyclohexane gel (Charlet and Delmas, 1984).

When we study the formation kinetics of polymorphs in polymers, the correlation between unit cell symmetry and crystal habit, as described in Section 1.2 of this Chapter, begs the following question: What occurs in single crystal habits when a semicrystalline polymer transforms from one crystal structure to another that has a different lattice symmetry? The answer depends on whether the transformation takes place in the form of a solid–solid transition or a solid–liquid–solid transition. For example, a solid–solid transformation occurs when the hexagonal phase of polyethylene is grown at an elevated pressure and transforms to the orthorhombic polyethylene crystal phase as the pressure and temperature are reduced. This transformation can give rise to twinned symmetries having three equivalent orientations 120° apart from each other (DiCorleto and Bassett, 1990), as shown in Fig. 5.3A. The morphology of the single crystal with high unit cell symmetry is retained in this solid–solid transition. In contrast, polyethylene spherulites formed at atmospheric pressure always grow with the *b* axis of the orthorhombic lattice in the radial direction, and no twin symmetry can be found (Keith and Padden, 1959a,b; Keller, 1959; Point, 1955; Price, 1959).

From a conformational point of view, solid–solid crystal transformation between two phases requires more attention. In some solid–solid crystal transformations, both crystalline phases possess the identical chain conformation. The transformations thus only require a positional displacement of the chains from their placement in the initial phase to their placement in the final phase. This type of transformation can be observed in the cases of polyethylene between the all *trans*-conformation orthorhombic and monoclinic phases (Seto *et al.*, 1968; Wittmann and Lotz, 1989), poly(ethylene-co-carbon monoxide) between the α- and β-phases (Klop *et al.*, 1995), and poly(vinylidene fluoride) between the α- to δ-phases (the tgtg$^-$ helical conformation, Lovinger, 1982). For other solid–solid crystal transformations, the initial and final phases involve different chain conformations. Examples include the transition in isotactic poly-1-butene from phase II with a 11_3 helical conformation to phase I with a 3_1 helical conformation (Holland and Miller, 1964; Lotz *et al.*, 1998b) and the transformation in syndiotactic polypropylene from the low-temperature orthorhombic phase with an all *trans*-conformation to the high-temperature phases with a helical chain conformation (Chatani *et al.*, 1991; De Rosa *et al.*, 1998; Lotz *et al.*, 1998b).

On the other hand, if modifications of the chain positions and conformations must be on a relatively large scale, then solid–liquid–solid transformations are necessary to accomplish the transformation. In this case, memory of the original single crystal habit is lost in the transition to the liquid state. Only the latest structural symmetry adopted determines the newly formed single crystal habit.

The first interesting example of polymorphism is isotactic poly-1-butene. This polymer exhibits three crystal modifications in solution crystallization (Holland and Miller, 1964; Luciani *et al.*, 1988): phase I with a 3_1 helical conformation and a trigonal unit cell formed by the solid–solid crystal transformation from phase II (Natta *et al.*, 1960a) (also, a directly crystallized phase I′ exists with an identical crystal structure as phase I but with lower stability); phase II with a 11_3 helix conformation and a tetragonal unit cell (Petraccone *et al.*, 1976); and phase III with a 4_1 helix conformation and an orthorhombic unit cell with a $P2_12_12_1$ space group (Cojazzi *et al.*, 1976; Dorset *et al.*, 1994). These polymorphs possess different stabilities in the temperature region between room temperature and their melting temperatures. Note that both left- and right-handed helices with an equal conformational energy exist, such that poly-1-butene, like other vinyl polymers, has a chiral center but exists in a racemic form. In the trigonal phase I and tetragonal phase II of isotactic poly-1-butene, the crystals require antichiral chain packing, where the right-handed helix must be surrounded by left-handed helices and vice versa. However, in phase III, all the helical chains have the identical handedness, resulting in isochiral packing. It is particularly interesting that phase III can form scrolled single crystals (Holland and Miller, 1964; Lotz and Cheng, 2006).

Much attention has been paid to the transformation between these polymorphic phases in isotactic poly-1-butene. For example, phase II can spontaneously transform into the phase I at room temperature via a solid–solid crystal transition (Chau *et al.*, 1986; Holland and Miller, 1964; Kopp *et al.*, 1994a; Lotz *et al.*, 1998b; Luciani *et al.*, 1988; Tosaka *et al.*, 2000). On the other hand, phases I′ and III can transform to phase II through annealing below their melting temperatures (~90 °C), probably via melt-recrystallization processes (Armeniades and Baer, 1967; Geacintov *et al.*, 1963; Woodward and Morrow, 1968). In crystallization of isotactic poly-1-butene from the isotropic melt, the metastable phase II is usually formed because of its favorable kinetics rather than its thermodynamic stability. After annealing at room temperature for a long time, a phase II crystal can transform to phase I. In a recent experiment concerning thin film epitaxial crystallization, all three phases were produced by using the appropriate substrate (Kopp *et al.*, 1994b,c; Mathieu *et al.*, 2001). A solvent-assisted method to produce a large population of phase III in the melt was also developed (Lotz and Thierry, 2003). Figure 5.18 shows an optical microscopy image of isotactic poly-1-butene crystallized utilizing this newly

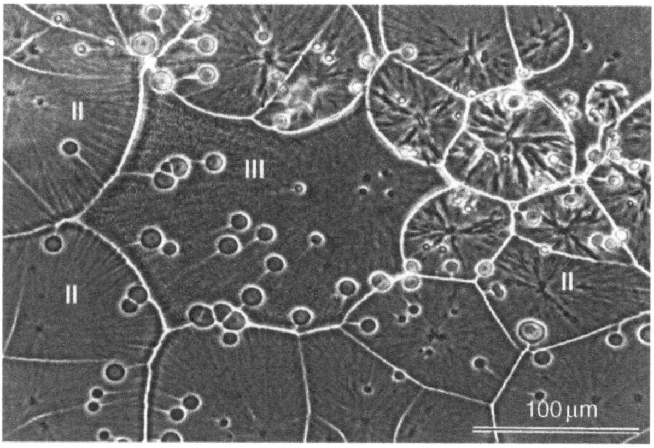

Figure 5.18 Optical microscopy image of isotactic poly-1-butene crystallized from the melt utilizing a newly developed solvent-assisted method. The center spherulite is the phase III with extinction rings and the surrounding spherulites are the phase II. The encapsulated small spherulites are the phase I (Courtesy of B. Lotz).

developed method. At the center of this image, the area exhibiting weak birefringence is a phase III spherulite, while the surrounding spherulites are in phase II (or perhaps they have transformed to phase I). Phase III spherulites possess concentric extinction rings with a periodicity of 5 μm, indicating cooperative lamellar twisting. In Fig. 5.19, a bright field transmission electron microscopy image clearly reveals the cooperative twisting of phase III lamellae in the spherulite (Lotz and Thierry, 2003).

At the crystallization temperature used in Fig. 5.18, the small phase I spherulites are encapsulated by the advancing growth front of phase III, as well as by phase II spherulite growth. Recent results showed that the ratio of the II and I′ phase growth

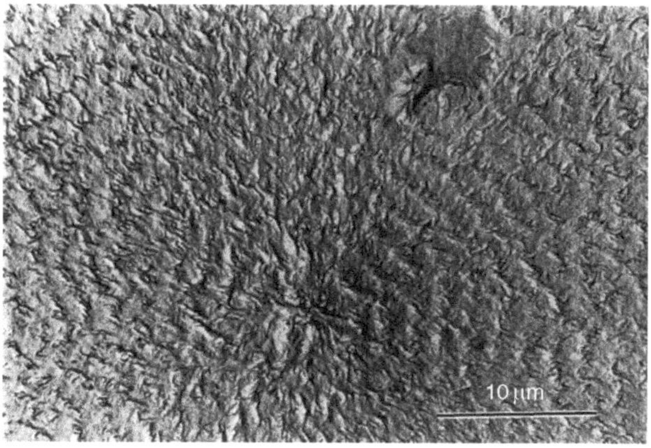

Figure 5.19 Bright field transmission electron microscopy image of isotactic poly-1-butene crystallized from the melt utilizing a solvent-assisted method. The phase III spherulite possesses concentric extinction rings with a periodicity of 5 μm, indicating lamellar cooperative twisting [reprinted from Lotz and Thierry (2003), with permission].

rates is greater than two orders of magnitude (Yamashita *et al.*, 2007). This experimental evidence further indicates that appearance of the phase II tetragonal phase is not determined by its ultimate phase stability, but rather from the height of the nucleation barrier.

In addition to structural information on the polymorphs, detailed growth kinetics analysis can be carried out. Poly(ether ketone ketone) serves to illustrate this concept. It possesses two different orthorhombic crystal phases (phases I and II). Both the unit cells contain two chains and have the identical symmetry ($Pbcn\text{-}D_{2h}^{14}$). The phase I crystal ($a = 0.786$ nm, $b = 0.575$ nm, and $c = 1.016$ nm) (Avakian *et al.*, 1990; Gardner *et al.*, 1992, 1994) is thermodynamically more stable than the phase II crystal ($a = 0.417$ nm, $b = 1.134$ nm, and $c = 1.008$ nm) (Blundell and Newton, 1991; Ho *et al.*, 1994a, 1995b). Figure 5.20 shows single crystals grown from the melt of these two phases in addition to their [001] zone electron diffraction patterns and molecular packing in the *ab* plane. The diffraction results provide decisive evidence that both the orthorhombic phases are two-chain unit cells with one chain at the center of the unit cell but with very different unit cell dimensions (as shown in Fig. 5.20C).

These two phases prefer to grow in different temperature regions. Figure 5.21 shows two sets of wide angle x-ray powder diffraction patterns for both these phases formed at different crystallization temperatures. Note that the diffraction at $2\theta = 15.7°$ (d-spacing of 0.565 nm) is identified as the (020) of the phase II crystals. Figure 5.21A is for crystallization from the isotropic liquid and Fig. 5.21B represents crystallization from the quenched amorphous glass. It is evident that phase I grows faster at high temperatures, while the phase II grows faster at low temperatures. These observations reflect that the heights of the nucleation barriers of these two phases in poly(ether ketone ketone) possess distinct supercooling dependencies.

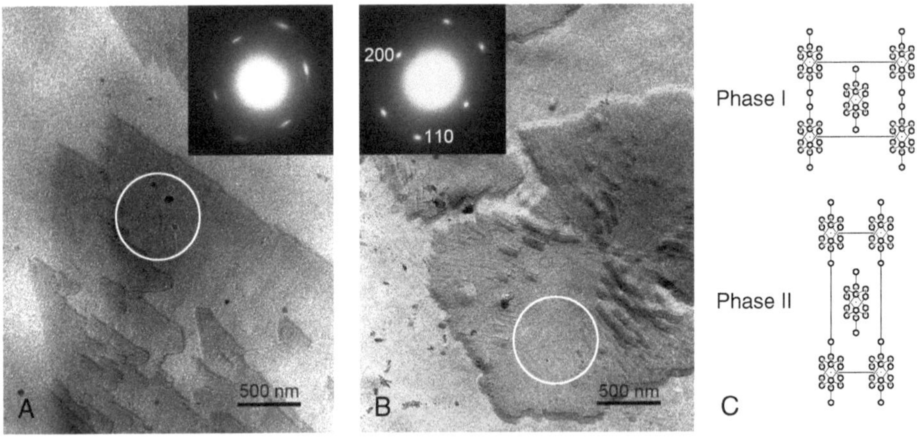

Figure 5.20 Transmission electron microscopic bright field images of poly(ether ketone ketone) single crystals grown from the melt and their corresponding [001] electron diffraction patterns for phases I (A) and II (B). Illustrations of the chain packing in the *ab* plane of the unit cells for these two phases (C) [reprinted from Ho *et al.* (1995b), with permission].

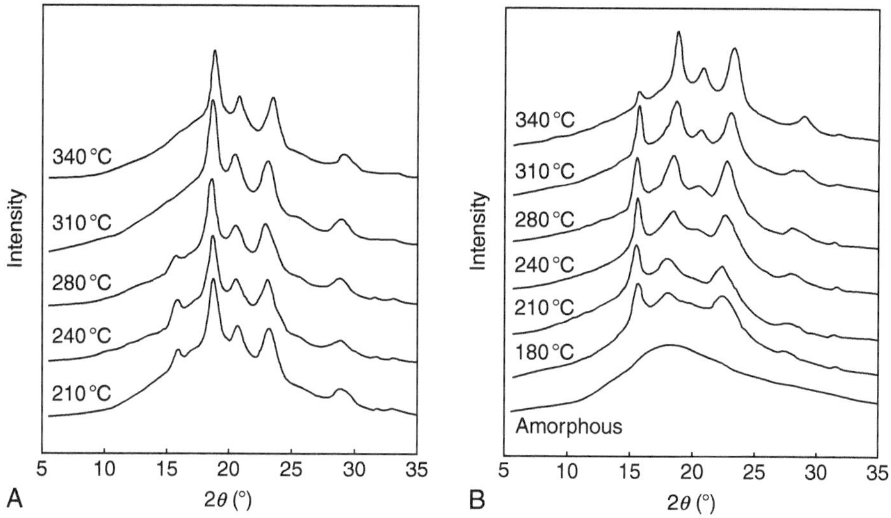

Figure 5.21 Two sets of one-dimensional wide angle x-ray diffraction powder patterns for poly (ether ketone ketone) crystallized from the isotropic melt (A) and from the amorphous glassy state (B). The (020) diffraction peak at $2\theta = 15.7°$ is representative of the phase II crystals [replotted from Ho *et al.* (1994a), with permission].

Namely, at high crystallization temperatures, the barrier height of phase I is lower than that of phase II, and this situation is reversed at low crystallization temperatures.

Another interesting example is syndiotactic polystyrene. Extensive studies have been reported on its complex polymorphic phase behavior (Cartier *et al.*, 1998; De Rosa, 1996; De Rosa *et al.*, 1992; Guerra *et al.*, 1990, 1991; Pradere and Thomas, 1990; Tosaka *et al.*, 1999). Focusing on polymorph formation in the melt, the trigonal α-phase has been identified as a supermolecular frustrated form (Cartier *et al.*, 1998), and the orthorhombic β-phase is more stable then the α-phase. Wide angle x-ray diffraction and infrared spectroscopic observations show that when the syndiotactic polystyrene melt is below 300 °C, isothermal crystallization leads exclusively to the formation of the modified trigonal α'-phase, which is less ordered than the α-phase but maintains the identical unit cell lattice. When the melt is above 340 °C, the modified orthorhombic β'-phase grows. When the melt is in between these two temperatures, a mixture of the α'- and β'-crystals forms grows. Note that all these isothermal crystallization experiments were carried out at temperatures ranging between 235 and 250 °C. Figure 5.22A shows two wide angle x-ray diffraction powder patterns. In one, the melt was held at 290 °C; in the other, the melt was held at 340 °C, but both were isothermally crystallized at 250 °C. Figure 5.22B shows the two corresponding infrared spectra that distinguish the α'- and β'-phases (Ho *et al.*, 2000).

These experimental observations imply that the nucleation barriers of these polymorphs are dependent upon the residence temperature of the melt. However, the α' phase initially grows during isothermal crystallization when the sample is quenched from the melt below 300 °C and gradually transforms to the β'-phase after

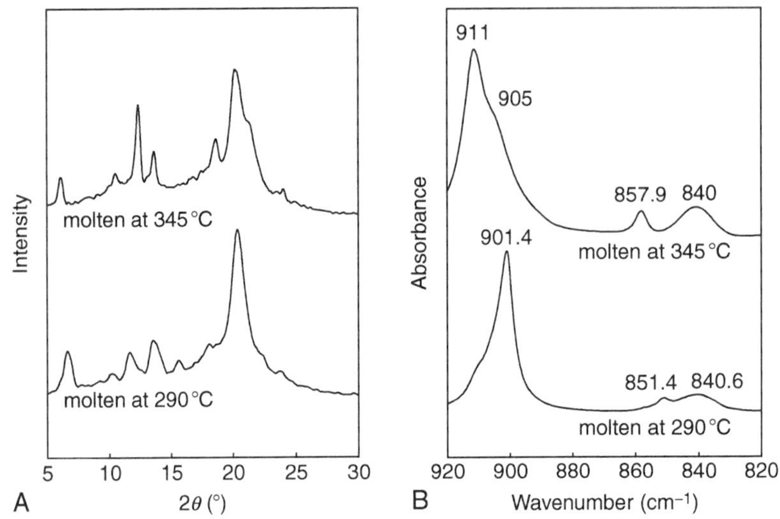

Figure 5.22 Set of wide angle x-ray diffraction powder patterns for the α′- and β′-phases of syndiotactic polystyrene crystallized at 250 °C after they were quenched from 290 and 340 °C, respectively (A), and the corresponding infrared spectra of these two phases under the same crystallization conditions (B) [replotted from Ho *et al.* (2000), with permission].

prolonged annealing at the isothermal crystallization temperature (Ho *et al.*, 2000, 2001). This transformation reflects a change of metastability via phase transformation. Furthermore, their isothermal crystallization behaviors have different Avrami exponents and other kinetic parameters (Woo *et al.*, 2001) when the samples are quenched from the melt to below 300 °C or above 340 °C. More interestingly, the single crystal habits of these two phases are also different. There is a hexagonal-shaped single crystal for the trigonal α′-phase and a truncated lozenge-shaped single crystal for the orthorhombic β′-phase (Ho *et al.*, 2001). The remaining issue is to understand how the less stable α′-phase transforms to the more stable β′-phase, particularly whether the lamellar thickening of the α′-phase triggers this transformation into the more stable β′-phase. If the lamellar thickening triggers this solid–solid phase transition, the hexagonal-shaped single crystal should at least be partially retained after the formation of the β′-phase. Further study is clearly required. Recent studies have also reported that the solution-grown δ-phase can transform to the γ-phase via solvent evaporation, indicating a stability difference between these two phases (Yashioka and Tashiro, 2003).

In terms of polymer polymorphism, one of the most well known and technologically important examples is poly(vinylidene fluoride). This polymer exhibits four polymorphs. Among them, the α- and γ-phases are commonly observed when crystallization occurs from the isotropic melt. The α-phase has an orthorhombic unit cell with $a = 0.966$ nm, $b = 0.496$ nm, and $c = 0.464$ nm (Lando *et al.*, 1966). The γ-phase possesses a monoclinic unit cell with $a = 0.496$ nm, $b = 0.967$ nm, $c = 0.920$ nm, and $\beta = 93°$ (Lovinger, 1981b). These two phases grow spherulitic aggregates with different textures. The α-phase spherulites possess high birefringence

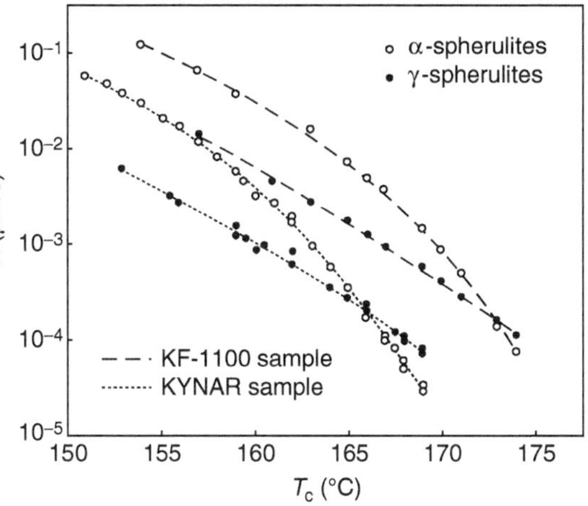

Figure 5.23 Linear growth rates of the α- and γ-phase spherulites crystallized from the melt for two poly(vinylidene fluoride) samples at different temperatures. The two samples (KF-1100 and KYNAR) possess different head-to-head and tail-to-tail compositions. In both cases, the growth rate inversion can be observed [replotted from Lovinger (1980), with permission].

and tight bending, while the birefringence of the γ-phase spherulites is lower (Lovinger, 1982). Although the growth rates of these two phases vary due to different head-to-head and tail-to-tail compositions in the polymers, the general trend of the growth rates for these two phases is what is shown in Fig. 5.23 (Lovinger, 1980). In this figure, the spherulitic crystal growth rates of two different poly(vinylidene fluoride) samples are shown.

Generally, the growth rates of the γ-phase are several times slower than those of the α-phase in the lower crystallization temperature region. At high crystallization temperatures, however, the growth rates are reversed, with the γ-phase growing faster. This growth rate inversion indicates that the heights of the nucleation barriers for these two phases are inverted. Although the barriers are dependent upon supercooling, the divergence of these two growth rates with increasing crystallization temperature above the crossover temperature indicates that the height of these two nucleation barriers must be affected by other factors in addition to supercooling.

Isotactic polypropylene, which has four different crystal modifications (Brückner and Meille, 1989; Dorset *et al.*, 1998; Lotz and Wittmann, 1986; Lotz *et al.*, 1986, 1991, 1996; Natta *et al.*, 1959; Stocker *et al.*, 1998; Turner-Jones *et al.*, 1964), has become the most illustrative example to explain the role of metastable states during phase transformation. Among the polymorphs in this polymer, the γ-phase may be enhanced by crystallizing under high pressure (Campbell *et al.*, 1993; Pae *et al.*, 1966); by using low molecular weight polypropylene (Addink and Beintema, 1961; Morrow and Newman, 1968); by growing it epitaxially on the α-phase (Lotz *et al.*, 1986, 1991); by the presence of chain defects or chemical heterogeneity (Alamo *et al.*, 1999; Turner-Jones, 1971); or by the inclusion of ethylene comonomer units in the chains (Alamo *et al.*, 2003; De Rosa *et al.*, 2002; Laihonen *et al.*, 1997; Mezghani and Phillips, 1995, 1998; Morrow and Newman, 1968; Zimmermann, 1993).

Both the α- (monoclinic) and β- (trigonal) phases grow in a wide range of crystallization temperatures at atmospheric pressure. The chain conformation in these two phases is the three-fold helix. Therefore, the structural growth of these two phases is completely determined by the stem packing schemes, as described in Section 1.5 of Chapter 4. Figure 5.24 illustrates the structural relationship between these two phases. In the α-phase, the handedness of the stems alternates layer-by-layer, forming an antichiral packing along the *b* axis [the (040) planes of the α-phase]. Furthermore, the tips of the triangular helices (or the flats of the triangular helices) are pointed to the neighboring two triangular helices (or face each other), as shown in this figure. The β-phase possesses a frustrated packing (Lotz *et al.*, 1994). This phase grows along the {110} planes with isochiral packing, yet the tips of the helices are oriented up, down, and down (as in Fig. 5.24). Between these two phases, the β-phase is a thermodynamically metastable phase compared with the α-phase.

The formation of both the α- and β-phases in isotactic polypropylene is determined by nucleation processes. The spherulitic morphology of these two phases is very distinct. Spherulites of the α-phase generally exhibit weak birefringence due to their crosshatched lamellar growth (Lotz and Wittmann, 1986), while spherulites of the β-phase have strong birefringence under polarized optical microscopy. The underlying morphological characteristics can be observed in Fig. 5.25, which is a transmission electron microscopy bright field image. In α-phase spherulites with low birefringence, the lamellar cross-hatching can be clearly seen with lamellae oriented along the radial and tangential directions of the spherulite due to self-epitaxial growth. In β-phase spherulites with high birefringence, only one lamellar population can be seen with a uniform orientation (in this figure, they are flat-on type lamellae) (Fillon *et al.*, 1993).

The growth rates for each phase in isotactic polypropylene can be experimentally determined using polarized optical microscopy. Extensive growth rate data has been collected over the last 30 years (Fillon *et al.*, 1993; Lotz, 1998; Lovinger *et al.*, 1977; Varga, 1989). Using directional solidification for isotactic polypropylene crystallized

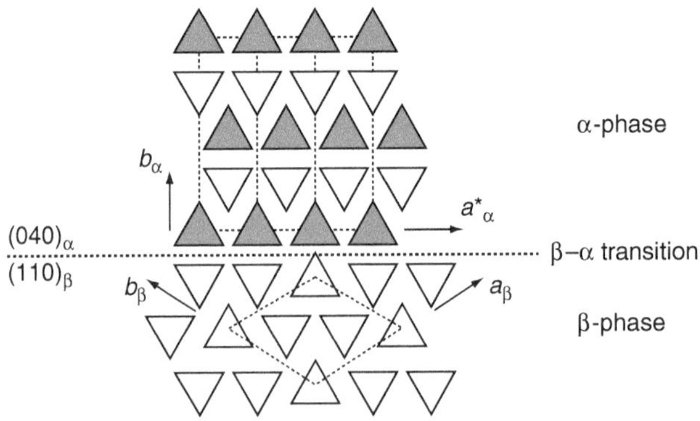

Figure 5.24 Illustration of the structural relationship between the α- and β-phases in isotactic polypropylene. The densest packing of the α-phase is in the (040) plane normal to the *b* axis, while that of the β-phase is in the (110) plane [replotted from Lotz (1998), with permission].

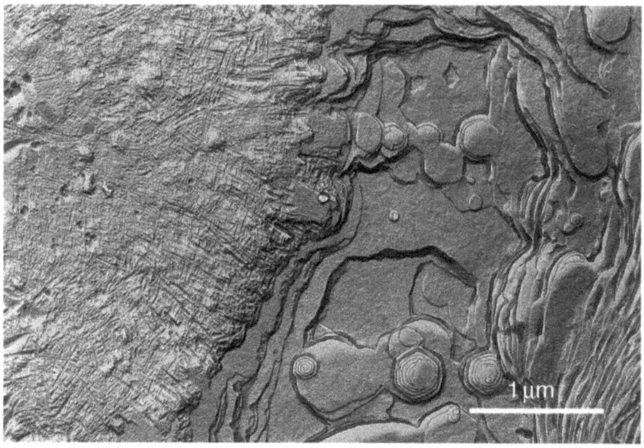

Figure 5.25 Transmission electron microscopy bright field image of isotactic polypropylene crystallized in the melt. Both crosshatched lamellae of the α-phase spherulite (left) and flat-on uniform lamellae of the β-phase spherulite (right) can be identified [reprinted from Fillon *et al.* (1993), with permission].

in a temperature gradient (Lovinger *et al.*, 1977) or specific nucleation agents (Fillon *et al.*, 1993; Lotz, 1998; Varga, 1989), the β-phase can reach almost 100%. In the quiescent crystallization, although the β-phase is metastable, its growth rate is up to 70% faster than that of the α-phase across a wide range of crystallization temperatures. Figure 5.26 illustrates experimental growth rate data measured at different temperatures for both the α- and β-phases (Lotz, 1998). When the crystallization temperature exceeds 141 °C, the growth rates invert and the growth rate of the α-phase again becomes faster than that of the β-phase. Very interestingly, at lower crystallization temperatures between 100 and 105 °C, the growth rates invert again and the growth of the α-phase becomes faster than that of the β-phase. Between 100

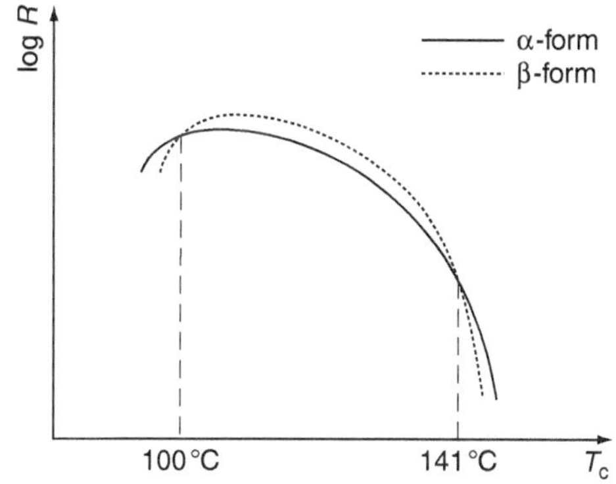

Figure 5.26 Crystal growth rates of both the α- and β-polymorphs at different temperatures. It is evident that there are two crossover growth rates between these two phases at about 100 and 141 °C, serving as an example of growth rate inversion [replotted from Lotz (1998), with permission].

and 141 °C, the growth rate of the β-phase is faster. These two growth rate inversions lead to a "growth rate reentry" and thus, a "phase reentry," which is similar to the case in liquid crystals described in Section 1.4 of Chapter 2. In this case, the growth rates of both phases can be experimentally observed, indicating that the heights of the nucleation barriers for the formation of both phases are not too different, yet possess a "barrier height reentry." The appearance of a metastable state is therefore solely dependent upon the nucleation barriers that may critically rely on the chain packing in these two phases.

2.2. Phase stability changes in polymorphs at high pressures and temperatures

Phase stability can also be investigated by isothermally changing pressure. The most well-known example is that of the polyethylene polymorph that appears at elevated pressures. The hexagonally packed extended chain crystal of polyethylene with a thickness on the order of micrometers was first grown under elevated pressures in the melt (Geil *et al.*, 1964; Prime and Wunderlich, 1969; Rees and Bassett, 1968, 1971; Wunderlich and Arakawa, 1964; Wunderlich and Davidson, 1969; Wunderlich and Melillo, 1968). Figure 5.27 shows a transmission electron microscope bright field image of polyethylene extended chain crystals. The different lamellar thicknesses in this figure may be due to molecular segregation during the

Figure 5.27 Transmission electron microscopy bright field image of polyethylene extended chain lamellar crystals. The sample was melt-crystallized isothermally at a pressure of 4.8 kbar and at 220 °C for 20 h. The scale bar represents 2 μm [reprinted from Prime and Wunderlich (1969), with permission].

crystallization of the sample, which had a polydispersity of 18 and a weight average molecular weight of 150 kg/mol (Prime and Wunderlich, 1969).

The formation mechanism of polyethylene extended chain crystals was experimentally shown to be associated with the hexagonal phase grown at elevated pressures. This hexagonal phase was first observed in experiments by Bassett and Turner (1972, 1974a,b) and Yasuniwa *et al.* (1973, 1976). Figure 5.28 shows a temperature–pressure phase diagram for polyethylene (Bassett and Turner, 1974a,b). In this phase diagram, polyethylene can exist in two ordered polymorphs, the orthorhombic and the hexagonal phases. This phase diagram ignores, for present purposes, the mechanically or epitaxially induced monoclinic polymorph. In the orthorhombic structure, the chains form a regular crystal, while the hexagonal structure is a columnar phase with large molecular mobility along the chain direction. The triple point, Q, in this figure indicates the coexistence of these three phases.

A detailed structural study utilizing wide angle x-ray diffraction at elevated pressures demonstrates that the formation of this hexagonal phase includes a discontinuous 0.1 nm expansion of the orthorhombic crystal unit cell *a* axis to 0.84 nm at the transition temperature and pressure; while, the *c* axis contracts along the chain direction, indicating a disordering of the all-*trans* polyethylene conformation (Bassett and Turner, 1974a,b; Yamamoto *et al.*, 1977; Yasuniwa *et al.*, 1976). Figure 5.29 shows the discontinuous changes in the *a* and *c* axes at a pressure around 5 kbar and a temperature around 240 °C, revealing a first-order phase transition.

As a consequence of this distinction in the polymorphs, the orthorhombic phase has the structure of the folded chain crystals of polyethylene, while the hexagonal phase is for extended chain crystals. In the former case, the crystals grow only laterally at a fixed supercooling with a determined lamellar thickness; while in the latter case, the crystals also keep growing in the thickness direction through sliding motion along the chain, terminated only by crystal impingement (Hikosaka *et al.*, 1992; Wunderlich and Melillo, 1968). As shown in Fig. 5.28, the growth of this

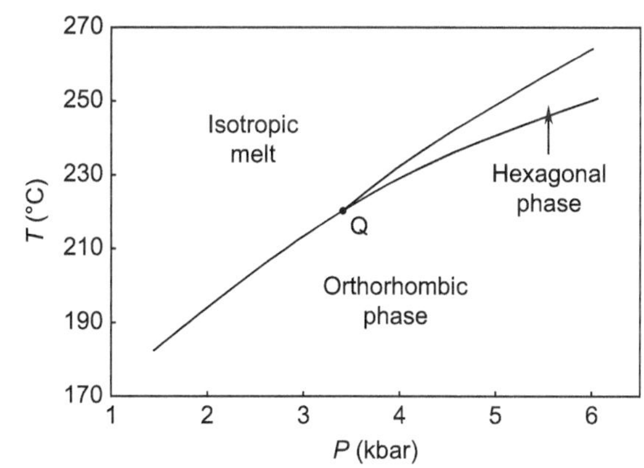

Figure 5.28 Temperature–pressure phase diagram of polyethylene. There are three phases seen in this diagram: the orthorhombic phase, the hexagonal phase, and the isotropic melt. The triple point is presented as Q [replotted from Bassett and Turner (1974b), with permission].

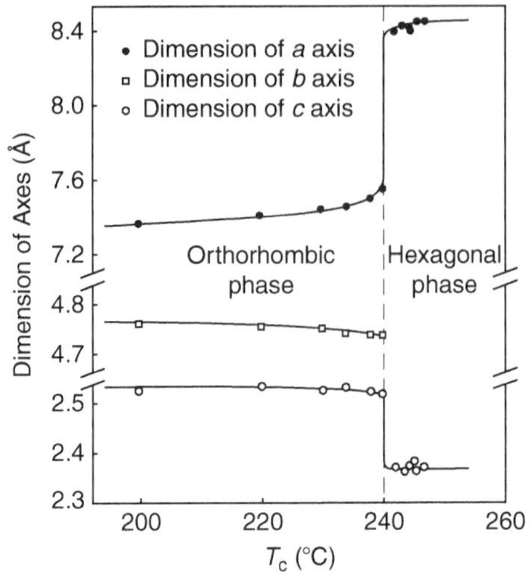

Figure 5.29 Discontinuous dimensional changes of the polyethylene *a* and *c* axes at 5 kbar and 240 °C as determined by wide angle x-ray diffraction [replotted from Yasuniwa *et al.* (1976), with permission].

hexagonal phase starts at the point Q shared by all three states. Figure 5.28 gives rise to the phase boundaries for each of these three phases. It is evident that outside their stable temperature–pressure regions, these phases become metastable. Therefore, the orthorhombic phase is stable at low pressures and temperatures, while the hexagonal phase is stable at high pressures and temperatures.

Polyethylene is not the only example of polymorphism changes with pressure. Another important example is poly(vinylidene fluoride). It was reported that this polymer could grow the γ-phase at sufficiently high pressures and temperatures (Hasegawa *et al.*, 1970, 1972a,b, Kobayashi *et al.*, 1975). However, further experimental results found that the high-pressure phase in poly(vinylidene fluoride) was the β-phase (Matsushige and Takemura, 1978, 1980), which is technologically important due to its piezoelectric and pyroelectric properties. The extended chain crystals were observed to have a lamellar thickness of 150–200 nm. Later, wide angle x-ray diffraction results at elevated pressures and temperatures showed that formation of the extended chain crystals in poly(vinylidene fluoride) is also due to the existence of a hexagonal phase above 0.3 kbar and 300 °C (Hattori *et al.*, 1996, 1997).

Another form of extended chain crystal occurs with annealing at elevated temperature but at atmospheric pressure. The existence of the high-temperature hexagonal polymorph can be observed in polytetrafluoroethylene. It is known that polytetrafluoroethylene possesses four different crystalline polymorphs; they are the triclinic phase II with a 13_6 helix chain conformation, the trigonal phase IV with a 15_7 helix chain conformation, the hexagonal phase I with irregular helices, and the orthorhombic phase III with a 2_1 planar zigzag chain conformation, which only appears at high pressures (Clark, 1967; Clark and Muus, 1962; Corradini and Guerra, 1977; Flack, 1972). Again, the hexagonal phase I is a columnar phase with large molecular mobility along the chain direction. Figure 5.30 shows a temperature–pressure phase diagram

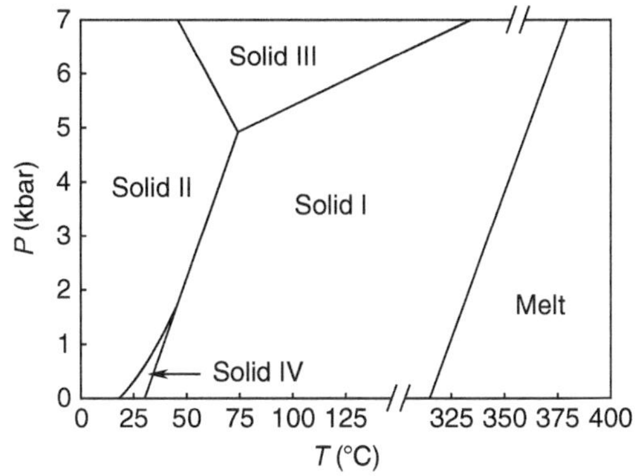

Figure 5.30 Temperature–pressure phase diagram of polytetrafluoroethylene with four different polymorphs [replotted from Wunderlich *et al.* (1988), with permission].

for these four phases. At atmospheric pressure, phase II is stable in the lowest temperature region, and it transforms to phase IV at 19 °C. Phase IV converts to phase I around 33 °C. Therefore, phase IV is only stable in a very narrow temperature and pressure region. On the other hand, phase I is stable across a broad range of temperatures and pressures (Starkweather *et al.*, 1982; Wunderlich, 1980; Wunderlich *et al.*, 1988).

Within the hexagonal phase temperature region, the growth of polytetrafluoroethylene extended chain crystals at high temperatures even at atmospheric pressure was observed in microscopy experiments (Melillo and Wunderlich, 1972). Figure 5.31 shows the extended chain crystals of this polymer with a molecular weight of 500 kg/mol after slowly cooling from the phase I transition temperature. As shown in Fig. 5.30, there are two triple points in this phase diagram. At these two points, three phases coexist. Although polytetrafluoroethylene is a good candidate for investigating the role of metastable states in phase transitions between these polymorphs, little effort has been made in this area. A few other examples of extended chain crystals having this formation mechanism include polychlorotrifluoroethylene (Miyamoto *et al.*, 1972) and *trans*-1,4-polybutadiene (Finter and Wegner, 1981; Natta and Corradini, 1960; Rastogi and Ungar, 1992; Suehiro and Takayanagi, 1970).

These experimental observations raise a question: Can the existence of the hexagonal phase at high pressures and/or temperatures be generally associated with the growth of extended chain lamellar crystals in aliphatic semicrystalline polymers? Efforts to answer this question have been made by Wunderlich *et al.* (1988) and Ungar (1993). From the structural perspective, this subset of hexagonal phases should be classified as a columnar phase, as described in Section 1.4 of Chapter 2. All of the polymers that can grow extended chain crystals at elevated pressures and/or high temperatures do so from a hexagonal phase because this subset always includes conformational disorder with some molecular rotational and translational mobility. This disorder and mobility have been detected by spectroscopic experimental methods such as nuclear magnetic resonance (see, e.g., Bovey *et al.*, 1988; Brown and Spiess, 2001; Tonelli, 1989), as well as infrared and Raman vibrational spectroscopy

Figure 5.31 Transmission electron microscopy bright field image of polytetrafluoroethylene extended chain lamellar crystals. The sample was slowly cooled at a rate of 4.6 °C/h from above the highest phase I transition temperature. The scale bar represents 0.5 μm [reprinted from Melillo and Wunderlich (1972), with permission].

(see, e.g., Colthup *et al.*, 1990; Wunderlich *et al.*, 1988). A special feature of these hexagonal phases is that their dynamic shear moduli are close to or even lower than that of their corresponding isotropic melt (Nagata *et al.*, 1980). Figure 5.32 shows that in polytetrafluoroethylene, the minimum shear stress in the hexagonal phase appears around 270 °C (Starkweather, 1979). Upon entering the isotropic melt, the shear stress increases. The hexagonal phases in these polymers play an important role in investigating the phase stability and size effects of metastable states on phase transformations (see Section 1.1 of Chapter 6). A similar viscosity minimum was later observed in polyethylene (Kolnaar and Keller, 1994, 1995; Waddon and Keller, 1990, 1992).

There is another group of polymers that can grow extended chain crystals at elevated pressures and temperatures. It was recently reported that poly(ethylene terephthalate) can form very thick lamellar crystals. The formation of thick lamellar crystals was determined to be accompanied by *trans*-esterification chemical reactions (Lu *et al.*, 2006). No reports have indicated that this thick lamellar crystal grows from a hexagonal phase. Several other polymers also exhibit similar behavior.

One recent study on the phase stability dependence on pressure involved syndiotactic polypropylene. This polymer possesses a rich range of polymorphs. Now, we know that at atmospheric pressure, the most stable crystal phase is phase I with an orthorhombic unit cell where $a = 1.450$ nm, $b = 1.120$ nm, and $c = 0.740$ nm, with an *Ibca* symmetry group and an antichiral chain packing (De Rosa *et al.*, 1997; Lotz *et al.*, 1988; Lovinger *et al.*, 1990, 1991, 1993). Phase II with an orthorhombic unit cell where $a = 1.450$ nm, $b = 0.560$ nm, and $c = 0.740$ nm, with a $C222_1$ symmetry group and an isochiral chain packing (Corradini *et al.*, 1967;

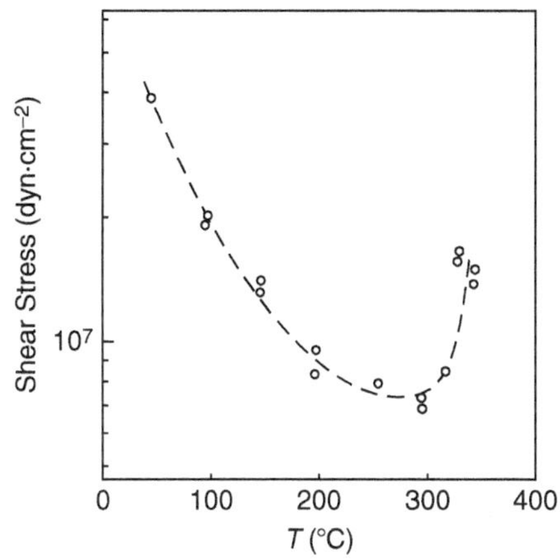

Figure 5.32 Polytetrafluoroethylene shear-stress dependence on temperature in its hexagonal phase. The data was obtained using a capillary rheometer with a diameter of 1.111 mm and a length of 19.0 mm. The apparent shear rate was 8.9/sec [replotted from Starkweather (1979), with permission].

De Rosa *et al.*, 1998; Natta *et al.*, 1960b) is metastable. The chains in both phases are in a s(2/1)2 ttgg helical conformation. The melting temperature of this isochiral packed phase II is always lower than that of the antichiral packed phase I when the crystal sizes are identical. Phase II is usually obtained by uniaxially deforming samples with either low or high stereoregularity (Auriemma *et al.*, 1995; De Rosa and Corradini, 1993; De Rosa *et al.*, 1998; Lotz *et al.*, 1998b). Figure 5.33 illustrates the chain molecular packing schemes of the orthorhombic unit cells of these two phases and their schematic corresponding [001] electron diffraction patterns. It is evident that the size of the *b* axis in phase II is half of that of phase I. Their [001] electron diffraction patterns are very different due to the different lattice symmetries. This structural difference serves as a reference for identifying these two polymorphs in wide angle x-ray diffraction experiments at elevated pressures.

Real-time wide angle x-ray diffraction experiments at varying pressures and temperatures showed that even though the antichiral-packed phase I of syndiotactic polypropylene can normally be obtained upon cooling from the isotropic melt at atmospheric pressure up to around 1.5 kbar, the isochiral-packed phase II can be obtained upon cooling from the isotropic melt at high pressures above 1.5 kbar. Furthermore, the melting temperature of this isochiral-packed phase II is higher than that of the antichiral-packed phase I, indicating a phase stability inversion. These findings suggest the existence of a thermodynamically stable region for the isochiral-packed phase II at high pressures and temperatures in the pressure–temperature phase diagram, as schematically illustrated in Fig. 5.34. The question then becomes: How do elevated pressures cause phase stability inversion? The first possible reason is a density inversion. Note that these two phases have almost identical densities because of the fact that the volume of two phase II unit cells is equivalent to that of a single phase I unit cell, as illustrated in Fig. 5.33. However, at elevated pressures (~3.2 kbar), the phase I unit cell deforms and expands in volume,

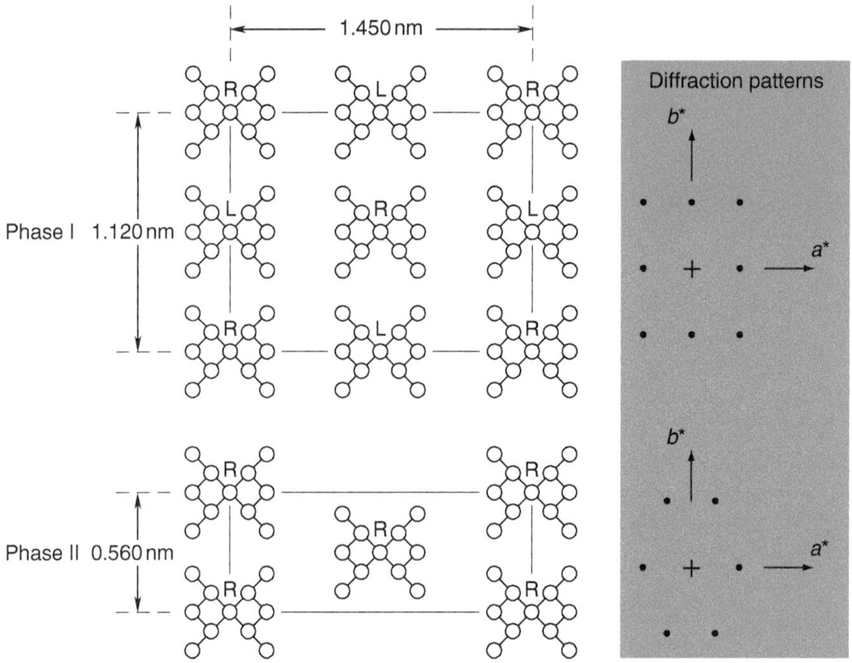

Figure 5.33 Chain packing schemes in the stable phase I and metastable phase II of syndiotactic polypropylene at atmospheric pressure. Schematics corresponding to the [001] electron diffraction patterns are also included [replotted from Lovinger *et al.* (1991), with permission].

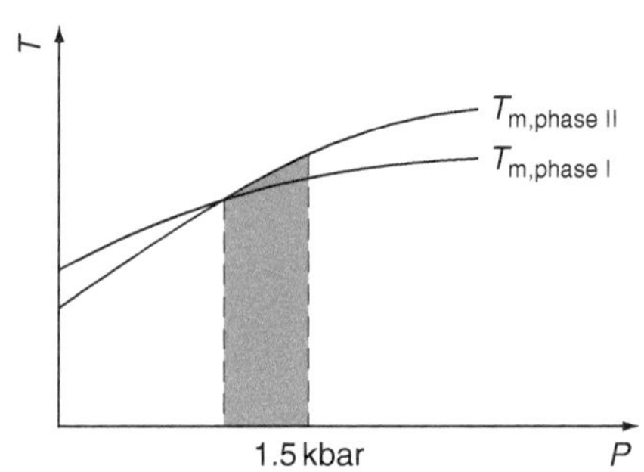

Figure 5.34 Schematic illustration of a pressure–temperature phase diagram based on the wide angle x-ray diffraction experimental results. Below around 1.5 kbar, the antichiral-packed phase I crystals are thermodynamically most stable; while, above this pressure, the isochiral-packed phase II crystals become most stable, a clear phase stability inversion with changing pressure [replotted from Rastogi *et al.* (2001), with permission].

while the phase II unit cell dimensions remain unchanged based on wide angle x-ray diffraction results (Rastogi *et al.*, 2001). The isochiral-packed helical conformation in phase II thus seems to be the thermodynamically more favorable phase at elevated pressures, compared with the antichiral-packed phase I. In this study, only experimentally observed melting temperatures are taken into account and folded chain lamellar crystals themselves are metastable phases, even though they were grown at elevated pressures and high temperatures. Therefore, strictly speaking, Fig. 5.34 is not quantitatively a thermodynamic equilibrium phase diagram, but only a qualitative illustration (Rastogi *et al.*, 2001). If we use kinetic terminology to describe this phase stability inversion, at elevated pressures above 1.5 kbar, the barrier to form phase II becomes lower than that to form phase I; while, below that pressure, the relative barrier heights are inverted.

2.3. External field-induced polymorphs

As discussed above, it is well-documented that the phase stability of polymorphs can be affected by temperature and pressure. This section focuses on elongational and shear flow in melts and mechanical deformation in solids that also induce new phases and cause phase transformations.

Polymer crystallization in oriented liquids can be achieved using a moving temperature gradient. Polymer crystal lamellae were found to orient parallel to the direction of solidification. Therefore, the polymer samples, despite retaining their initial morphology after crystallization, are highly anisotropic (Lovinger, 1978a,b; Lovinger and Gryte, 1976a,b; Lovinger *et al.*, 1977). Several polymers were investigated using oriented crystallization from the melt. Some of them possessed polymorphs such as isotactic polypropylene (Lovinger *et al.*, 1977) and poly(vinylidene fluoride) (Lovinger, 1982; Lovinger and Wang, 1979). The formation kinetics of these metastable polymer states are affected by the oriented melt, indicating that the moving temperature gradient is a dominant factor influencing the barrier height for structure formation.

For isotactic polypropylene crystallized in a temperature gradient, α-phase-oriented spherulites grew via nucleation, while the β-phase did not significantly develop. However, the β-phase growth could be easily initiated via transformation and growth along the oriented front of the α-phase. Furthermore, because the β-phase crystallizes faster than the α-phase in that temperature range, locations where the α-phase transforms to the β-phase were dispersed across the sample, interrupting the growth of the oriented α-phase. This phase transformation is a good example of an orientational melt induction of metastable β-phase (Lovinger *et al.*, 1977).

It has also been recently reported that mechanical shearing induces different polymorphisms in the case of isotactic polypropylene, as shown in Fig. 5.35 (Huo *et al.*, 2004). Wide angle x-ray diffraction experiments showed that with increasing shear rate up to 10 reciprocal seconds, the population of the metastable β-phase rapidly increases and reaches a broad maximum at a shear rate of 20 reciprocal seconds. The population then gradually decreases. On the other hand, the population of the stable α-phase correspondingly drops when the shear rate increases up to 10 reciprocal seconds and reaches a broad minimum at 20 reciprocal seconds. The α-phase population gradually increases with a further increase of the shear rate and undergoes an inversion with the β-phase population at a shear rate of 55 reciprocal seconds. This implies that the nucleation barriers of both the α- and β-phases in isotactic polypropylene respond differently to mechanical shear.

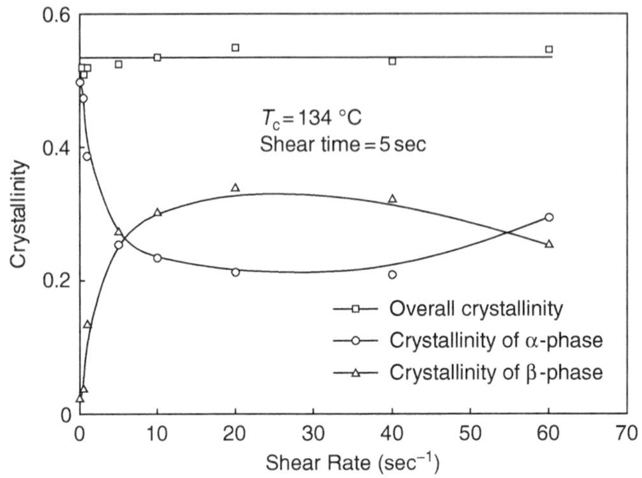

Figure 5.35 Overall and α- and β-phase crystallinity dependencies on shear rate in an isotactic polypropylene sample. The overall crystallinity of the sample is almost constant; however, the populations of the α- and β-phases are sensitive to the shear rate [replotted from Huo *et al.* (2004), with permission].

It is quite common in crystalline polymers that solid state deformation leads to the development of new phases that normally are not observed in quiescent crystallization. One of the most important examples of obtaining a pure metastable phase via solid-state deformation is the β-phase of poly(vinylidene fluoride). The structure of this phase was introduced in Section 2.1 of this Chapter. In industrial production, this phase is routinely obtained by mechanical deformation of melt-crystallized films. During the deformation, other phases of poly(vinylidene fluoride), such as α- and γ-phases, easily transform to the β-phase (Lovinger, 1982), indicating that the barriers of transformation of these phases to the β-phase are relatively low and easily overcome.

Other examples include such polymers as syndiotactic polypropylene. As described in Section 2.2 of this Chapter, the most stable phase is the orthorhombic phase I with antichiral packing. However, when the polymer undergoes mechanical deformation, the metastable orthorhombic phase II with isochiral packing can be formed largely by transformation via annealing to release the stress from phase III that contains an all-*trans* planar conformation in this polymer. With further annealing at high temperatures, the deformed fiber samples having phase II or phase III crystals give rise to fibers with a mixture of phase I and phase II crystals (De Rosa *et al.*, 1998). This example indicates that the formation barrier of the metastable phase II from phase III is lower than that of phase I from phase III with mechanical deformation, even though phase I is thermodynamically most stable. It is also a typical example that illustrates Ostwald's stage rule (1897).

Another example is poly(butylene terephthalate) that exhibits two modifications, the α- and β-phases. Both phases have triclinic lattices (Yokouchi *et al.*, 1976). Under a uniaxial mechanical deformation, the α-phase undergoes a reversible transition to the β-phase (Gomez *et al.*, 1988). Without deformation, the α-phase is thermodynamically most stable. Again, this example indicates that mechanical deformation helps to stabilize the metastable β-phase and forces the transformation from the stable α-phase.

3. Monotropic Phase Transitions in Polymers

3.1. Crystallization kinetics enhanced by a preordered mesophase

If we consider two ordered phases, one of which is crystalline and the other is liquid crystalline, the competition between the formations of these two phases can be clearly understood when the liquid crystal is monotropic, as described in Section 3 of Chapter 3. In this case, the liquid crystal–isotropic melt transition temperature is lower than the crystal–isotropic melt transition temperature, and it can be detected only during cooling at a rate fast enough to bypass crystallization, which requires supercooling. This type of phase transition behavior allows studies of crystal formation either from the isotropic melt or from the liquid crystalline phase. Specifically, as shown in Fig. 3.8, an interesting aspect of monotropic phase behavior is that in temperature region II, slightly lower than the liquid crystal–isotropic melt transition temperature which is near the crossover temperature, the crystals can grow from either the isotropic melt or the liquid crystalline phase with comparable barrier heights. It is thus possible to investigate the competing crystal growths from these two distinct initial states. This overall topic also falls into the broader concept of polymorphs indicated in Section 3 of Chapter 3 (Cheng and Keller, 1998; Keller and Cheng, 1998; Keller *et al.*, 1990; Percec and Keller, 1990).

Examples of this type of phase transition behavior have been extensively studied and reported. One example is a series of polymers synthesized by coupling 1-(4-hydroxyphenyl)-2-(2-methyl-4-hydroxyphenyl)-ethane with α,ω-dibromoalkanes and its derivatives (Cheng *et al.*, 1992; Jing *et al.*, 2002; Percec and Yourd, 1989a,b; Ungar *et al.*, 1990; Yandrasits *et al.*, 1992). In this series of polyethers, monotropic liquid crystal phases were experimentally observed. Figure 5.36 shows the thermal behavior of a representative molecule of this series and its chemical structure. In Fig. 5.36, differential scanning calorimetric results during cooling (diagram A) show that an isotropic to a liquid crystal phase transition at a isotropic melt–liquid crystal transition temperature (T_d) is observed before the polyether started crystallizing. The second exothermic process is crystallization from the liquid crystalline phase. The subsequent heating (diagram B) shown in Fig. 5.36 indicates that only one endothermic process can be found higher than the isotropic melt–liquid crystal transition temperature (T_d) during cooling, which is the crystal melting temperature (T_m) (Yandrasits *et al.*, 1992).

In order to prove there is a transformation between the metastable liquid crystalline phase and the stable crystal phase, an experiment was specifically conducted where the sample was cooled to a temperature between these two exothermic peaks and isothermally annealed there for different lengths of time. The sample was then heated to above its melting temperature without further cooling, as shown in Fig. 5.37. The heating diagram without annealing recovered almost all of the

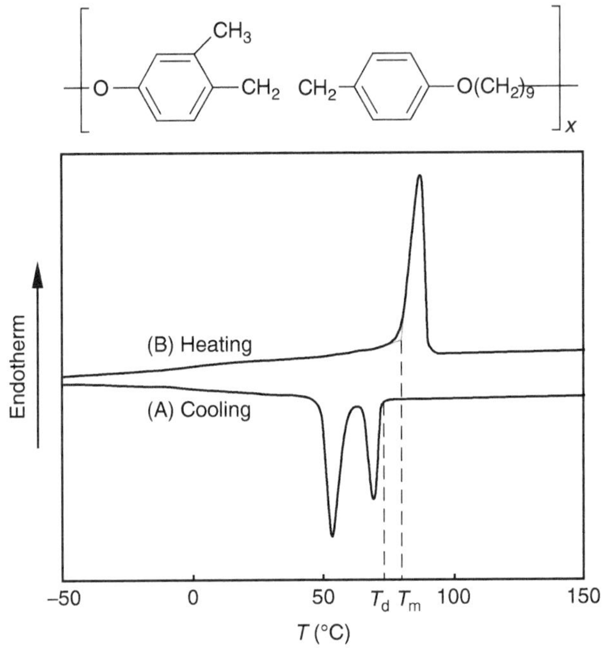

Figure 5.36 Experimentally observed monotropic liquid crystal phase in a polyether synthesized by coupling 1-(4-hydroxyphenyl)-2-(2-methyl-4-hydroxyphenyl)-ethane and α,ω-dibromoalkanes with nine methylene units during cooling at a rate of 10 °C/min (A) and subsequent heating at the same rate (B) [replotted from Yandrasits *et al.* (1992), with permission].

enthalpy change of the liquid crystalline phase at the liquid crystal–isotropic melt transition temperature (T_d). After only 6 min of annealing, the enthalpy change of the liquid crystalline phase disappeared, and the crystal melting dominated at the crystal–isotropic melting transition temperature (Yandrasits *et al.*, 1992). These results indicate that with short annealing times, the system transforms from the metastable liquid crystal phase to the stable crystal phase. The remaining liquid crystalline phase in the sample (since polymers rarely reach 100% crystallinity) does not undergo the isotropization shown in Fig. 5.36, possibly due to the confinement imposed by the newly formed crystals, resulting in a superheated liquid crystalline phase. Wide angle x-ray diffraction experiments indicate the presence of the monotropic nematic liquid crystal phase and the crystal phase (Jing *et al.*, 2002; Yandrasits *et al.*, 1992). Molecular mobility changes as a result of passing through these phase transitions were also monitored by carbon-13 solid state nuclear magnetic resonance for this series of polyethers (Cheng *et al.*, 1992).

Another example is a series of poly(ester imide)s synthesized via *N*-[4-(chloroformyl)-phenyl]-4-(choroformyl)phthalimide and diols with different methylene units (Kricheldorf *et al.*, 1991a,b; Pardey *et al.*, 1992, 1993). Its chemical structure is included in Fig. 5.38. All of these polymers form crystal phases. A specific structural feature of these crystal phases is that the crystals also possess a strong supermolecular layer order, called a "crystalline smectic state" (Kricheldorf *et al.*, 1991a,b).

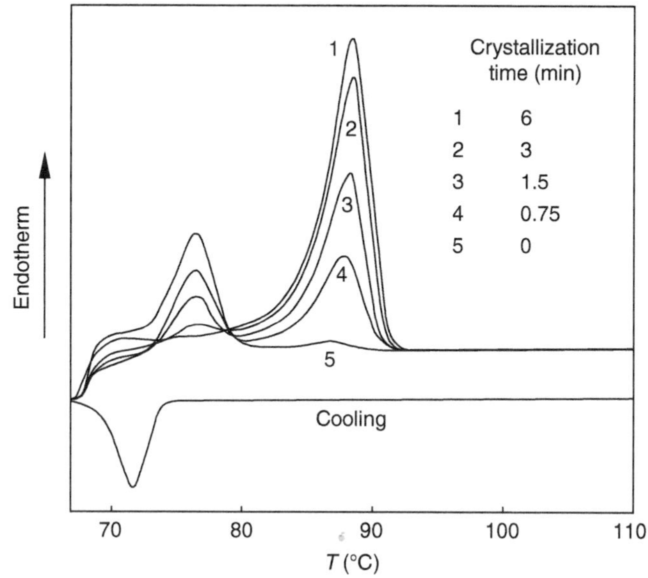

Figure 5.37 Specifically designed experiment with differential scanning calorimetry by cooling the same sample in Fig. 5.36 at a rate of 10 °C/min to 67 °C, which is just at the end of the first exothermic peak shown in Fig. 5.36A and then staying there for different time periods (0–6 min) before the sample was heated at 10 °C/min to above the melting temperature [replotted from Yandrasits *et al.* (1992), with permission].

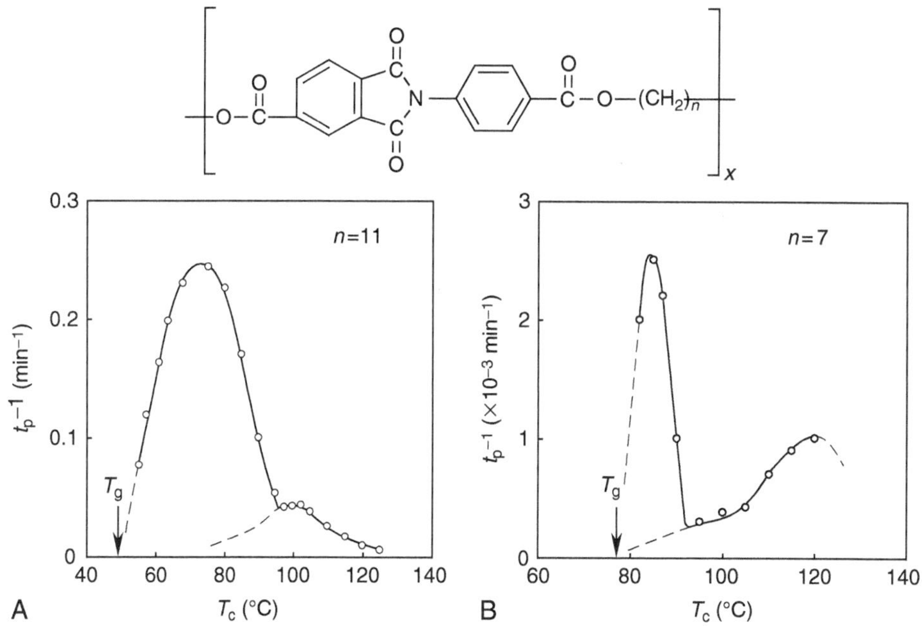

Figure 5.38 Overall crystallization rates for poly(ester imide) synthesized from *N*-[4-(chloroformyl)-phenyl]-4-(choroformyl)phthalimide and diols with 11 (A) and 7 (B) methylene units [replotted from Pardey *et al.* (1994), with permission].

Monotropic smectic liquid crystal phases were later found in this series of poly(ester imide)s (Pardey *et al.*, 1992, 1993). We are specifically interested in the transition kinetics of the crystalline phases. The advantage of this series is that we can study the phase transition kinetics across the whole temperature range from the glass transition temperature to the melting temperature. Figure 5.38A and B shows the overall crystallization rates at different isothermal temperatures of two of these monotropic poly(ester imide)s containing eleven and seven methylene units. Based on the wide angle x-ray diffraction results, the crystal structures formed above and below the liquid crystalline phase transition temperatures are identical for this series of polymers with odd-numbered methylene units (Pardey *et al.*, 1994).

It is evident that the overall crystallization rate from the isotropic melt and the liquid crystalline phase possesses a symmetric bell-shaped relationship with respect to isothermal crystallization temperature, as shown in Fig. 5.38A and B. However, the existence of the liquid crystalline phase, which has at least long-range stem orientational order, substantially enhances the overall crystallization rate of the polymer compared with crystallization from the corresponding isotropic melt. In Fig. 5.38A, the polymer with eleven methylene units possesses a monotropic liquid crystalline transition temperature at 97 °C, which is 48 °C above the glass transition temperature ($T_g = 49$ °C). Both overall crystallization kinetics from the isotropic melt and the liquid crystal phase can be monitored, which is shown in this figure. The fastest rate of crystallization from the liquid crystalline phase is about seven times higher than when grown from the isotropic melt. On the other hand, for the poly(ester imide) containing seven methylene units, Fig. 5.38B shows that the liquid crystalline transition temperature for this polymer is only 15 °C higher than its glass transition temperature at 77 °C. As soon as the liquid crystalline phase forms below this transition temperature, the overall crystallization rate, however, is still enhanced. The fastest rate is about four times greater than that from the isotropic melt. In this case, the bell-shaped rate curve is squeezed to fit into the 15 °C temperature range. Note that at the glass transition, large-scale translational mobility ceases, stopping the crystallization process (Pardey *et al.*, 1994).

Next, the question arises as to whether this liquid crystalline phase speeds up the primary nucleation rate or enhances the crystal growth (surface nucleation) rate or both. A recent study on a polyether synthesized from 1-(4-hydroxyphenyl)-2-(2-methyl-4-hydroxyphenyl)-ethane and α,ω-dibromoalkanes with nine methylene units reported that both the overall crystallization (Fig. 5.39A) and the crystal growth rates (Fig. 5.39B) were enhanced. The liquid crystal–isotropic melt transition temperature is at 75.5 °C (Jing *et al.*, 2002). As shown in this figure, the enhancement of both the nucleation and growth rates in the preordered state slightly below this transition temperature implies that the height of the nucleation barriers are reduced by introducing stem orientational order with the low-ordered liquid crystalline phase. This rate enhancement also indicates that during crystal growth from the isotropic melt, the chain orientation near and at the growth front cannot be in a preordered mesomorphic phase with orientational long-range or quasi-long-range order.

If we carefully examine Fig. 5.39B, two growth rates can be observed between 73 and 75.5 °C where the sudden increase in growth rate occurs. This example suggests

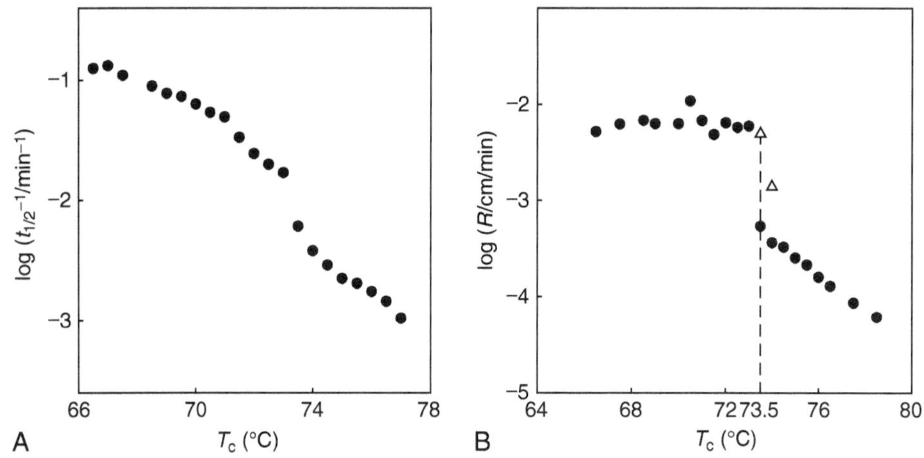

Figure 5.39 Overall crystallization rates (A) and growth rates (B) of a polyether synthesized from 1-(4-hydroxyphenyl)-2-(2-methyl-4-hydroxyphenyl)ethane and α,ω-dibromoalkanes with nine methylene units at different crystallization temperatures [replotted from Jing *et al.* (2002), with permission].

that two growth rates coexist at a single temperature (see the open triangle and solid circle in Fig. 5.39B). In this temperature region, the growth rates of the liquid crystal phase are still slightly lower than the crystal growth rates (see Fig. 3.8 in region II). It can be speculated that one of the crystal growth rates must be attributed to the growth from the liquid crystalline phase, while the other growth rate may represent direct growth from the isotropic melt (Jing *et al.*, 2002). The final but most interesting question is: Can we find the crossover temperature of region II where the liquid crystal formation rate is identical to the crystal growth rate, as shown in Fig. 3.8? Unless the phase structures and/or morphologies of crystals grown from the liquid crystalline phase are different from crystals grown from the isotropic melt, this crossover temperature is very difficult to identify. Because in at least one case this identification is possible, we will discuss this crossover temperature in the following section.

3.2. Change of phase transition sequence due to existence of a preordered state

A more complicated case is when a monotropic liquid crystalline phase induces a crystal phase that is structurally different from that grown directly from the isotropic melt. An example of this behavior comes from a series of poly(ester imide)s synthesized from *N*-[4-(chloroformyl)-phenyl]-4-(choroformyl)-phthalimide and diols with even-numbered methylene units (Kricheldorf *et al.*, 1991a,b; Pardey *et al.*, 1992, 1993). We focus on the phase structures and transition behaviors of the poly(ester imide)s with eight, ten, and twelve methylene units. It was found that above the liquid crystal–isotropic melt transition, the crystals that grow directly from the isotropic melt are monoclinic. However, when the crystallization takes place below this transition temperature, the ordered phases have a hexagonal structure (Pardey *et al.*, 1994).

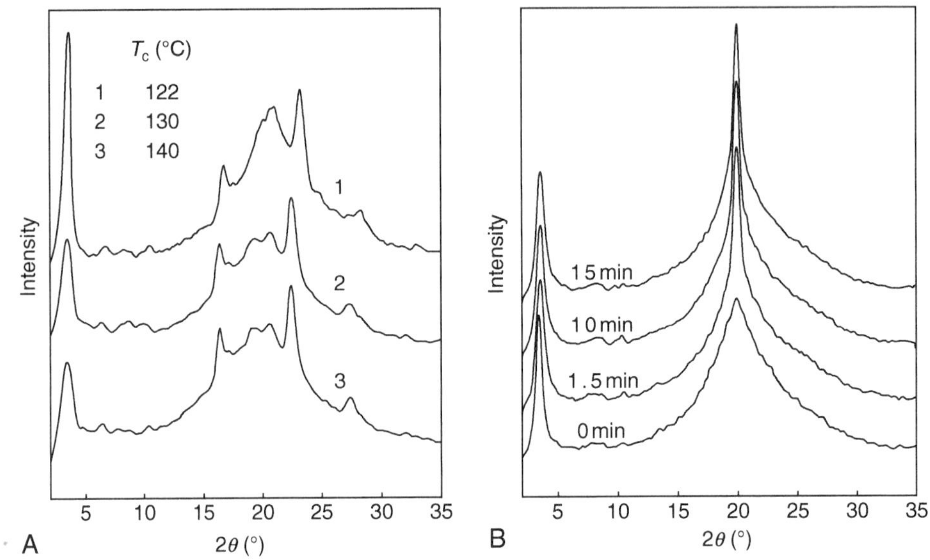

Figure 5.40 Sets of wide angle x-ray diffraction powder patterns for a poly(ester imide) with twelve methylene units isothermally crystallized at different temperatures above (A) and below (B) the liquid crystal–isotropic melt transition temperature of 124 °C. Note that in (A), the isothermal crystallization temperature of 122 °C is 2 °C lower than the liquid crystal–isotropic melt transition temperature, while in (B), the isothermal temperature is at 100 °C [replotted from Pardey *et al.* (1994), with permission].

Figure 5.40 shows wide angle x-ray diffraction results for the poly(ester imide) with twelve methylene units. Figure 5.40A is the wide angle x-ray diffraction of this polymer isothermally crystallized at two different temperatures, 130 °C and 140 °C (patterns 2 and 3 in this figure), both of which are above the liquid crystal–isotropic melt transition temperature at 124 °C. They exhibit monoclinic crystalline order. On the other hand, Fig. 5.40B contains the real-time wide angle x-ray diffraction patterns for the polymer isothermally crystallizing at 100 °C. Right after quenching to 100 °C at the zero minute, the x-ray pattern possesses smectic liquid crystalline characteristics with a layer structure and a short-range lateral packing order. With increasing time, a single diffraction peak gradually develops at $2\theta = 20.0°$ with a substantial increase in intensity, indicating the formation of a hexagonal structure. In the low angle region, the diffraction peak at $2\theta = 3.05°$ represents the layer spacing of the smectic phase. Therefore, Fig. 5.40A and B shows two structurally different crystal phases. If we carefully examine the diffraction pattern taken at the 122 °C isothermal crystallization temperature in Fig. 5.40A (pattern 1), which is 2 °C below the liquid crystal–isotropic melt transition temperature, this pattern is in fact a mixture of both the monoclinic and hexagonal structures. The low angle diffraction is shared by both structures (Pardey *et al.*, 1994).

The question is: How do we explain these experimental observations? The lower melting temperature indicates that the hexagonal phase in Fig. 5.40B is metastable with respect to the monoclinic phase. Yet, the hexagonal phase is more stable than

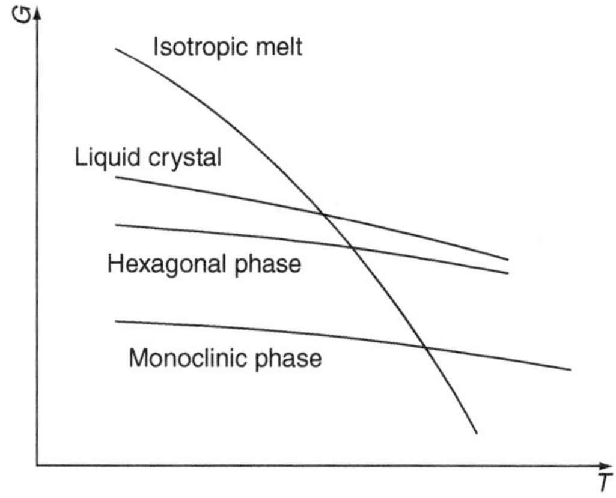

Figure 5.41 Schematic drawing of the four relationships between Gibbs free energy and temperature for the monoclinic phase, isotropic melt, liquid crystal phase, and hexagonal phase. Among all these four phases, both the liquid crystal and hexagonal phases are metastable and monotropic.

the liquid crystalline phase; hence, this phase is induced by the formation of the liquid crystalline phase. The stability of this hexagonal phase is thus in between that of the monoclinic phase and the liquid crystalline phase. This stability relationship can be illustrated by plotting Gibbs free energy against temperature, as shown in Fig. 5.41. In this figure, four Gibbs free energy versus temperature relationships are given for each of the following phases: monoclinic, isotropic melt, liquid crystalline, and hexagonal. For the purpose of convenience in our discussion, we are presently ignoring the metastabilities associated with phase size, such as lamellar thickness. It is then evident that among these four phases, the liquid crystalline and hexagonal phases are metastable across the entire temperature range studied. Both of them are monotropic phases with respect to the monoclinic crystalline phase, and the liquid crystalline phase is also a monotropic phase with respect to the hexagonal phase. These observations are a perfect illustration of Ostwald's stage rule (1897), but now, we have two metastable stages to go through before reaching the final, stable monoclinic phase. The appearance of these two monotropic, metastable phases is critically dependent on the experimental conditions.

The deepest supercooling of the melt one can reach to form the monoclinic phase is located at the liquid crystal–isotropic melt transition temperature. This lower limit of supercooling is estimated because the formation of the hexagonal phase also requires some supercooling, and the hexagonal phase–isotropic melt transition temperature is too close to the liquid crystal–isotropic melt transition temperature. The closeness of these two transition temperatures does not provide sufficient supercooling for the hexagonal phase to form. As a result, only after the liquid crystalline phase forms at and below the liquid crystal–isotropic melt transition temperature does the existence of this preordered phase induce the formation of the hexagonal phase. The observation of this transition is not only because the Gibbs free energy line of the hexagonal phase is just beneath that of the liquid crystalline phase, but also because the barrier for the transformation between the liquid crystal and hexagonal phases is lower than that between the liquid crystalline and the monoclinic phases. Now the question is: Can we ever reach the most stable monoclinic phase? The answer is critically dependent upon the chain mobility in the hexagonal phase, the height of the transition barrier between the hexagonal and monoclinic phases, and the amount of experimental time available. Note that this hexagonal phase is not a columnar phase because it possesses the long-range ordered layer structure. Therefore, we may need a very long period for the Gibbs free energy to drop from the hexagonal to the monoclinic phase, or the hexagonal phase may be permanent for kinetic reasons rather than thermodynamic ones.

Another example of a liquid crystal-induced metastable crystal phase can be found again in the case of a polyether synthesized from 1-(4-hydroxyphenyl)-2-(2-methyl-4-hydroxyphenyl)-ethane and α,ω-dibromoalkanes with nine methylene units, as shown in Fig. 5.39. In fact, this polymer also possesses multiple crystalline phases. When the crystal is grown directly from the isotropic melt, it forms an orthorhombic phase. When the isothermal crystallization is carried out between 48 and 72 °C, a triclinic phase forms, as observed by wide angle x-ray diffraction. The triclinic phase is metastable compared with the orthorhombic phase, as indicated by its lower melting temperature (Jing *et al.*, 2002). Therefore, a similar explanation can be given using the Gibbs free energy versus temperature relationship for both the metastable liquid crystal and triclinic phases illustrated in Fig. 5.41.

The question is why the crystal structure change does not occur at the liquid crystal–isotropic melt transition temperature (75.5 °C), rather than at 72 °C. In order to investigate this point, a method to distinguish these two phases was found. The spherulitic morphologies grown in these two different temperature regions have different birefringence and textures, as shown in Fig. 5.42. Figure 5.42A is the spherulite grown from the isotropic melt at 77.5 °C. This type of orthorhombic phase spherulite morphology was observed down to 72 °C. While

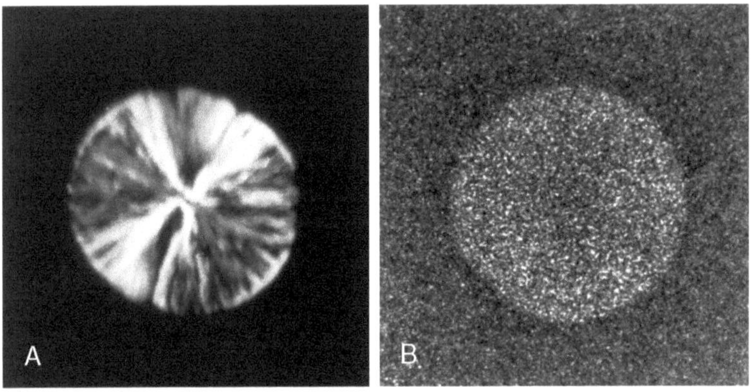

Figure 5.42 Two spherulitic morphologies of a polyether synthesized from 1-(4-hydroxyphenyl)-2-(2-methyl-4-hydroxyphenyl)ethane and α,ω-dibromoalkanes with nine methylene units under polarized optical microscopy at crystallization temperatures of 77.5 °C for the growth of the orthorhombic phase (A) and 69.5 °C for the growth of the triclinic phase (B). The crystals are around 30 μm [reprinted from Jing *et al.* (2002), with permission].

in Fig. 5.42B, the spherulite grown from the liquid crystalline phase at 69.5 °C had a triclinic structure. This morphology was observed up to 72 °C. It is speculated that the crossover temperature at which the rate of forming the orthorhombic phase and the rate of forming the liquid crystal are identical is near 72 °C. Above 72 °C, the orthorhombic nuclei form first. The liquid crystalline phase only assists the further growth of the orthorhombic phase structure and speeds up the crystal growth kinetics. This acceleration occurs because the crystallization rate in this temperature region is still slightly faster than the liquid crystal formation rate (see Fig. 3.8). However, below 72 °C, the liquid crystalline phase forms first and the triclinic crystal grows from the liquid crystalline phase, indicating that the barrier for this triclinic phase is lower than for the orthorhombic phase from the liquid crystalline phase. The metastable triclinic phase may be permanent due to the difficulty of transforming from the triclinic phase to the orthorhombic phase (Jing *et al.*, 2002).

4. Surface- and Interface-Induced Metastable Phases

4.1. Surface-induced metastable polymorphs

As discussed in Section 2.3 of Chapter 4, polymer lamellar crystals are metastable; thus, with decreasing thickness, their melting temperatures decrease. As the phase size shrinks, the surface area-to-volume ratio grows, and both the atomic and molecular interactions at surfaces are different from those in the bulk state. Different amounts of surface and bulk interactions may lead to a change in the location of the Gibbs free energy minimum and, thus, lead to a change of transition temperature or even a change in molecular arrangement and packing, which sometimes induces new phases.

In small molecules, a well-known example of a surface-induced phase is surface-stabilized ferroelectric liquid crystals (Clark and Lagerwall, 1980). Ferroelectricity was first predicted by Meyer *et al.* (1975) and is defined for materials with a nonvanishing polarization component whose direction can be reversed by an electric field (Goodby *et al.*, 1994). In the chiral smectic C phase, each smectic layer has a nonvanishing component of the polarization vector along the C_2 two-fold axis parallel to the layer surface and normal to the tilting direction. However, the net polarization component in each layer cannot accumulate because the helical structure of this chiral smectic C phase does not allow uniform directional orientation in a monodomain. Therefore, the chiral smectic C phase is not ferroelectric. In order to exhibit ferroelectricity, chiral smectic C liquid crystals have to be placed in a cell with a thickness ranging between 0.5 and 2 μm. Because the thickness of the cells is smaller than the pitch length of the chiral smectic C phase helix, the helix is unwound, and the molecules are forced to lie down parallel to the surface of the cell, causing the liquid crystal to become ferroelectric. By changing the direction of the electric field, the liquid crystals can be switched between two equally stable states (Galerne and Liebert, 1991; Sirota *et al.*, 1987; Tweet *et al.*, 1990).

Generally speaking, the melting temperature in near two-dimensional systems, such as in thin films, is lower than the bulk melting temperature due to the reduced number of interactions of the surface atoms and molecules (see, e.g., Dosch, 1992;

Faraday, 1860). However, in the case of short *n*-alkanes and alcohols, a surface freezing phenomenon can exist where there is an ordered monolayer on the surface above the bulk melting temperature (Gang *et al.*, 1998; Ocko *et al.*, 1997; Wu *et al.*, 1993). When the short *n*-alkanes are grafted onto polymer backbones, these short alkyl chains aggregate to the surface to form ordered monolayers. Sharp order–disorder transitions in these monolayers were experimentally detected above the bulk melting temperatures (Gautam and Dhinojwala, 2002; Gautam *et al.*, 2003; Prasad *et al.*, 2005). The stability changes, with respect to the bulk state, reflect the interaction changes in these near two-dimensional systems. A recent report also showed that *n*-alkanes, such as *n*-nonadecane, crystallized in confined microcapsules with a diameter of about 5 μm, can form a frozen surface monolayer before the bulk crystallization. This monolayer induces a metastable rotator phase (Xie *et al.*, 2006).

In polymers, a few examples of unique surface-induced phase structures with different thermodynamic stabilities have been reported. One example is in a series of polyethers synthesized from 1-(4-hydroxy-4′-biphenyl)-2-(4-hydroxy-phenyl)propane and α,ω-dibromoalkanes. An equilibrium phase diagram of this series of polymers with an odd number of methylene units is shown in Fig. 5.43 (Yoon *et al.*, 1996a,b). The determination of each phase in this figure was carried out via wide angle x-ray and electron diffraction. Morphological identification involved polarized light and transmission electron microscopy. Some smectic liquid crystals and smectic crystal phases are only stable in narrow temperature regions. No metastable states are given in this phase diagram, and all of these phases are thermodynamically stable in each of their own temperature regions.

The concept of metastability can be used to explain the phases and phase transition behaviors in the thin films of the polymer with seven methylene units on surfaces with different chemistries and different physical environments. As shown in Fig. 5.43, the bulk polymer shows three liquid crystalline phases: a nematic phase, a highly ordered smectic F phase, and a smectic G phase. The latter two phases possess hexagonal lateral packing (Yoon *et al.*, 1996a). The smectic H phase, which possesses an orthorhombic lateral packing lattice, was observed in members of this series with an odd-number of methylene units larger than nine, but it was not found in the polymer with seven methylene units. From Fig. 5.43, it is speculated that the smectic H phase in the bulk polymer with seven methylene units must be below the glass transition temperature, which sets the lower temperature limit where the molecules are sufficiently mobile to form the stable smectic G phase.

However, in the case of thin films with thicknesses ranging between 10 and 100 nm, the constrained environment may provide additional stability for the smectic H phase above the glass transition temperature; thus, this phase may become observable. The detailed structure and morphology of the polymer with seven methylene units in thin films was studied by electron diffraction and transmission electron microscopy on three different types of substrates: silane-grafted, amorphous-carbon-coated, and clean glass surfaces. The development of homeotropic molecular alignment in liquid crystal monodomains was achieved by using substrates with silane-grafted and amorphous-carbon-coated surfaces. Both surfaces can induce the smectic H structural ordering of this polymer and form an orthorhombic lateral packing that is not observed in the bulk and fiber samples of this

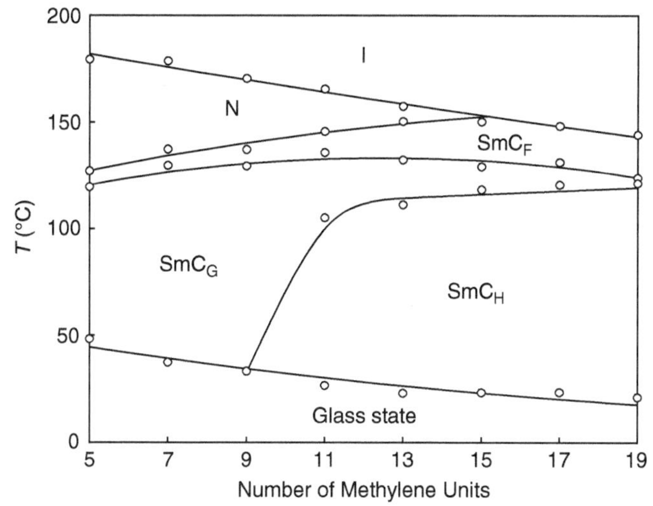

Figure 5.43 Phase diagram of a series of polyethers synthesized from 1-(4-hydroxy-4′-biphenyl)-2-(4-hydroxy-phenyl) propane and α,ω-dibromoalkanes with odd-numbered methylene units. In this figure, the abbreviations I, N, SmC_F, SmC_G, and SmC_H represent the isotropic melt, nematic, smectic F, smectic G, and smectic H phases, as defined in Fig. 1.3 of Section 1.2 in Chapter 1 [replotted from Yoon *et al.* (1996a), with permission].

polymer, but was found in this series of polymers when the number of methylene units was above nine. It was also found that the monodomain morphology of the highly ordered smectic crystal phases with homeotropic molecular alignment depends strongly on the structural symmetry. These results are shown in Fig. 5.44. The smectic G phase monodomain with hexagonal lateral packing exhibited a circular habit, while the orthorhombic packing of the smectic H phase exhibited an elongated ellipsoid habit (Ho *et al.*, 1996). This example indicates that metastability can be extended to understand the constraints of size and dimensionality on polymer systems. Because the film thickness is so small, the surfaces of the films become increasingly important in the determination of the thermodynamic stability of the whole system. The detailed mechanism by which the surfaces stabilize metastable phases is not well understood at the present time.

A more familiar approach for inducing new phases and phase transitions in semicrystalline polymers is crystallizing polymers on specifically generated and defined crystal surfaces utilizing epitaxy. In the epitaxial growth of crystals, the key issue is that the newly developed crystals must have a crystal growth plane that lattice matches the surface lattice plane of the crystalline substrate. Furthermore, epitaxial growth adds a new element of symmetry over and above the symmetry that either component originally possesses. These types of crystal surfaces are generated using small molecules or different polymers, and the surfaces possess specific (hkl) crystalline planes. Almost all the experimental evidence indicates that the driving force for stabilizing the metastable state via epitaxial growth exists because the epitaxially grown polymer crystal matches up and interacts with the substrate crystal along two lattice periodicities, even though the crystal phase is thermodynamically

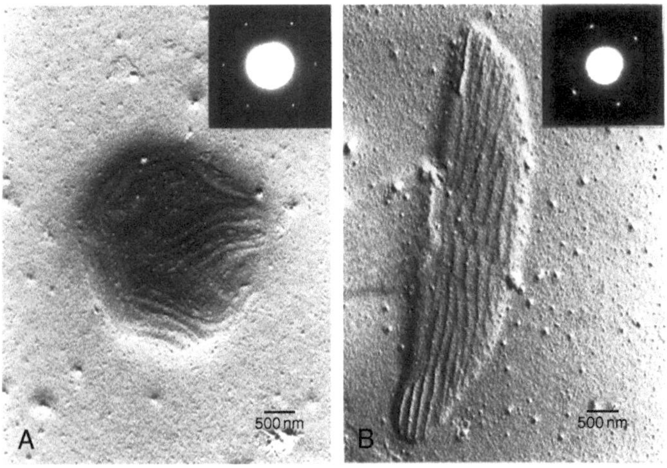

Figure 5.44 Monodomain morphologies and electron diffraction patterns (inserted) of a polyether synthesized from 1-(4-hydroxy-4′-biphenyl)-2-(4-hydroxy-phenyl)-propane and α,ω-dibromoalkanes with seven methylene units. In this figure, (A) is a hexagonal packing in the smectic G phase and (B) is an orthorhombic packing in the surface-induced smectic H phase [reprinted from Ho *et al.* (1996), with permission].

less stable when it is grown quiescently. In this sense, the epitaxy technique can grow surface-induced and stabilized metastable phases based on crystallographic lattice matching. Namely, epitaxy decreases the nucleation barrier and increases the growth kinetics of the metastable state, thus allowing it to appear. On the other hand, this metastable state induced by epitaxy may be expected to transform back to the stable state. How long the metastable state will exist is dependent upon how stable this metastable state is and the difficulty in transitioning to the stable state. In some cases, the metastable phase can only develop in a thin film and then transform to the stable state when the film thickness increases, where as in other cases, the metastable state can grow into the bulk and exist for a long time, depending, again, on the transition barrier from the metastable to the stable states.

A number of examples in which a metastable phase is stabilized by epitaxial growth on small molecule and polymer surfaces have been reported in the past half-century for many semicrystalline polymers. Perhaps the most well-known report on epitaxial growth in polymer crystals is for isotactic polypropylene. We know that the β-phase of isotactic polypropylene is metastable to the α-phase, although in the 100–141 °C temperature range, the growth rate of the β-phase is faster than that of the α-phase, as described in Section 2.1 of this Chapter. The observation of the β-phase was first reported by Keith *et al.* (1959); while, its crystal structure was correctly solved 35 years later (Lotz *et al.*, 1994; Meille *et al.*, 1994). One of the reasons for this delay was the difficulty in obtaining the pure β-phase from quiescent crystallization because both the α- and β-phases coexist in this temperature region. However, the metastable β-phase can be exclusively crystallized epitaxially on two specific small molecular crystals, γ-quinacridone and dicyclohexyl-terephthalamide, forming thin films that yield a biaxially oriented β-phase, as shown in Fig. 5.45. Large rectangular

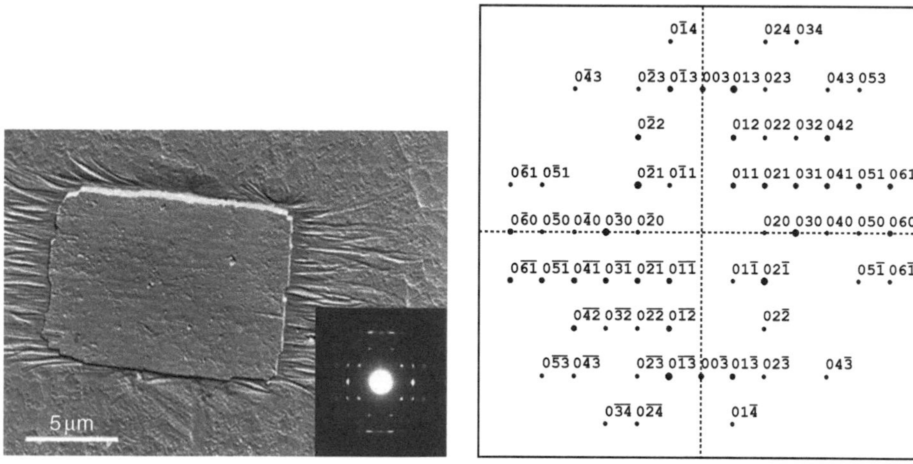

Figure 5.45 Bright field image of a β-phase crystal of isotactic polypropylene grown on a crystal substrate of dicyclohexylterephthalamide. The isotactic polypropylene film was melted and recrystallized. The dicyclohexylterephthalamide crystal lying on the top of the isotactic polypropylene film was redissolved, and the sample was shadowed with Pt/C. The electron diffraction pattern was from the single crystal shown in this figure and indices are obtained using the Cerius^2 program from Molecular Simulations [reprinted from Stocker *et al.* (1998), with permission].

single crystals of the β-phase can be observed in bright field images under transmission electron microscopy. The electron diffraction pattern shown in Fig. 5.45 is for the β-phase of isotactic polypropylene after the dicyclohexylterephthalamide was completely dissolved and removed. The trigonal (hkl) indices shown in this figure indicate that the (110) plane of the β-phase is the contact face involved in this epitaxial growth. Furthermore, the relative intensities of the (013) and (003) spots reveal that the weak (003) diffraction on the vertical axis of the electron diffraction pattern results from a c/6 shift of two helices in the unit cell, which translates into an opposition of phases with respect to the (003). This is a specific feature of a frustrated structure (Cartier *et al.*, 1996; Lotz *et al.*, 1994). Figure 5.46A and B shows two high resolution atomic force microscopy images with a chain packing model to illustrate the lateral helical packing periodicity of 1.9 nm on the (110) contact face. This 1.9 nm corresponds to the distance between three chains perpendicular to the lateral packing, as indicated in Fig. 5.46C. This periodicity results from the different azimuthal positions of the chains and is an indicator of the frustrated packing structure of the helices in the β-phase of isotactic polypropylene. The reasons for the epitaxial growth of the pure β-phase are lattice matching along the *c* axis in the β-phase (0.65 nm), one prominent periodicity of dicyclohexylterephthalamide and the orthogonal geometry of the contact face (Stocker *et al.*, 1998). In this case, dicyclohexylterephthalamide can serve as a nucleation agent for β-phase growth, and the β-phase of isotactic polypropylene can exist for a very long time. A similar analysis can also be deduced in the case of γ-quinacridone. The microporous morphology of the β-phase after deformation is important for technological applications.

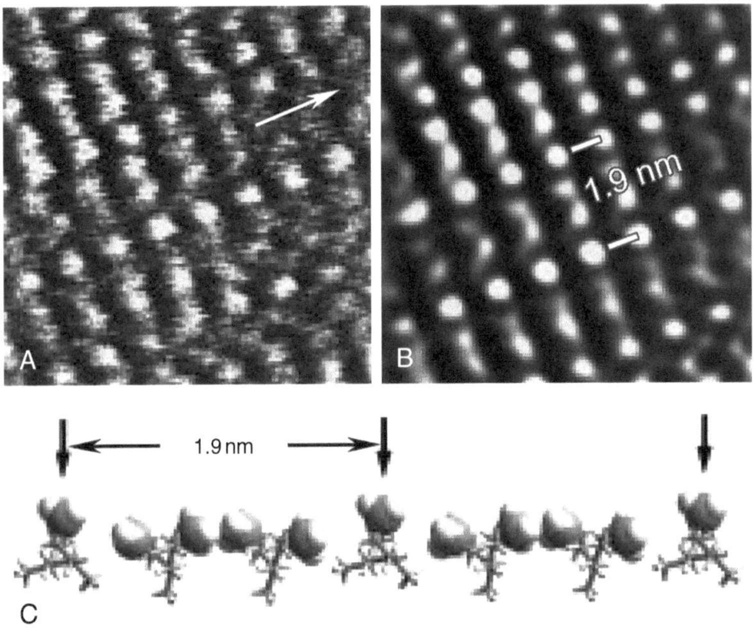

Figure 5.46 High resolution atomic force microscopy images for the β-phase as shown in Fig. 5.45: (A) unfiltered and (B) Fourier-filtered images. The chain direction is indicated by the arrow. The periodicity of 1.9 nm corresponds to the distance between two neighboring chains that are pointed north. This occurs in every third chain packing along the (110) plane. The chain direction is perpendicular to the printing paper in panel (C) [reprinted from Stocker *et al.* (1998), with permission].

Another example of isolating and stabilizing metastable phases via epitaxial growth is the case of isotactic poly-1-butene. All three phases of poly-1-butene can be isolated by epitaxial growth on appropriate substrates. Among them, as described in Section 2.1 of this Chapter, phases II and III are metastable states compared with phase I. These three crystal phases differ in their helix conformation and unit cell parameters, and they can be isolated by crystallization via epitaxy on the appropriate organic substrate (in particular, aromatic acids or salts). The phase III epitaxy relies mainly on matching the chain axis periodicity, while phase II is an example of the epitaxy of irrational helices in which the interturn distance plays a major role (Kopp *et al.*, 1994b). Based on composite electron diffraction patterns, the epitaxial relationship between phases III and II was established. However, isolating phase III from phase II was not successful, and the best way to obtain the pure phase III is by the solvent-assisted method, as described in Section 2.1 of this Chapter (Lotz and Thierry, 2003). In addition, the thermodynamically most stable hexagonal phase I′ (the same structure as phase I but grown directly from the melt with a lower melting temperature, presumably due to a reduced ability to reorganize) was also epitaxially crystallized on 4-bromo- and 4-chlorobenzoic acids, their potassium salts, and the hemi-acid, potassium hydrogen 4-chlorobenzoate. The contact plane in all cases is the (110) plane. In phase I′, the successive (110) planes alternately contain only left-handed or only right-handed helices. The observed epitaxy can, therefore, be differentiated according to the chirality of the helices interacting with the substrate. In this case, epitaxy is governed by a one-dimensional periodicity match-up, which involves the interturn distance, namely the distance between successive outer helical paths (Kopp *et al.*, 1994c). Based on electron diffraction and atomic force microscopy observations, all three phases can be induced on a 3-fluorobenzoic acid substrate, depending on the crystallization temperature. This example is a rare one of versatility in the epitaxial crystallization of polymers (Mathieu *et al.*, 2001).

In yet another example, a new crystal modification, in addition to the stable orthorhombic α-crystal modification of poly(L-lactic acid) and the racemate of poly(L-lactic acid) and poly(D-lactic acid), can be grown epitaxially on a hexamethylbenzene crystal substrate (Zwiers *et al.*, 1983). In this epitaxial study, the α-crystal modification of the optically active polymer, based on the 10_3 helix conformation of poly(D-lactic acid) or the 10_7 helical conformation for poly(L-lactic acid), was obtained when the sample was crystallized near 155 °C. However, a new crystal modification was produced by epitaxial crystallization at a slightly lower temperature of around 140 °C. The crystal structure of this new modification was established by electron diffraction and packing energy analysis. Two antiparallel helices are packed in an orthorhombic unit cell with parameters of $a = 0.995$ nm, $b = 0.625$ nm, and $c = 0.88$ nm (Cartier *et al.*, 2000). It seems that this modification is metastable to the α-crystal modification. The racemate of poly(L-lactic acid) and poly(D-lactic acid) was also crystallized epitaxially near 165 °C on a hexamethylbenzene substrate (Cartier *et al.*, 2000).

In polyethylene, the monoclinic phase can be epitaxially grown on alkali halides and several other organic substrates (Wittmann and Lotz, 1989). The monoclinic phase in polyethylene is metastable compared to its orthorhombic phase. It is generally understood that the monoclinic phase can be obtained by application of stress. In this study, it was found that the monoclinic phase is dominant when the layer thickness, in which growth proceeds, does not exceed approximately 10 nm. This finding reveals that the monoclinic phase, probably through a crystal growth transition, reverts back to the orthorhombic phase after only about 20 growth layers. This transformation takes place at the boundary planes between the (010) plane of the orthorhombic phase and the (210) plane of the monoclinic phase. Therefore, the epitaxially induced monoclinic phase of polyethylene is only stabilized for a short distance (Wittmann and Lotz, 1989).

Syndiotactic polypropylene has its stable (phase I) and metastable (phase II) phases at atmospheric pressure, as described in Section 2.2 of the Chapter. The metastable, isochiral phase II can be epitaxially induced on the substrate of 2-quinoxalinol in the low crystallization temperature region (Zhang *et al.*, 2001). The (110) plane of phase II is in contact with the (001) plane of the 2-quinoxalinol substrate. However, phase II can only grow to a thickness of around 50 nm. Beyond this thickness, a transformation takes place to the disordered, antichiral phase I. In other words, one cannot use a nucleating agent to induce a pure phase II in the bulk. This observation is different from the high-pressure crystallization of syndiotactic polypropylene described in Section 2.2 of this Chapter, where phase II possesses a stability inversion at high pressures; thus, the pure phase II can grow in the bulk (Rastogi *et al.*, 2001).

Because the lattice matching between the polymer crystals and substrate enables the growth of metastable crystals that cannot normally be isolated, the lattice matching must significantly lower the nucleation barrier to grow metastable crystals.

Another type of epitaxial growth is that of polymer crystals grown on their own or on another polymer crystal surface. A unique example is the lamellar cross-hatching in the monoclinic α-phase of isotactic polypropylene, of which two lamellae grow across each other with a fixed angle of 80° or 100° (Khoury, 1966). The mechanism of this lamellar cross-hatching is attributed to the fact that one lamella epitaxially grown occurs on the lateral (010) crystallographic plane of another lamella by a satisfactory interdigitation of the methyl groups of facing planes. This condition arises from the near identical dimensions of the *a* and *c* axes in the unit cell. From a molecular point of view, chains that deposit onto the (010) plane for the initiation of this epitaxy have the same helical handedness as chains in the (010) plane but with an 80° or 100° angle to obtain favorable interactions of the methyl groups on the helices (Lotz and Wittmann, 1986).

A well-known example of polymer crystals grown on their own crystal surface is the shish-kebab structure. The systematic study of this structure was initiated by Pennings and Kiel (1965). When a polymer solution is stirred, such as in a 5% polyethylene solution of *p*-xylene at a temperature of 104 °C and at a stirring rate of 510 rpm, a shish-kebab structure forms as shown in Fig. 5.47 (Pennings *et al.*, 1970). The structure clearly has two components: the kebab structure is made of metastable lamellar crystals that grow along the chain direction of the shish single crystals and,

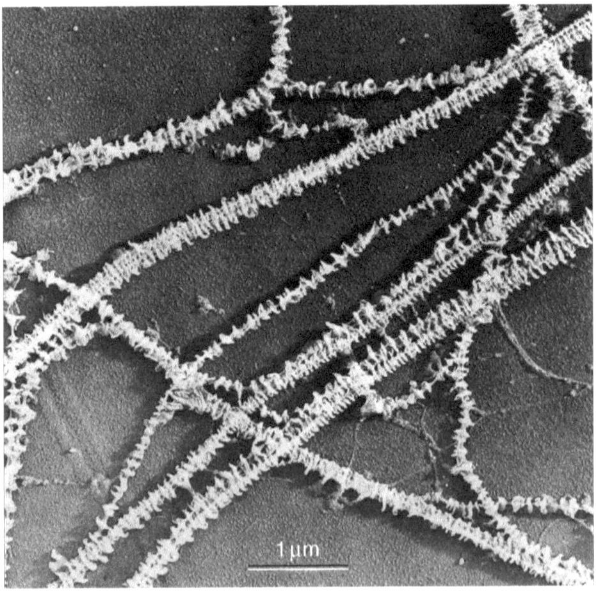

Figure 5.47 Bright field transmission electron microscopy image of shish-kebab structures of 5% polyethylene solution in *p*-xylene. The structures were obtained by stirrer crystallization at 104.5 °C and 510 rpm stirrer speed [reprinted from Pennings *et al.* (1970), with permission].

thus, forming the epitaxial growth of a folded chain polymer on its own extended chain crystals (Keller and Kolnaar, 1997).

Very recently, micro-shish-kebab structures have been observed in flow-induced crystallization. Systematic studies of the shish-kebab structures concerning how they are affected by chain stretching, dynamics, critical orientation and molecular weight have been reported in polyethylene and isotactic polypropylene (Hsiao *et al.*, 2005; Somani *et al.*, 2005). It has been concluded that the shish structure consists of stretched chains and the kebab structure results from the crystallization of unstretched chains. In the ultrahigh molecular weight polyethylene, multiple shishs can be observed in the shish-kebab structure, as shown in Fig. 5.48 (Hsiao *et al.*, 2005).

Polymer crystals can also be epitaxially grown on other ordered polymer surfaces. For example, the α-phase of isotactic polypropylene can be epitaxially grown on an oriented polytetrafluoroethylene film surface (Yan *et al.*, 2000), and a variety of semicrystalline polymers can be grown on carbon nanotubes to form heterogeneous shish-kebab structures (Li *et al.*, 2005, 2006). Semicrystalline polymers can also grow on inorganic surfaces, which may be recognized as a kind of morphological epitaxy (Wittmann and Lotz, 1990). However, there are few reports on stabilizing a metastable phase using this type of epitaxial growth.

A soft epitaxy technique, by which the epitaxial growth relies only on the molecular orientational order of surface material, has also been developed. This is the lamellar decoration technique (Wittmann and Lotz, 1982, 1985). The historical importance of this technique was in providing the first direct experimental evidence of how chains fold on the lamellar crystal surface and of chain-folding in polymer crystals. The detailed experimental procedure includes heating a linear polyethylene to the point of degradation in vacuum. The degraded polyethylene molecules have a molecular weight of about 1 kg/mol. These molecules then evaporate and deposit themselves on the folded chain crystal surface where they crystallize. If the chain folds are oriented along one (hkl) plane, the rod-like polyethylene crystals are oriented with their *c* axis parallel to the chain-folded direction (along the {hkl}

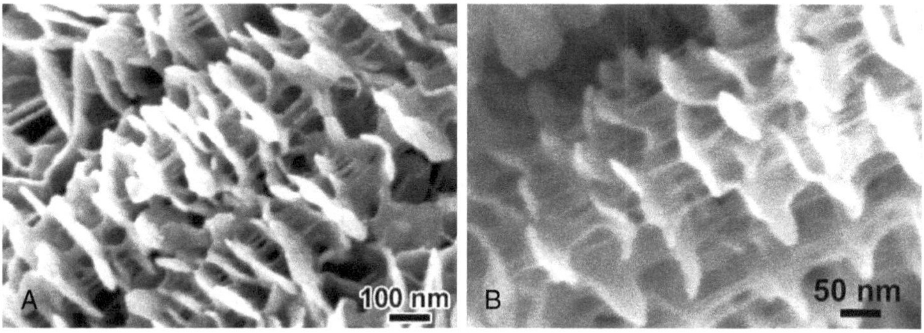

Figure 5.48 Field-emission scanning electron microscopy images of multiple micro-shish-kebab structures of a blend consisting of an ultrahigh molecular weight polyethylene and a low molecular weight amorphous ethylene copolymer at two different magnifications. These structures were formed after the sample was mechanically sheared [reprinted from Hsiao *et al.* (2005), with permission].

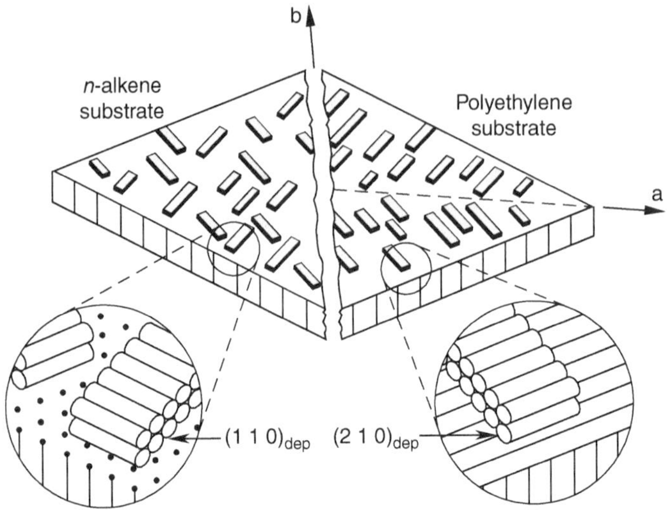

Figure 5.49 Schematic illustration of the mechanism of the lamellar decoration method. The right half represents the decoration of the chain folded polyethylene single crystals. The left half is the decoration of the chain-end surface of an *n*-alkane single crystal [redrawn from Wittmann and Lotz (1985), with permission].

plane). Figure 5.49 is an illustration of the mechanism of how the degraded polyethylene chains align along the chain-folded direction of a polyethylene single crystal. In this figure, the orientation of the rod-like polyethylene crystals is absent when they are decorated on an *n*-alkane single crystal without chain-folding. A bright field image shows four sectors identified by the degraded polyethylene chains on a polyethylene single crystal grown in solution; thus, the long axis of the rod-like polyethylene crystals are oriented perpendicular to the (110) planes of the growth faces, as shown in Fig. 5.50 (Wittmann and Lotz, 1982, 1985).

This lamellar decoration method has also been utilized on other polymer crystals such as isotactic polypropylene (Wittmann and Lotz, 1985), poly(ethylene oxide) (Chen *et al.*, 1995), a series of chiral nonracemic polyesters (Li *et al.*, 1999a,b, 2000), and others. This lamellar decoration technique can sensitively detect the surface topology down to a thickness of around 1 nm (Wittmann and Lotz, 1985). Recently, this technique has been utilized to probe the orientation of rubbed polymer alignment layers used to orient liquid crystals (Ge *et al.*, 2001). This technique is also used to identify the molecular orientation in columnar and other supramolecular phases (Xue *et al.*, 2004).

The mechanism of soft epitaxy is not lattice matching but relies on surface topological constraints. In most cases, the surface constraints are associated with topological orientation, such as chain folds and molecular shapes, which provide stable nucleation sites for decorating materials. Therefore, the length scale probed using this technique is larger than the lattice scale and is usually on the range of one to a few nanometers.

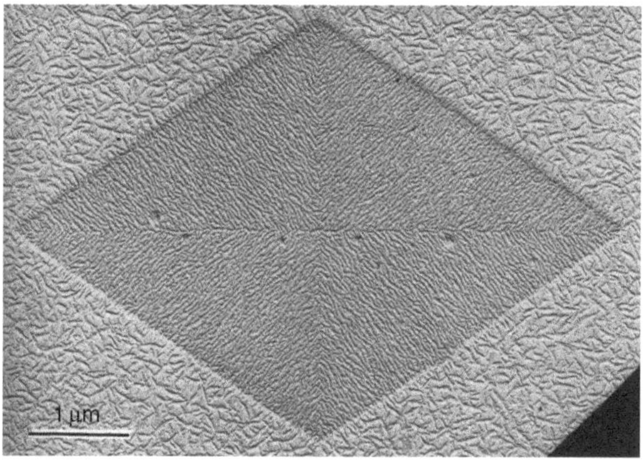

Figure 5.50 Bright field transmission electron microscopy image for the decoration of a polyethylene single crystal grown in solution [reprinted from Wittmann and Lotz (1985), with permission].

4.2. Metastable states introduced by unbalanced surface stresses caused by chain folding

There is another kind of metastable state in polymer crystals on the morphological scale rather than on the scale of the structural lattice. Experimental observations showed that lamellar crystals are not always flat. The scrolling of lamellae is frequently observed in small molecules, as in the case of asbestos chrysotile. This scrolling was explained based on the crystal structure of chrysotile. Chrysotile crystals are made of two different sheets. One is a network of linked SiO_4 tetrahedral sections (SiO_5); the other sheet is a brucite-type octahedral layer. These two sheets are connected by covalent bonds. Two-thirds of the hydroxyl ions at the base of the brucite layers are substituted by the oxygen at the apices of the Si–O tetrahedral links. The brucite and SiO_4 tetrahedral links now have different dimensions in the plane of the sheet, resulting in a layer that is scrolled with the "tighter" sheet inside to release the surface stress (Monkman, 1979).

Different from the scrolled crystals in small molecules, polymer chains form lamellar single crystals with chain-folding that originates from molecular connectivity. The noncrystalline chain folds located at crystal surfaces distinguish the polymer single crystals from small molecule single crystals and also play an important role in determining the habits of polymer single crystals. For example, if the occupied volumes of chain-folds at two opposite lamellar surfaces of a single crystal are different, unbalanced surface stress is created. This stress may result in curved or scrolled single crystals.

Curved polymer crystals have been observed to form in dilute polymer solutions. These observations can be traced back to the early 1970s in poly-4-methyl-1-pentene (Khoury and Barnes, 1972), polyoxymethylene (Khoury and Barnes, 1974a), and polychlorotrifluoroethylene (Khoury and Barnes, 1974b). In particular, the curve is more pronounced (they have a smaller radius of curvature) when the crystallization temperature decreases. It is believed that individual fold domains are nonplanar. A simple four- or six-sectored crystal is actually slightly conical in overall shape. Therefore, a hypothetical model of the origin of curved crystals in polymers has been proposed.

Scrolled single crystals have also been observed. One example is the phase III (orthorhombic) crystal of isotactic poly-1-butene (Holland and Miller, 1964; Lotz and Cheng, 2006). Figure 5.51A shows a transmission electron microscope bright field image of a scrolled single crystal in phase III, grown in amyl acetate dilute solution at 50 °C. Because the single crystals were dried during sample preparation for the transmission electron microscope, they collapsed onto the carbon-coated surfaces (Lotz and Cheng, 2006). The attached electron diffraction pattern of the scrolled single crystal in this figure indicates that first, the scrolled crystals possess a phase III cell and the *c* axes are perpendicular to the basal planes. Second, the scroll axis is approximately along the (110) normal. These conclusions can be further supported by the morphology of the initially developed scrolled crystals in Fig. 5.51b. In this figure, the scrolls were not yet fully developed due to unbalanced surface stress generated by different fold "encumbrance" with asymmetric azimuthal setting of the chain folds (Lotz and Cheng, 2005, 2006). However, when the crystallization temperature is increased to 55 °C, only flat single crystals of phase III isotactic poly-1-butene can be observed. Therefore, there is an issue of stability for the single crystals: scrolled versus flat.

Another even more striking example is for scrolled single crystals of nylon 66. The first report of scrolled single crystals of nylon 66 appeared almost a half-century ago (Geil, 1960). This observation can be understood based on the existence of two different fold types. Polyamides are known to form dense hydrogen-bonded sheets, and the folds are most likely located in the more flexible aliphatic part of the repeat unit. For nylon 66, there remain two possibilities. Folding may take place in the acidic or in the amide part of the repeat unit, and thus the fold could be made from

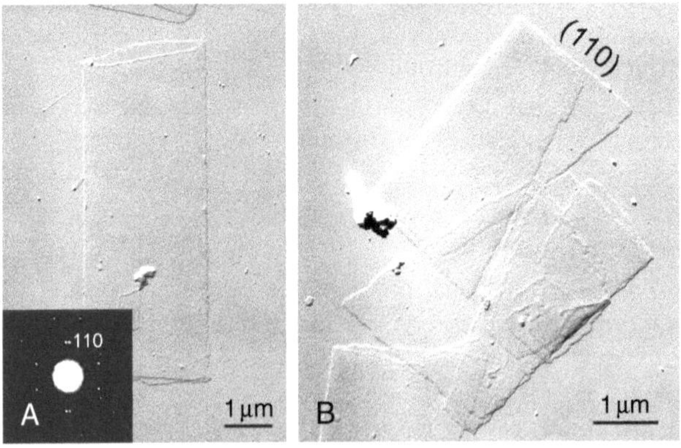

Figure 5.51 Bright field image of a scrolled isotactic poly-1-butene phase III single crystal after collapsing onto a carbon film (A) and isotactic poly-1-butene phase III single crystals grown to the initial stage of scrolling [reprinted from Lotz and Cheng (2006), with permission].

four or six atoms of carbon, respectively. The existence of two potential folds introduces the possibility of a difference in fold stress.

Recently, a set of unexpected results was reported for nylon 66 single crystals produced in solution after self-seeding. The self-seeding temperature was increased, while the crystallization temperature of the nylon 66 was maintained at 172 °C. When the self-seeding temperature was 202 °C, the nylon 66 crystals produced at the crystallization temperature of 172 °C were flat. When the self-seeding temperature was increased to 206 °C, all of the single lamellar crystals became scrolled. With a further increase in the self-seeding temperature to 208 °C, the crystals became flat again (Cai *et al.*, 2004). Figure 5.52 shows a set of single crystal morphologies of nylon 66 to illustrate the "flat–scrolled–flat" sequence.

It is known that when using the self-seeding technique, the lamellar single crystal thickness is controlled by the thickness of the seeds. At increased self-seeding temperatures, the seed thickness increases. At the lower and upper self-seeding temperatures of 202 °C and 208 °C, both fold surfaces could be acid or amine folds, with the lamellar thickness differing by one full chemical repeat unit of one diacid and one diamine. Each of the acid folds contains four methylene units, while the amide folds contain six methylene units (Fig. 5.52A and C). However, for the intermediate self-seeding temperature at 206 °C, it is possible that the thickness is only half of a repeat unit larger (either the acid part or the amine part). In that case, both the acid and amine parts must be involved in the folds (Fig. 5.52B). Because of hydrogen bonding in the crystal, these two different folds are, by necessity, segregated to opposite surfaces of the lamellar crystal; thus, they introduce a difference in fold-surface stress. The observed flat–scrolled–flat sequence in the lamellae, as shown in Fig. 5.52, lends strong support to the mechanism of the unbalanced surface stresses due to the different pressures experienced by the folds on opposite lamellar surfaces (Cai *et al.*, 2004).

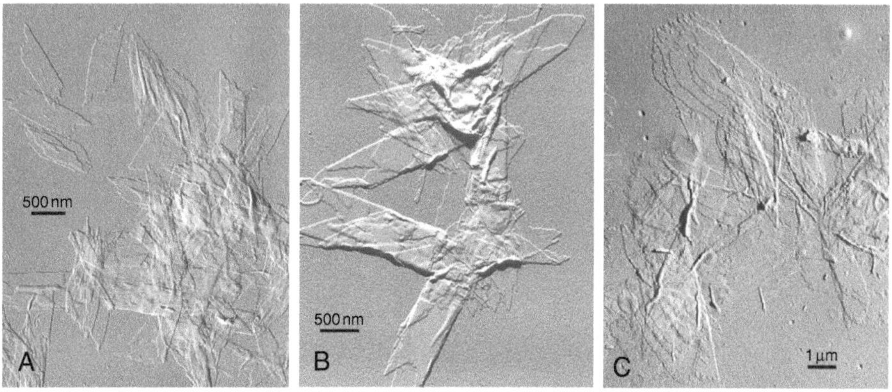

Figure 5.52 Transmission electron microscope bright field images of nylon 66 single crystals grown at 172 °C in dilute solution at self-seeding temperatures of (A) 202 °C, (B) 206 °C, and (C) 208 °C. This set of images illustrates the flat-scrolled-flat sequence [reprinted from Cai *et al.* (2004), with permission].

Although all the details have not yet been worked out for nylon 66 single crystals (such as the origin of the scroll axis orientation), this analysis appears to be applicable to the bulk crystallization of polyamides. Several polyamides display a curious sequence of spherulite birefringence when the crystallization temperature varies. In particular, one observes a shift from positive to negative birefringence in spherulites. Because the hydrogen-bonding direction is by far the fastest growth direction, the changes in birefringence would make sense if under specific crystallization conditions, the lamellae become scrolled: the fastest growth direction is radial for the "flat" lamellae and normal or nearly normal to the radius of the scrolled lamellae (Lotz and Cheng, 2005).

Scrolled crystals can also be observed from melt crystallization. A well-known example is poly(vinylidene fluoride) lamellar crystals, although they are not strictly single crystals. The lamellar structures within the poly(vinylidene fluoride) spherulites in the γ-phase have a novel architecture in which the lamellae adopt a highly curved, "scroll-like" morphology with the scroll axis parallel to the spherulite radius, as shown in Fig. 5.53A and B (Vaughan, 1993). It has been suggested that

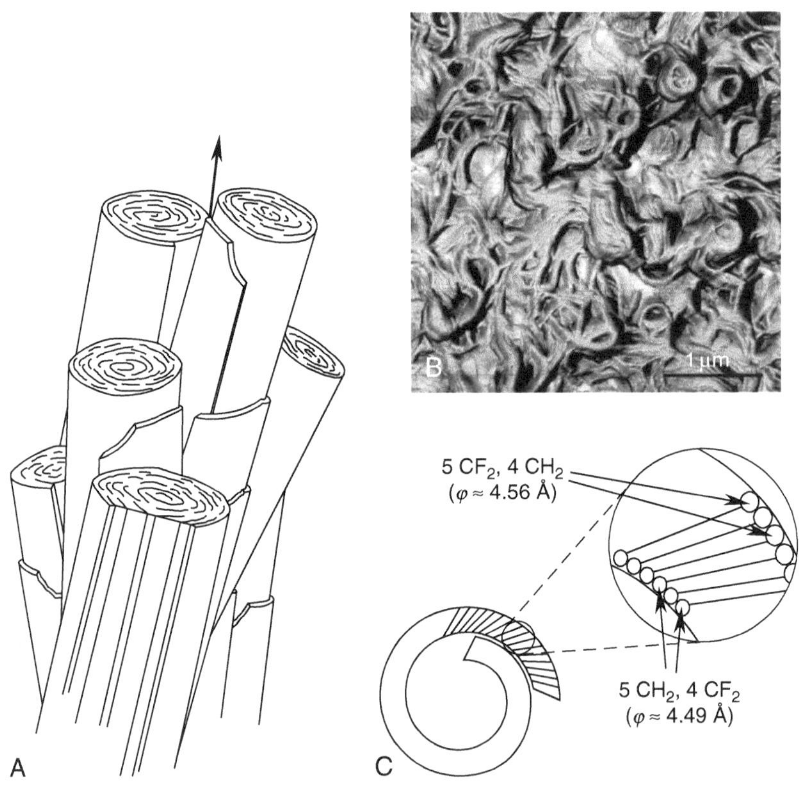

Figure 5.53 Schematic illustration of the scrolled lamellar crystals in γ-phase spherulites of poly (vinylidene fluoride) [redrawn from Vaughan (1993), with permission] (A) atomic force microscopy image of a poly(vinylidene fluoride) γ-phase spherulite with a view along the radius of the spherulite (B); and schematic illustration of stem splay in the γ-phase lamellar crystal with different fold volumes on opposite fold surfaces (C) [reprinted from and redrawn from Lotz and Cheng (2005), with permission].

this scroll-like morphology results from stresses associated with an imbalance of the fold composition on opposite fold surfaces in the γ-phase lamellar crystals. This imbalance reflects the polarity and steric constraints on the fold conformation outside of the crystals, which can deform the lamellar crystal core. Lotz *et al.* (1998a) proposed that these scrolled lamellar crystals are due to the $t_3gt_3g^-$ crankshaft conformation with the stem tilting in a synclinic (tilting in the opposite direction) or isoclinic (tilting in the same direction) arrangement with the polar folds on opposite surfaces. The polar behavior is as follows: on one fold surface of the lamellar crystal, every fold starts and ends with CH_2 groups, and on the opposite fold surface, every fold starts and ends with CF_2 groups. As a result, the folds contain an odd number of carbon atoms. However, on one side, they have five CH_2 and four CF_2 groups, while on another, four CH_2 and five CF_2 groups. This compositional distinction in these two types of folds leads to a difference in fold volume (1 nm^3), thus affecting the fold diameter, as shown in Fig. 5.53C, which causes the scrolled lamellar morphology (Lotz *et al.*, 1998a).

The concept of unbalanced surface stresses was proposed to be a major reason for all of the observed curved and scrolled single crystals. It is known that a curved lamellar single crystal introduces a bending energy that adds a positive free energy. This bend is caused by lattice expansion on one side and contraction on another side (Keith and Padden, 1984). Therefore, scrolling a single crystal must destabilize the crystal. Compared with flat single crystals of an identical thickness, these curved or scrolled single crystals are metastable. Yet, the exclusive observation of these curved or scrolled single crystals implies that the energy barrier to form these crystals is lower than that which forms the corresponding flat crystals at certain crystallization conditions.

In particular, in the case of nylon 66, the molecules prefer to grow scrolled single crystals instead of slightly increasing their lamellar thickness to form the flat ones at a constant crystallization temperature. This case indicates that the bending energy must be smaller than the new surface free energy that is required to increase the lamellar crystal thickness constrained by the thickness of the seeds. In the case of the γ-phase in poly(vinylidene fluoride), the unbalanced surface stress originates from the partitioning of the CF_2 and CH_2 groups in the nine carbon atoms of the folds. The molecules choose to form scrolled crystals rather than to increase the size of the amorphous folds by one more carbon atom. Therefore, decreasing the maximum crystallinity of the γ-phase in poly(vinylidene fluoride) would cost more free energy than introducing the bending energy to generate the scrolled crystals.

Polymer crystals can also form twists. The simplest physical model for lamellar twisting was proposed by Keith and Padden (1984), as shown in Fig. 5.54. This model consists of a single lamella made up of two half-lamellae that scroll in opposite directions. The cause of scrolling in the half-lamellae is attributed to the unbalanced surface stresses due to differences in the fold surface structures. Because these two half-lamellae must form a whole lamella, a lamellar twist develops in order to release the surface stresses.

The most well-known example is cooperative crystal twisting in polyethylene spherulites, shown in Fig. 5.55A. A simple illustration given uses the model schematically drawn in Fig. 5.55B to provide the construction of the banded spherulites

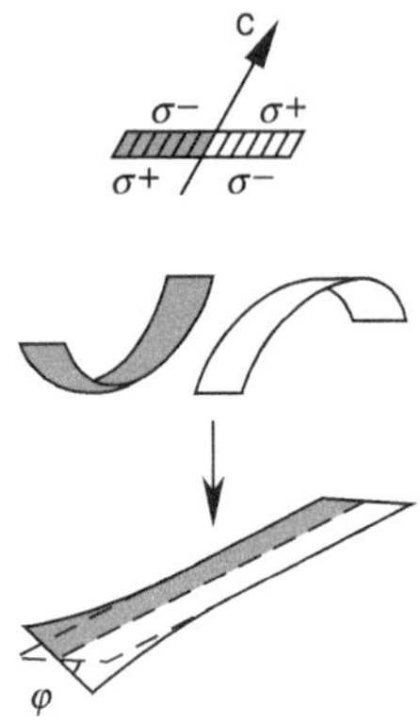

Figure 5.54 Schematics of the origin of lamellar twisting. On the top, the lamella is seen along the growth direction with a chain tilted arrangement. This asymmetry generates different fold stresses on the opposite fold surfaces of the lamella. In the middle, if exerted on half-lamellae split along their growth direction, the lamellar curvatures are in opposite directions of these two half-lamellae, and unbalanced surface stresses are introduced. On the bottom, because these two half-lamellae are attached together, they generate a twist along the whole lamella to relieve the surface stresses [reprinted from Keith and Padden (1984), with permission].

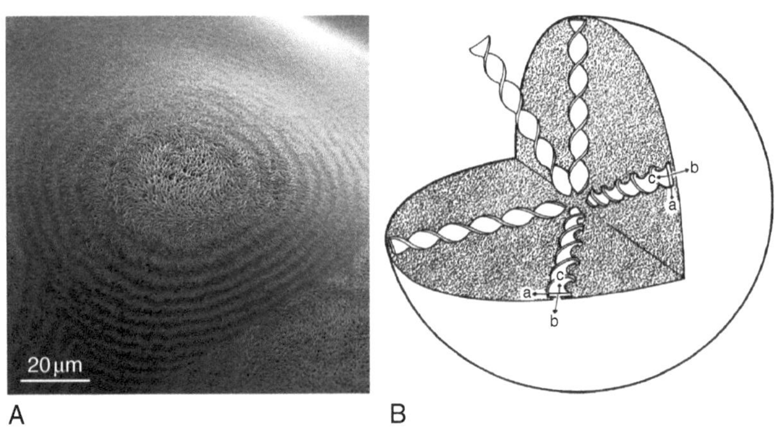

Figure 5.55 Scanning electron microscopy image of the surface of a polyethylene spherulite as the twisting lamellae reach that surface (the spherulite center is beneath the surface). The concentric rings indicate that the twist is cooperative, in phase, and corresponds to regions where the lamellae are seen nearly flat-on. The lamellar profiles between these rings are C-shaped or (in the present case) inverted Cs when seen from the spherulite center outward. These profiles correspond to the intersection of the twisting lamellae and the spherulite surface. The polyethylene sample used (Sclair) is known for its low nucleation density, which makes it suited to illustrate the present effect. Average distance between concentric bands is four micrometers (A) [reprinted from Lustiger *et al.* (1989), with permission]. Schematic representation of the lamellar twist in polyethylene spherulites (B) [replotted from Barham and Keller (1977), with permission].

[Barham and Keller, 1977, for a recent review, see Lotz and Cheng (2005)]. It is known that in polyethylene spherulites grown from the melt, the radial growth direction is along the *b* axis. The stems in the crystals are tilted in the *ac* plane with respect to the lamellar normal. Tilting angles range from 18° to 35°. In some cases at high crystallization temperatures, the tilting angle can reach 45°. The fold surfaces are thus not the (001) plane but rather the (101), (201), or even the (301) plane when the tilting angle is 45°. The polyethylene spherulites generally exhibit a banded texture, implying that the lamellae adopt a twisting scheme along the *b* axis (Keith and Padden, 1959a,b, Keller, 1959; Point, 1955; Price, 1959). The twisting has been identified as being either "right-handed" or "left-handed" (Lustiger *et al.*, 1989).

It took a long time to verify the existence of twisted lamellae in polyethylene. Kunz *et al.* (1995) reported the observation of polyethylene twisted lamellar single crystals when they investigated the crystalline morphology of ultrahigh molecular weight polyethylene physical gels in decalin, as shown in Fig. 5.56.

The most spectacular example of a twisted lamellar single crystal is, however, from a series of nonracemic chiral polyesters that were synthesized from (R)-(−)-4′-{ω-[2-(*p*-hydroxy-*o*-nitrophenyloxy)-1-propyloxy]-1-nonyloxy}-4-biphenyl carboxylic acid. This series of polyesters has an atomic chiral center attached to an aliphatic dialcohol, with a different number of methylene units ranging from seven to eleven. This chemical structure is included in Fig. 5.57. The single crystals can be either "curved flat" or twisted with identical crystal structures, as determined by electron diffraction. In the "curved flat" single crystals, it was determined by utilizing dark field transmission electron microscopy that the curvature is along the

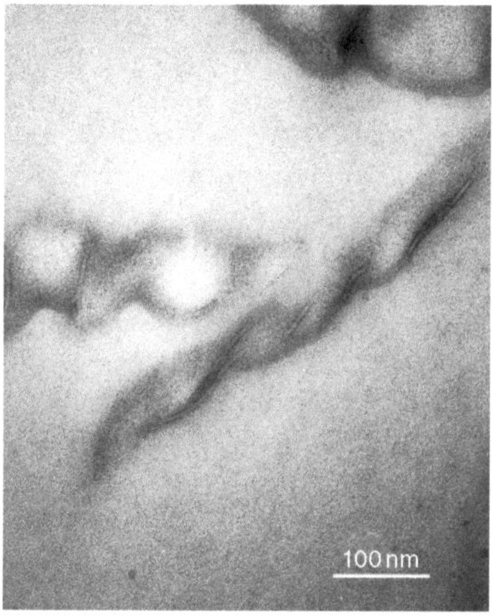

Figure 5.56 Transmission electron microscope bright field image of a twisted single lamella of ultrahigh molecular weight polyethylene in a decalin solution. Decalin was replaced by methacrylate that was polymerized to yield crystals of the polyethylene crystal embedded in a solid poly (methacrylate) matrix, while its original morphology was retained. The sample was microtomed and stained with ruthenium tetraoxide [reprinted from Kunz *et al.* (1995), with permission].

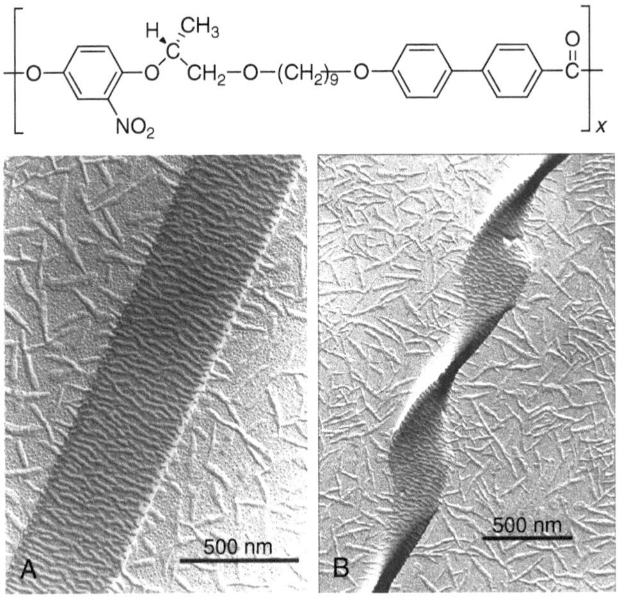

Figure 5.57 Transmission electron microscope bright field images of flat (A) and twisted helical single crystals (B) for a polyester synthesized from (R)-(−)-4′-{ω-[2-(*p*-hydroxy-*o*-nitrophenyloxy)-1-propyloxy]-1-nonyloxy}-4-biphenyl carboxylic acid with a right-handed chiral center and nine methylene units. The crystals were grown at 145 °C from the smectic A phase in the bulk. Polyethylene lamellae decorate the crystal surface and indicate the chain folding direction of the single crystals [reprinted from Li *et al.* (1999a,b, 2000), with permission].

short axis of the single crystals (Li *et al.*, 2000). In the twisted single crystals, the stems in the twisted helical single crystals are double twisted based on dark field transmission electron microscopy (Li *et al.*, 1999a,b, 2000). Based on experimental results from the lamellar decoration method described in Section 4.1 of this Chapter, the polymer chain-folding direction in both flat and helical lamellar crystals is also identical and always along the long axis of the lamellar crystals. Figure 5.57 shows that both the "curved flat" and twisted single crystals for a polyester with a right-handed chiral center and nine methylene units observed in bright field transmission electron microscopy (Li *et al.*, 1999a,b, 2000). All the twisted helical single crystals for this polymer possess identical handedness.

When the chiral center was changed from right-handed to left-handed, the handedness of the twisted helical single crystal changed accordingly. The most striking observation was that these polyesters with the identical handedness of the chiral centers, but a different number of the methylene units, displayed either right- or left-handed helical crystals. Specifically, when the polyesters had an odd number of methylene units, the helical single crystals possessed a right-handed twist, while those with an even number of methylene units had a left-handed twist. This unexpected feature was summarized in a paper titled "Left or right, it is a matter of one methylene unit" (Li *et al.*, 2001). These results indicate that the chirality apparent in the lamellar crystal is the ultimate outcome of a cascade of chiralities that exist at different length scales: the chiral center (bonds attached to the chiral atom), conformational chirality (helical handedness), helical lamellar single crystals (or monodomains), and object chiralities (clustering of crystals or aggregating domains). Chirality transfer from one length scale to another is neither automatic nor necessary. It is critically dependent upon the packing scheme of the twisted helical building blocks to transfer the chirality to a larger length scale (Li *et al.*, 2002). Because the mechanism of these double-twisted helical single crystals must be associated with these transfer processes, a complete understanding of the mechanism is still awaited.

One specifically interesting observation is that the chain folds located at two opposite sides of the "curved flat" lamellar single crystals behave differently, resulting in dissimilar patterns of polyethylene decoration (Weng *et al.*, 2002). It is understood that a bending energy is needed to generate this type of "curved flat" single crystals (Fig. 5.57A). On the other hand, for the twisted single crystals, the chain folds on both sides of the lamellae are identical, as evidenced by the same polyethylene decoration patterns on both fold surfaces of the lamellae (Fig. 5.57B). Based on the experimental observations, the helical crystals are always observed to be in the majority, while the "curved flat" single crystals are always in the minority. This evidence indicates that the chains prefer to twist rather than to bend when packing into these crystals, independent of the ultimate stability. Using kinetic language, the determining factor is the nucleation barrier which favors growth of the helical crystals. Quantitative explanations for this observation need to be associated with the microscopic molecular packing and interactions, as well as the growth environment.

REFERENCES AND FURTHER READING

Addink, E. J., and Beintema, J. (1961). Polymorphism of crystalline polypropylene. *Polymer* **2**, 185–193.

Alamo, R. G., Kim, M. H., Galante, M. J., Isasi, J. R., and Mandelken, L. (1999). Structural and kinetic factors governing the formation of the γ polymorph of isotactic polypropylene. *Macromolecules* **32**, 4050–4064.

Alamo, R. G., Blanco, J. A., Agarwal, P. K., and Randall, J. C. (2003). Crystallization rates of matched fractions of $MgCl_2$-supported Ziegler Natta and metallocene isotactic poly(propylene)s. 1. The role of chain microstructure. *Macromolecules* **36**, 1559–1571.

Arlie, J. P., Spegt, P., and Skoulios, A. (1965). Variation discontinue du nombre de repliement des chaînes d'un polyoxyéthylène crystallize en masse. *C. R. Hebd. Séances Acad. Sci.* **260**, 5774–5777.

Arlie, J. P., Spegt, P. A., and Skoulios, A. (1966). Etude de la cristallisation des polymères. I. Structure lamellaire de polyoxyéthylène de faible masse moléculaire. *Die Makromolekulare Chemie* **99**, 160–174.

Arlie, J. P., Spegt, P. A., and Skoulios, A. (1967). Etude de la cristallisation des polymères II. Structure lamellaire et repliement des chaines du polyoxyéthylène. *Die Makromolekulare Chemie* **104**, 212–229.

Armeniades, C. D., and Baer, E. (1967). Effect of pressure on the polymorphism of melt crystallized poly-1-butene. *J. Macromol. Sci. Part B: Phys.* **1**, 309–334.

Auriemma, F., Lewis, R. H., Spiess, H. W., and De Rosa, C. (1995). Phase transition from a C-centered to a B-centered orthorhombic crystalline form of syndiotactic poly(propylene). *Macromol. Chem. Phys.* **196**, 4011–4024.

Avakian, P., Gardner, K. H., and Matheson, R. P. Jr. (1990). A comment on crystallization in PEEK and PEEK resins. *J. Polym. Sci. Polym. Letter Ed.* **28**, 243–246.
Barham, P. J., and Keller, A. (1977). The problem of thermal expansion in polyethylene spherulites. *J. Mater. Sci.* **12**, 2141–2148.
Bassett, D. C. (1964). Moire patterns in the electron microscopy of polymer crystals. *Philos. Mag.* **10**, 595–615.
Bassett, D. C. (1981). "Principles of Polymer Morphology." Cambridge University Press, Cambridge.
Bassett, D. C., and Turner, B. (1972). New high-pressure phase in chain-extended crystallization of polyethylene. *Nature* **240**, 146–148.
Bassett, D. C., and Turner, B. (1974a). On chain-extended and chainfolded crystallization of polyethylene. *Philos. Mag.* **29**, 285–307.
Bassett, D. C., and Turner, B. (1974b). On the phenomenology of chain-extended crystallization in polyethylene. *Philos. Mag.* **29**, 925–955.
Bassett, D. C., Dammont, F. R., and Salovey, R. (1964). On the morphology of polymer crystals. *Polymer* **5**, 579–588.
Bassett, D. C., Olley, R. H., and Al Raheil, I. A. M. (1988). On isolated lamellae of melt-crystallized polyethylene. *Polymer* **29**, 1539–1543.
Beekmans, L. G. M., and Vancso, G. J. (2000). Real-time crystallization study of poly(ϵ-caprolactone) by hot-stage atomic force microscopy. *Polymer* **41**, 8975–8981.
Bidd, I., and Whiting, M. C. (1985). The synthesis of pure n-paraffins with chain-lengths between one and four hundred. *J. Chem. Soc. Chem. Commun.* no volume 543–544.
Bidd, I., Holdup, D. W., and Whiting, M. C. (1987). Studies on the synthesis of linear aliphatic-compounds. Part 3. The synthesis of paraffins with very long chains. *J. Chem. Soc. Perkin Transactions 1: Organic and Bio-Organic Chem.* **11**, 2455–2463.
Blundell, D. J., and Newton, A. B. (1991). Variations in the crystal lattice of PEEK and related para-substituted aromatic polymers. 2. Effect of sequence and proportion of ether and ketone links. *Polymer* **32**, 308–313.
Boda, E., Ungar, G., Brooke, G. M., Burnett, S., Mohammed, S., Proctor, D., and Whiting, M. C. (1997). Crystallization rate minima in a series of n-alkanes from $C_{194}H_{390}$ to $C_{294}H_{590}$. *Macromolecules* **30**, 4674–4678.
Bovey, F. A., Mirau, P. A., and Gutowsky, H. S. (eds.) (1988). "Nuclear Magnetic Resonance Spectroscopy," 2nd edn. Academic Press, San Diego.
Briber, R. M., and Khoury, F. A. (1993). The morphology of poly(vinylidene fluoride) crystallized from blends of poly(vinylidene fluoride) and poly(ethyl acrylate). *J. Polym. Sci. Polym. Phys. Ed.* **31**, 1253–1272.
Brizzolara, D., Cantow, H.-J., Diederichs, K., Keller, E., and Domb, A. J. (1996). Mechanism of the stereocomplex formation between enantiomeric poly(lactide)s. *Macromolecules* **29**, 191–197.
Brown, S. P., and Spiess, H. W. (2001). Advanced solid-state NMR methods for the elucidation of structure and dynamics of molecular, macromolecular, and supramolecular systems. *Chem. Rev.* **101**, 4125–4155.
Brückner, S., and Meille, S. V. (1989). Non-parallel chains in crystalline γ-isotactic polypropylene. *Nature* **340**, 455–457.
Bu, Z., Yoon, Y., Ho, R.-M., Zhou, W., Jangchud, I., Eby, R. K., Cheng, S. Z. D., Hsieh, E. T., Johnson, T. W., Geerts, R. G., Palackal, S. J., Hawley, G. R., *et al.* (1996). Crystallization, melting, and morphology of syndiotactic polypropylene fractions. 3. Lamellar single crystals and chain folding. *Macromolecules* **29**, 6575–6581.
Cai, W., Li, C. Y., Li, L., Lotz, B., Keating, M., and Marks, D. (2004). Submicrometer scroll/tubular lamellar crystals of nylon 66. *Adv. Mater.* **16**, 600–605.
Campbell, R. A., Phillips, P. J., and Lin, J. S. (1993). The gamma phase of high-molecular-weight polypropylene: 1. Morphological aspects. *Polymer* **34**, 4809–4816.
Cartier, L., Spassky, N., and Lotz, B. (1996). Frustrated structures of chiral polymers. *C. R. Acad. Sci. Serie II B: Méc. Phys. Chim., Astron.* **322**, 429–435.
Cartier, L., Okihara, T., and Lotz, B. (1997). Triangular polymer single crystals: Stereocomplexes, twins, and frustrated structures. *Macromolecules* **30**, 6313–6322.

Cartier, L., Okihara, T., and Lotz, B. (1998). The α′ "superstructure" of syndiotactic polystyrene: A frustrated structure. *Macromolecules* **31**, 3303–3310.

Cartier, L., Okihara, T., Ikada, Y., Tsuji, H., Puiggali, J., and Lotz, B. (2000). Epitaxial crystallization and crystalline polymorphism of polylactides. *Polymer* **41**, 8909–8919.

Charlet, G., and Delmas, G. (1984). Effect of solvent on the polymorphism of poly(4-methylpentene-1). 2. Crystallization in semi-dilute solutions. *Polymer* **25**, 1619–1625.

Chatani, Y., Maruyama, H., Asanuma, T., and Shiomura, T. (1991). Structure of a new crystalline phase of syndiotactic polypropylene. *J. Polym. Sci. Polym. Phys. Ed.* **29**, 1649–1652.

Chau, K. W., Yang, Y. C., and Geil, P. H. (1986). Tetragonal → twinned hexagonal crystal phase transformation in polybutene-1. *J. Mater. Sci.* **21**, 3002–3014.

Chen, J. H., Cheng, S. Z. D., Wu, S. S., Lotz, B., and Wittmann, J.-C. (1995). Polymer decoration study in chain folding behavior of solution-grown poly(ethylene oxide) crystals. *J. Polym. Sci. Polym. Phys. Ed.* **33**, 1851–1855.

Cheng, J., Jin, Y., Wunderlich, B., Cheng, S. Z. D., Yandrasits, M. A., Zhang, A., and Percec, V. (1992). Solid-state ^{13}C NMR studies of molecular motion in MBPE-9 and MBPE-5. *Macromolecules* **25**, 5991–5999.

Cheng, S. Z. D., and Chen, J. H. (1991). Nonintegral and integral folding crystal growth in low-molecular mass poly(ethylene oxide) fractions. III. Linear crystal growth rates and crystal morphology. *J. Polym. Sci. Polym. Phys. Ed.* **29**, 311–327.

Cheng, S. Z. D., and Keller, A. (1998). The role of metastable states in polymer phase transitions: Concepts, principles, and experimental observations. *Annu. Rev. Mater. Sci.* **28**, 533–562.

Cheng, S. Z. D., and Li, C. Y. (2002). Structure and formation of polymer single crystal textures. *Mater. Sci. Forum* **408**, 25–37.

Cheng, S. Z. D., Barley, J. S., and Von Meerwall, E. D. (1991a). Self-diffusion of poly(ethylene oxide) fractions and its influence on the crystalline texture. *J. Polym. Sci. Polym. Phys. Ed.* **29**, 515–525.

Cheng, S. Z. D., Chen, J. H., Zhang, A. Q., and Heberer, D. P. (1991b). Nonintegral and integral folding crystal growth in low-molecular mass poly(ethylene oxide) fractions. II. End-group effect: α,ω-methoxy-poly(ethylene oxide). *J. Polym. Sci. Polym. Phys. Ed.* **29**, 299–310.

Cheng, S. Z. D., Zhang, A. Q., Chen, J. H., and Heberer, D. P. (1991c). Nonintegral and integral folding crystal growth in low-molecular-mass poly(ethylene oxide) fractions. I. Isothermal lamellar thickening and thinning. *J. Polym. Sci. Polym. Phys. Ed.* **29**, 287–297.

Cheng, S. Z. D., Zhang, A., Barley, J. S., Chen, J., Habenschuss, A., and Zschack, P. R. (1991d). Isothermal thickening and thinning processes in low molecular weight poly(ethylene oxide) fractions. 1. From nonintegral-folding to integral-folding chain crystal transitions. *Macromolecules* **24**, 3937–3944.

Cheng, S. Z. D., Chen, J., Barley, J. S., Zhang, A., Habenschuss, A., and Zschack, P. R. (1992a). Isothermal thickening and thinning processes in low molecular weight poly(ethyleneoxide) fractions crystallized from the melt. 3. Molecular weight dependence. *Macromolecules* **25**, 1453–1460.

Cheng, S. Z. D., Chen, J. H., Zhang, A. Q., Barley, J. S., Habenschuss, A., and Zschack, P. R. (1992b). Isothermal thickening and thinning processes in low molecular weight poly(ethylene oxide) fractions crystallized from the melt. 2. Crystals involving more than one fold. *Polymer* **33**, 1140–1149.

Cheng, S. Z. D., Wu, S. S., Chen, J., Zhuo, Q., Quirk, R. P., von Meerwall, E. D., Hsiao, B. S., Habenschuss, A., and Zschack, P. R. (1993). Isothermal thickening and thinning processes in low molecular weight poly(ethylene oxide) fractions crystallized from the melt. 4. End-group dependence. *Macromolecules* **26**, 5105–5117.

Clark, E. S. (1967). Molecular motion in poly(tetrafluoroethylene) at cryogenic temperatures. *J. Macromol. Sci. Part B: Phys.* **1**, 795–800.

Clark, E. S., and Muus, L. T. (1962). Partial disordering and crystal transition in polytetrafluoroethylene. *Zeitschrift für Kristallographie* **117**, 119–127.

Clark, N. A., and Lagerwall, S. T. (1980). Submicrosecond bistable electro-optic switching in liquid crystals. *Appl. Phys. Lett.* **36**, 899–901.

Cojazzi, G., Malta, V., Celotti, G., and Zannetti, R. (1976). Crystal structure of form III of isotactic poly-1-butene. *Die Makromolekulare Chemie* **177**, 915–926.

Colthup, N. B., Daly, L. H., and Wiberley, S. E. (1990). *In* "Introduction to Infrared and Raman Spectroscopy," 3rd edn. Acadamic Press, San Diego.

Corradini, P., and Guerra, G. (1977). The chain conformation of poly(tetrafluoroethylene) in the crystalline modification above 30 °C. *Macromolecules* **10**, 1410–1413.
Corradini, P., Natta, G., Ganis, P., and Temussi, P. A. (1967). Crystal structure of syndiotactic polypropylene. *J. Polym. Sci. Polym. Symp.* **16**, 2477–2484.
Davis, G. T., McKinney, J. E., Broadhurst, M. G., and Roth, S. C. (1978). Electric-field-induced phase changes in poly(vinylidene fluoride). *J. Appl. Phys.* **49**, 4998–5002.
De Rosa, C. (1996). Crystal structure of the trigonal modification (α-form) of syndiotactic polystyrene. *Macromolecules* **29**, 8460–8465.
De Rosa, C., and Corradini, P. (1993). Crystal structure of syndiotactic polypropylene. *Macromolecules* **26**, 5711–5718.
De Rosa, C., Auriemma, F., and Vinti, V. (1997). Disordered polymorphic modifications of form I of syndiotactic polypropylene. *Macromolecules* **30**, 4137–4146.
De Rosa, C., Auriemma, F., and Vinti, V. (1998). On the form II of syndiotactic polypropylene. *Macromolecules* **31**, 7430–7435.
De Rosa, C., Auriemma, F., Circelli, T., and Waymouth, R. M. (2002). Crystallization of the α and γ forms of isotactic polypropylene as a tool to test the degree of segregation of defects in the polymer chains. *Macromolecules* **35**, 3622–3629.
De Rosa, C., Rapacciuolo, M., Guerra, G., Petraccone, V., and Corradini, P. (1992). On the crystal structure of the orthorhombic form of syndiotactic polystyrene. *Polymer* **33**, 1423–1428.
DiCorleto, J. A., and Bassett, D. C. (1990). On circular crystals of polyethylene. *Polymer* **31**, 1971–1977.
Dorset, D. L. (1995). "Structural Electron Crystallography." Plenum Press, New York.
Dorset, D. L., McCourt, M. P., Kopp, S., Wittmann, J. C., and Lotz, B. (1994). Direct determination of polymer crystal structures by electron crystallography-isotactic poly(1-butene), form (III). *Acta Crystallogr. Sect. B: Struct. Sci.* **50**, 201–208.
Dorset, D. L., McCourt, M. P., Kopp, S., Schumacher, M., Okihara, T., and Lotz, B. (1998). Isotactic polypropylene, β-phase: A study in frustration. *Polymer* **39**, 6331–6337.
Dosch, H. (1992). "Critical Phenomena at Surfaces and Interfaces: Evanescent X-Ray and Neutron Scattering." Chapter 5, Springer-Verlag, Berlin.
Faraday, M. (1860). Note on regelation. *Proc. R. Soc. Lond.* **10**, 440–450.
Fillon, B., Thierry, A., Wittmann, J. C., and Lotz, B. (1993). Self-nucleation and recrystallization of polymers. Isotactic polypropylene, β phase: β-α conversion and β-α growth transitions. *J. Polym. Sci. Polym. Phys. Ed.* **31**, 1407–1424.
Finter, J., and Wegner, G. (1981). Relation between phase transition and crystallization behavior of 1,4-trans-poly(butadiene). *Die Makromolekulare Chemie* **182**, 1859–1874.
Fischer, E. W. (1957). Stufen- und spiralförmiges Kristallwachstum bei Hochpolymeren. *Zeitschrift für Naturforschung A: Astrophisik, Physik und Physikalische Chemie* **12**, 753–754.
Fischer, E. W., and Lorenz, R. (1963). Über fehlordnungen in polyäthylen-einkristallen. *Kolloid-Zeitschrift & Zeitschrift für Polymere* **189**, 97–110.
Flack, H. D. (1972). High-pressure phase of polytetrafluoroethylene. *J. Polym. Sci. Polym. Phys. Ed.* **10**, 1799–1809.
Frank, F. C., Keller, A., and O'Connor, A. (1959). Observations on single crystals of an isotactic polyolefin. Morphology and chain packing in poly-4-methyl-pentene-1. *Philos. Mag.* **4**, 200–214.
Galerne, Y., and Liebert, L. (1991). Antiferroelectric chiral smectic-O* liquid crystal. *Phys. Rev. Lett.* **66**, 2891–2894.
Gang, O., Wu, X. Z., Ocko, B. M., Sirota, E. B., and Deutsch, M. (1998). Surface freezing in chain molecules. II. Neat and hydrated alcohols. *Phys. Rev. E* **58**, 6086–6100.
Gardner, K. C. H., Hsiao, B. S., Matheson, R. R. Jr., and Wood, B. A. (1992). Structure, crystallization and morphology of poly(aryl ether ketone ketone). *Polymer* **33**, 2483–2495.
Gardner, K. H., Hsiao, B. S., and Faron, K. L. (1994). Polymorphism in poly(aryl ether ketone)s. *Polymer* **35**, 2290–2295.
Gautam, K. S., and Dhinojwala, A. (2002). Melting at alkyl side chain comb polymer interfaces. *Phys. Rev. Lett.* **88**, 145501.1–145501.4.
Gautam, K. S., Kumar, S., Wermeille, D., Robinson, D., and Dhinojwala, A. (2003). Observation of novel liquid-crystalline phase above the bulk-melting temperature. *Phys. Rev. Lett.* **90**, 215501.1–215501.4.

Ge, J. J., Li, C. Y., Xue, G., Mann, I. K., Zhang, D., Wang, S.-Y., Harris, F. W., Cheng, S. Z. D., Hong, S.-C., Zhuang, X., and Shen, Y. R. (2001). Rubbing-induced molecular reorientation on an alignment surface of an aromatic polyimide containing cyanobiphenyl side chains. *J. Am. Chem. Soc.* **123**, 5768–5776.

Geacintov, C., Schotland, R. S., and Miles, R. B. (1963). Phase transition of crystalline poly-1-butene in form III. *J. Polym. Sci. Polym. Lett. Ed.* **1**, 587–591.

Geil, P. H. Jr. (1960). Nylon single crystals. *J. Polym. Sci.* **44** 449–458.

Geil, P. H. Jr. (1963). "Polymer Single Crystals." Wiley-Interscience, New York.

Geil, P. H. Jr. Symons, N. K. J., and Scott, R. G. (1959). Solution grown crystals of an acetal resin. *J. Appl. Phys.* **30**, 1516–1517.

Geil, P. H. Jr., Anderson, F. R., Wunderlich, B., and Arakawa, T. (1964). Morphology of polyethylene crystallized from the melt under pressure. *J. Polym. Sci. Part A* **2**, 3707–3720.

Gomez, M. A., Cozine, M. H., and Tonelli, A. E. (1988). High-resolution solid-state 13CNMR study of the α and β crystalline forms of poly(butylene terephthalte). *Macromolecules* **21**, 388–392.

Goodby, J. W., Slaney, A. J., Booth, C. J., Nishiyama, I., Vuijk, J. D., Styring, P., and Toyne, K. J. (1994). Chirality and frustration in ordered fluids. *Mol. Cryst. Liq. Cryst.* **243**, 231–298.

Guerra, G., Vitagliano, V. M., De Rosa, C., Petraccone, V., and Corradini, P. (1990). Polymorphism in melt crystallized syndiotactic polystyrene samples. *Macromolecules* **23**, 1539–1544.

Guerra, G., De Rosa, C., Vitagliano, V. M., Petraccone, V., and Corradini, P. (1991). Effects of blending on the polymorphic behavior of melt-crystallized syndiotactic polystyrene. *J. Polym. Sci. Polym. Phys. Ed.* **29**, 265–271.

Hasegawa, R., Tanabe, Y., Kobayashi, M., Tadokoro, H., Sawaoka, A., and Hawai, N. (1970). Structural studies of pressure-crystallized polymers. I. Heat treatment of oriented polymers under high pressure. *J. Polym. Sci. Polym. Phys. Ed.* **8**, 1073–1087.

Hasegawa, R., Kobayashi, M., and Tadokoro, H. (1972a). Molecular conformation and packing of poly(vinylidene fluoride). Stability of three crystalline forms and the effect of high pressure. *Polym. J. (Japan)* **3**, 591–599.

Hasegawa, R., Takahashi, Y., Chatani, Y., and Tadokoro, H. (1972b). Crystal structures of three crystalline forms of poly(vinylidene fluoride). *Polym. J. (Japan)* **3**, 600–610.

Hattori, T., Hikosaka, M., and Ohigashi, H. (1996). The crystallization behavior and phase diagram of extended-chain crystals of poly(vinylidene fluoride) under high pressure. *Polymer* **37**, 85–91.

Hattori, T., Watanabe, T., Akama, S., Hikosaka, M., and Ohigashi, H. (1997). The high-pressure crystallization behaviors and piezoelectricity of extended chain lamellar crystals of vinylidene fluoride trifluoroethylene copolymers with high molar content of vinylidene fluoride. *Polymer* **38**, 3505–3511.

Hikosaka, M., Rastogi, S., Keller, A., and Kawabata, H. (1992). Investigations on the crystallization of polyethylene under high pressure: Role of mobile phases, lamellar thickening growth, phase transformations, and morphology. *J. Macromol. Sci. Part B: Phys.* **31**, 87–131.

Ho, R.-M., Cheng, S. Z. D., Hsiao, B. S., and Gardner, K. H. (1994a). Crystal morphology and phase identification in poly(aryl ether ketone)s and their copolymers. 1. Polymorphism in PEKK. *Macromolecules* **27**, 2136–2140.

Ho, R.-M., Cheng, S. Z. D., Fisher, H. P., Eby, R. K., Hsiao, B. S., and Gardner, K. H. (1994b). Crystal morphology and phase identification in poly(aryl ether ketone)s and their copolymers. 2. Poly(oxy-1,4-phenylenecarbonyl-1,3-phenylenecarbonyl-1,4-phenylene). *Macromolecules* **27**, 5787–5793.

Ho, R.-M., Cheng, S. Z. D., Hsiao, B. S., and Gardner, K. H. (1995a). Crystal morphology and phase identification in poly(aryl ether ketone)s and their copolymers. 3. Polymorphism in a polymer containing alternated terephthalic acid and isophthalic acid isomers. *Macromolecules* **28**, 1938–1945.

Ho, R.-M., Cheng, S. Z. D., Hsiao, B. S., and Gardner, K. H. (1995b). Crystal morphology and phase identification in poly(aryl ether ketone)s and their copolymers. 4. Morphological observations in PEKK with all p-phenylene linkages. *Macromolecules* **28**, 8855–8861.

Ho, R.-M., Yoon, Y., Leland, M., Cheng, S. Z. D., Yang, D., Percec, V., and Chu, P. (1996). Phase identification in a series of liquid crystalline TPP polyethers and copolyethers having highly ordered mesophase structures. 3. Thin film surface-induced ordering structure and morphology in TPP ($n = 7$). *Macromolecules* **29**, 4528–4535.

Ho, R.-M., Lin, C.-P., Tsai, H.-Y., and Woo, E.-M. (2000). Metastability studies of syndiotactic polystyrene polymorphism. *Macromolecules* **33**, 6517–6526.

Ho, R.-M., Lin, C.-P., Hseih, P.-Y., Chung, T.-M., and Tsai, H.-Y. (2001). Isothermal crystallization-induced phase transition of syndiotactic polystyrene polymorphism. *Macromolecules* **34**, 6727–6736.

Hobbs, J. K., Hill, M. J., and Barham, P. J. (2001). Crystallization and isothermal thickening of single crystals of $C_{246}H_{494}$ in dilute solution. *Polymer* **42**, 2167–2176.

Holland, V. F., and Miller, R. L. (1964). Isotactic polybutene-1 single crystals: Morphology. *J. Appl. Phys.* **35**, 3241–3248.

Hosier, I. L., Bassett, D. C., and Vaughan, A. S. (2000). Spherulitic growth and cellulation in dilute blends of monodisperse long n-alkanes. *Macromolecules* **33**, 8781–8790.

Hsiao, B. S., Yang, L., Somani, R. H., Avila-Orta, C. A., and Zhu, L. (2005). Unexpected shish-kebab structure in a sheared polyethylene melt. *Phys. Rev. Lett.* **94**, 117802.1–117802.4.

Huo, H., Jiang, S., An, L., and Feng, J. (2004). Influence of shear on crystallization behavior of the β phase in isotactic polypropylene with β-nucleating agent. *Macromolecules* **37**, 2478–2483.

Jaccodine, R. (1955). Observations of spiral growth steps in ethylene polymer. *Nature* **176**, 305–306.

Jing, A. J., Taikum, O., Li, C. Y., Harris, F. W., and Cheng, S. Z. D. (2002). Phase identification and monotropic transition behaviors in a thermotropic main-chain liquid crystalline polyether. *Polymer* **43**, 3431–3440.

Keith, H. D. (1964). On the relation between different morphological forms in high polymers. *J. Polym. Sci. Part A* **2**, 4339–4360.

Keith, H. D., and Padden, F. J. Jr. (1959a). The optical behavior of spherulites in crystalline polymers. Part I. Calculation of theoretical extinction patterns in spherulites with twisting crystalline orientation. *J. Polym. Sci.* **39** 101–122.

Keith, H. D., and Padden, F. J. Jr. (1959b). The optical behavior of spherulites in crystalline polymers. Part II. The growth and structure of the spherulites. *J. Polym. Sci.* **39** 123–138.

Keith, H. D., and Padden, F. J. Jr. (1984). Twisting orientation and the role of transient states in polymer crystallization. *Polymer* **25** 28–42.

Keith, H. D., and Padden, F. J. Jr. (1987). Spherulitic morphology in polyethylene and isotactic polystyrene: Influence of diffusion of segregated species. *J. Polym. Sci. Polym. Phys. Ed.* **25** 2371–2392.

Keith, H. D., Padden, F. J. Jr. Walter, N. W., and Wyckoff, H. W. (1959). Evidence for a second crystal form of polypropylene. *J. Appl. Phys.* **30**, 1485–1488.

Keith, H. D., Padden, F. J. Jr. Lotz, B., and Wittmann, J. C. (1989). Asymmetries of habit in polyethylene crystals grown from the melt. *Macromolecules* **22**, 2230–2238.

Keller, A. (1957). A note on single crystals in polymers: Evidence of a folded-chain configuration. *Philos. Mag.* **2**, 1171–1175.

Keller, A. (1959). Investigations on banded spherulites. *J. Polym. Sci.* **39**, 151–173.

Keller, A. (1968). Polymer crystals. *Rep. Prog. Phys.* **31**, 623–704.

Keller, A., and Cheng, S. Z. D. (1998). The role of metastability in polymer phase transitions. *Polymer* **39**, 4461–4487.

Keller, A., and Kolnaar, H. W. H. (1997). Flow-induced orientation and structure formation. *In* "Processing of Polymers" (H. E. H. Meijer, ed.), Chapter 4. VCH, Weinheim.

Keller, A., Ungar, G., and Percec, V. (1990). Liquid-crystalline polymers. A unifying thermodynamics-based scheme. *Am. Chem. Soc. Symposium Series* **435**, 308–334.

Khoury, F. (1966). The spherulitic crystallization of isotactic polypropylene from solution: Evolution of monoclinic spherulites from dendritic chain-folded crystal precursors. *J. Res. Natl. Bur. Stand. Sect. A: Physics and Chemistry* **70**, 29–61.

Khoury, F. (1979). Organization of macromolecules in the condensed phase: General discussion. *Faraday Discuss. Chem. Soc.* **68**, 404–405.

Khoury, F., and Barnes, J. D. (1972). Formation of curved polymer crystals. Poly(4-methyl-1-pentene). *J. Res. Natl. Bur. Stand. Sect. A: Phys. Chem.* **76**, 225–252.

Khoury, F., and Barnes, J. D. (1974a). Formation of curved polymer crystals. Poly(oxymethylene). *J. Res. Natl. Bur. Stand. Sect. A: Phys. Chem.* **78**, 95–128.

Khoury, F., and Barnes, J. D. (1974b). Formation of curved polymer crystals. Poly(chlorotrifluoroethylene). *J. Res. Natl. Bur. Stand. Sect. A: Phys. Chem.* **78**, 363–373.

Khoury, F., and Bolz, L. H. (1980). Scanning transmission electron microscopy of polyethylene crystals. *Proceedings—Annual Meeting, Electron Microscopy Society of American* **38**, 242–245.

Khoury, F., and Bolz, L. (1985). The lateral growth habits and sectored character of polyethylene crystals. *Bull. Am. Phys. Soc.* **30**, 493.

Khoury, F., and Passaglia, E. (1976). The morphology of crystalline synthetic polymers. *In* "Treatise on Solid State Chemistry. Volume 3. Crystalline and Noncrystalline Solids" (N. B. Hannay, ed.), Chapter 6. Plenum Press, New York.

Kim, I., and Krimm, S. (1996). Raman longitudinal acoustic mode studies of a poly(ethylene oxide) fraction during isothermal crystallization from the melt. *Macromolecules* **29**, 7186–7192.

Klop, E. A., Lommerts, B. J., Veurink, J., Aerts, J., and van Puijenbroek, R. R. (1995). Polymorphism in alternating polyketones studied by X-ray diffraction and calorimetry. *J. Polym. Sci. Polym. Phys. Ed.* **33**, 315–326.

Kobayashi, K. (1962). Koubunsi no Densikenbikyou Niyoru Kouzoukennkyuu (Transmission electron microscopy studies on the structure of polymers). *In* "Kobunshi no Bussei (Properties of Polymers)" (A. Nakajima, H. Tadokoro, T. Tsuruta, H. Yuki, and T. Ohtsu, eds.), Chapter 11. Kagakudojin, Kyoto.

Kobayashi, M., Tashiro, K., and Tadokoro, H. (1975). Molecular vibrations of three crystal forms of poly(vinylidene fluoride). *Macromolecules* **8**, 158–171.

Kolnaar, J. W. H., and Keller, A. (1994). A temperature window of reduced flow resistance in polyethylene with implications for melt flow rheology. 1. The basic effect and principle parameters. *Polymer* **35**, 3863–3874.

Kolnaar, J. W. H., and Keller, A. (1995). A temperature window of reduced flow resistance in polyethylene with implications for melt flow rheology. 2. Rheological investigations in the extrusion window. *Polymer* **36**, 821–836.

Kopp, S., Wittmann, J. C., and Lotz, B. (1994a). Phase II to phase I crystal transformation in polybutene-1 single crystals: A reinvestigation. *J. Mater. Sci.* **29**, 6159–6166.

Kopp, S., Wittmann, J. C., and Lotz, B. (1994b). Epitaxial crystallization and crystalline polymorphism of poly(1-butene): Forms II and III. *Polymer* **35**, 908–915.

Kopp, S., Wittmann, J. C., and Lotz, B. (1994c). Epitaxial crystallization and crystalline polymorphism of poly(1-butene): Form I′. *Polymer* **35**, 916–924.

Kovacs, A. J., and Gonthier, A. (1972). Crystallization and fusion of self-seeded polymers. II. Growth rate, morphology, and isothermal thickening of single crystals of low molecular weight poly (ethylene oxide) fractions. *Kolloid-Zeitschrift & Zeitschrift für Polymere* **250**, 530–551.

Kovacs, A. J., and Straupe, C. (1979). Isothermal growth, thickening, and melting of poly(ethylene oxide) single crystals in the bulk. Part 4. Dependence of pathological crystal habits on temperature and thermal history. *Faraday Discuss. Chem. Soc.* **68**, 225–238.

Kovacs, A. J., and Straupe, C. (1980). Isothermal growth, thickening, and melting of poly(ethylene oxide) single crystals in the bulk. III. Bilayer crystals and the effect of chain ends. *J. Cryst. Growth* **48**, 210–226.

Kovacs, A. J., Gonthier, A., and Straupe, C. (1975). Isothermal growth, thickening, and melting of poly(ethylene oxide) single crystals in the bulk. *J. Polym. Sci. Polym. Symp.* **50**, 283–325.

Kovacs, A. J., Straupe, C., and Gonthier, A. (1977). Isothermal growth, thickening, and melting of poly(ethylene oxide) single crystals in the bulk. II. *J. Polym. Sci. Polym. Symp.* **59**, 31–54.

Kricheldorf, H. R., Domschke, A., and Schwarz, G. (1991a). Liquid-crystalline polyimides. 3. Fully aromatic liquid-crystalline poly(ester imide)s derived from N-(4-carboxyphenyl) -trimellitimide and substituted hydroquinones. *Macromolecules* **24**, 1011–1016.

Kricheldorf, H. R., Schwarz, G., de Abajo, J., and de la Campa, J. G. (1991b). LC-polyimides. 5. Poly (ester imides) derived from N-(4-carboxyphenyl)trimellitimide and αω-dihydroxy- alkanes. *Polymer* **32**, 942–949.

Kunz, M., Drechsler, M., and Möller, M. (1995). On the structure of ultra-high molecular weight polyethylene gels. *Polymer* **36**, 1331–1339.

Laihonen, S., Gedde, U. W., Werner, P.-E., and Martinez-Salazar, J. (1997). Crystallization kinetics and morphology of poly(propylene-stat-ethylene) fractions. *Polymer* **38**, 361–369.

Lando, J. B., Olf, H. G., and Peterlin, A. (1966). Nuclear magnetic resonance and X-ray determination of the structure of poly(vinylidene fluoride). *J. Polym. Sci. Polym. Chem. Ed.* **4**, 941–951.

Lee, K. S., and Wegner, G. (1985). Linear and cyclic alkanes (C_n H_{2n+2}, C_n H_{2n}) with $n > 100$. Synthesis and evidence for chain-folding. *Die Makromolekulare Chemie Rapid Commun.* **6**, 203–208.

Li, C. Y., Yan, D., Cheng, S. Z. D., Bai, F., He, T., Chien, L.-C., Harris, F. W., and Lotz, B. (1999a). Double-twisted helical lamellar crystals in a synthetic main-chain chiral polyester similar to biological polymers. *Macromolecules* **32**, 524–527.

Li, C. Y., Cheng, S. Z. D., Ge, J. J., Bai, F., Zhang, J. Z., Mann, I. K., Harris, F. W., Chien, L.-C., Yan, D., He, T., and Lotz, B. (1999b). Double twist in helical polymer "soft" crystals. *Phys. Rev. Lett.* **83**, 4558–4561.

Li, C. Y., Cheng, S. Z. D., Ge, J. J., Bai, F., Zhang, J. Z., Mann, I. K., Chien, L.-C., Harris, F. W., and Lotz, B. (2000). Molecular orientations in flat-elongated and helical lamellar crystals of a main-chain nonracemic chiral polyester. *J. Am. Chem. Soc.* **122**, 72–79.

Li, C. Y., Cheng, S. Z. D., Weng, X., Ge, J. J., Bai, F., Zhang, J. Z., Calhoun, B. H., Harris, F. W., Chien, L.-C., and Lotz, B. (2001). Left or right, it is a matter of one methylene unit. *J. Am. Chem. Soc.* **123**, 2462–2463.

Li, C. Y., Jin, S., Weng, X., Ge, J. J., Zhang, D., Bai, F., Harris, F. W., Cheng, S. Z. D., Yan, D., He, T., Lotz, B., and Chien, L.-C. (2002). Liquid crystalline phases, microtwining in crystals and helical chirality transformations in a main-chain chiral polyester. *Macromolecules* **35**, 5475–5482.

Li, C. Y., Li, L., Cai, W., Kodjie, S. L., and Tenneti, K. K. (2005). Nanohybrid shish-kebab: Periodically functionalized carbon nanotubes. *Adv. Mater.* **17**, 1198–1202.

Li, L., Li, C. Y., and Ni, C. (2006). Polymer crystallization-driven, periodic patterning on carbon nanotubes. *J. Am. Chem. Soc.* **128**, 1692–1699.

Lotz, B. (1998). α and β phases of isotactic polypropylene: A case of growth kinetics 'phase reentrency' in polymer crystallization. *Polymer* **39**, 4561–4567.

Lotz, B., and Cheng, S. Z. D. (2005). A critical assessment of unbalanced surface stress as the mechanical origin of twisting and scrolling of polymer crystals. *Polymer* **46**, 577–610.

Lotz, B., and Cheng, S. Z. D. (2006). Comments on: 'A critical assessment of unbalanced surface stresses: Some complementary considerations'. by DC Bassett. *Polymer* **47**, 3267–3270.

Lotz, B., and Thierry, A. (2003). Spherulite morphology of form III isotactic poly(1-butene). *Macromolecules* **36**, 286–290.

Lotz, B., and Wittmann, J. C. (1986). The molecular origin of lamellar branching in the α (monoclinic) form of isotactic polypropylene. *J. Polym. Sci. Polym. Phys. Ed.* **24**, 1541–1558.

Lotz, B., Graff, S., and Wittmann, J. C. (1986). Crystal morphology of the γ (triclinic) phase of isotactic polypropylene and its relation to the α-phase. *J. Polym. Sci. Polym. Phys. Ed.* **24**, 2017–2032.

Lotz, B., Lovinger, A. J., and Cais, R. E. (1988). Crystal structure and morphology of syndiotactic polypropylene single crystals. *Macromolecules* **21**, 2375–2382.

Lotz, B., Graff, S., Straupe, C., and Wittmann, J. C. (1991). Single crystals of γ phase isotactic polypropylene: Combined diffraction and morphological support for a structure with nonparallel chains. *Polymer* **32**, 2902–2910.

Lotz, B., Kopp, S., and Dorset, D. (1994). Original crystal structure of polymers with ternary helices. *C. R. Acad. Sci. Ser. II B: Méc. Phys. Chim. Astron.* **319**, 187–192.

Lotz, B., Wittmann, J. C., and Lovinger, A. J. (1996). Structure and morphology of poly(propylene)s: A molecular analysis. *Polymer* **37**, 4979–4992.

Lotz, B., Thierry, A., and Schneider, S. (1998a). Molecular origin of the scroll-like morphology of lamellae in γ-PVDF spherulites. *C. R. Acad. Sci. Ser. II C: Chimie* **1**, 609–614.

Lotz, B., Mathieu, C., Thierry, A., Lovinger, A. J., De Rosa, C., de Ballesteros, O. R., and Auriemma, F. (1998b). Chirality constraints in crystal-crystal transformations: Isotactic poly(1-butene) versus syndiotactic polypropylene. *Macromolecules* **31**, 9253–9257.

Lovinger, A. J. (1978a). Crystallographic factors affecting the structure of polymeric spherulites. I. Morphology of directionally solidified polyamides. *J. Appl. Phys.* **49**, 5003–5013.

Lovinger, A. J. (1978b). Crystallographic factors affecting the structure of polymeric spherulites. II. X-ray diffraction analysis of directionally solidified polyamides and general conclusions. *J. Appl. Phys.* **49**, 5014–5028.

Lovinger, A. J. (1980). Crystallization and morphology of melt-solidified poly(vinylidene fluoride). *J. Polym. Sci. Polym. Phys. Ed.* **18**, 793–809.

Lovinger, A. J. (1981a). Crystallization of the β phase of poly(vinylidene fluoride) from the melt. *Polymer* **22**, 412–413.

Lovinger, A. J. (1981b). Unit cell of the γ phase of poly(vinylidene fluoride). *Macromolecules* **14**, 322–325.

Lovinger, A. J. (1982). Poly(vinylidene fluoride). *In* "Developments in Crystalline Polymers" (D. C. Bassett, ed.), Chapter 5. Elsevier Applied Science, Oxford.

Lovinger, A. J., and Gryte, C. C. (1976a). The morphology of directionally solidified poly(ethylene oxide) spherulites. *Macromolecules* **9**, 247–253.

Lovinger, A. J., and Gryte, C. C. (1976b). Model for the shape of polymer spherulites formed in a temperature gradient. *J. Appl. Phys.* **47**, 1999–2004.

Lovinger, A. J., and Wang, T. T. (1979). Investigation of the properties of directionally solidified poly (vinylidene fluoride). *Polymer* **20**, 725–732.

Lovinger, A. J., Chua, J. O., and Gryte, C. C. (1977). Studies on the α and β forms of isotactic polypropylene by crystallization in a temperature gradient. *J. Polym. Sci. Polym. Phys. Ed.* **15**, 641–656.

Lovinger, A. J., Lotz, B., and Davis, D. D. (1990). Interchain packing and unit cell of syndiotactic polypropylene. *Polymer* **31**, 2253–2259.

Lovinger, A. J., Davis, D. D., and Lotz, B. (1991). Temperature dependence of structure and morphology of syndiotactic polypropylene and epitaxial relationship with isotactic polypropylene. *Macromolecules* **24**, 552–560.

Lovinger, A. J., Lotz, B., Davis, D. D., and Padden, F. J. Jr. (1993). Structure and defects in fully syndiotactic polypropylene. *Macromolecules* **26** 3494–3503.

Lu, J., Huang, R., Chen, Y., and Li, L. B. (2006). Extended-chain crystals in high-pressure crystallized poly(ethylene terephthalate)/bisphenol A polycarbonate blends. *J. Polym. Sci. Polym. Phys. Ed.* **44**, 3148–3156.

Luciani, L., Seppälä, J., and Löfgren, B. (1988). Poly-1-butene: Its preparation, properties and challenges. *Prog. Polym. Sci.* **13**, 37–62.

Lustiger, A., Lotz, B., and Duff, T. S. (1989). The morphology of the spherulitic surface in polyethylene. *J. Polym. Sci. Polym. Phys. Ed.* **27**, 561–579.

Mareau, V. H., and Prud'homme, R. E. (2005). In-situ hot stage atomic force microscopy study of poly(ε-caprolactone) crystal growth in ultrathin films. *Macromolecules* **38**, 398–408.

Mathieu, C., Stocker, W., Thierry, A., Wittmann, J. C., and Lotz, B. (2001). Epitaxy of isotactic poly (1-butene): New substrates, impact and attempt at recognition of helix orientation in form I′ by AFM. *Polymer* **42**, 7033–7047.

Matsushige, K., and Takemura, T. (1978). Melting and crystallization of poly(vinylidene fluoride) under high pressure. *J. Polym. Sci. Polym. Phys. Ed.* **16**, 921–934.

Matsushige, K., and Takemura, T. (1980). Crystallization of macromolecules under high pressure. *J. Cryst. Growth* **48**, 343–354.

Meille, S. V., Ferro, D. R., Brückner, S., Lovinger, A. J., and Padden, F. J. Jr. (1994). Structure of β-isotactic polypropylene: A long-standing structural puzzle. *Macromolecules* **27** 2615–2622.

Melillo, L., and Wunderlich, B. (1972). Extended-chain crystals. VIII. Morphology of polytetrafluoroethylene. *Kolloid-Zeitschrift & Zeitschrift für Polymere* **250**, 417–425.

Meyer, R. B., Liébert, L., Strzelecki, L., and Keller, P. (1975). Ferroelectric liquid crystals. *J. Phys. Lettres* **36**, L69–L71.

Mezghani, K., and Phillips, P. J. (1995). γ-Phase in propylene copolymer at atmospheric pressure. *Polymer* **36**, 2407–2411.

Mezghani, K., and Phillips, P. J. (1998). The γ-phase of high molecular weight isotactic polypropylene: III. The equilibrium melting point and the phase diagram. *Polymer* **39**, 3735–3744.

Miyamoto, Y., Nakafuku, C., and Takemura, T. (1972). Crystallization of poly(chlorotrifluoroethylene). *Polym. J. (Japan)* **3**, 122–128.

Monkman, L. J. (1979). Asbestos—recent developments. *In* "Applied Fiber Science" (F. Happey, ed.), Vol. 3. Chapter 4. Academic Press, London.

Morgan, R. L., Barham, P. J., Hill, M. J., Keller, A., and Organ, S. J. (1998). The crystallization of the n-alkane $C_{294}H_{590}$ from solution: Inversion of crystallization rates, crystal thickening, and effects of supersaturation. *J. Macromol. Sci. Part B: Phys.* **37**, 319–338.

Morrow, D. R., and Newman, B. A. (1968). Crystallization of low-molecular-weight polypropylene fractions. *J. Appl. Phys.* **39**, 4944–4950.

Nagata, K., Tagashira, K., Taki, S., and Takemura, T. (1980). Ultrasonic study of high pressure phase in polyethylene. *Japn. J. Appl. Phys.* **19**, 985–990.

Natta, G., and Corradini, P. (1959). Conformation of linear chains and their mode of packing in the crystal state. *J. Polym. Sci.* **39**, 29–46.

Natta, G., and Corradini, P. (1960). General considerations on the structure of crystalline polyhydrocarbons. *Del Nuovo Cimento, Supplemento* **15**, 9–39.

Natta, G., Peraldo, M., and Corradini, P. (1959). Modificazione mesomorpha smettica del polipropilene isotattico. Atti della Accademia nazionale dei Lincei. *Rendiconti della Classe di scienze fisiche, matematiche e naturali* **26**, 14–17.

Natta, G., Corradini, P., and Bassi, I. W. (1960a). Crystal structure of isotactic poly-alpha-butene. *Del Nuovo Cimento, Supplemento* **15**, 52–67.

Natta, G., Corradini, P., and Bassi, I. W. (1960b). Crystal structure of poly-ortho-fluorostyrene. *Del Nuovo Cimento, Supplemento* **15**, 83–95.

Ocko, B. M., Wu, X. Z., Sirota, E. B., Sinha, S. K., Gang, O., and Deutsch, M. (1997). Surface freezing in chain molecules: Normal alkanes. *Phys. Rev. E* **55**, 3164–3182.

Okihara, T., Kawaguchi, A., Tsuji, H., Hyon, S. H., Ikada, Y., and Katayama, K. (1988). Lattice disorders in the stereocomplex of poly(L-lactide) and poly(D-lactide). *Bull. Inst. Chem. Res. Kyoto University* **66**, 271–282.

Okihara, T., Tsuji, M., Kawaguchi, A., Katayama, K.-, Tsuji, I. H., Hyon, S.-H., and Ikada, Y. (1991). Crystal structure of stereocomplex of poly(L-lactide) and poly(D-lactide). *J. Macromol. Sci. Part B: Phys.* **30**, 119–140.

Organ, S. J., and Keller, A. (1985a). Solution crystallization of polyethylene at high temperatures. Part 1. Lateral crystal habits. *J. Mater. Sci.* **20**, 1571–1585.

Organ, S. J., and Keller, A. (1985b). Solution crystallization of polyethylene at high temperatures. Part 2. Three-dimensional crystal morphology and melting behavior. *J. Mater. Sci.* **20**, 1586–1601.

Organ, S. J., and Keller, A. (1985c). Solution crystallization of polyethylene at high temperatures. Part 3. The fold lengths. *J. Mater. Sci.* **20**, 1602–1615.

Organ, S. J., Ungar, G., and Keller, A. (1989). Rate minimum in solution crystallization of long paraffins. *Macromolecules* **22**, 1995–2000.

Organ, S. J., Barham, P. J., Hill, M. J., Keller, A., and Morgan, R. L. (1997). A study of the crystallization of the n-alkane $C_{246}H_{494}$ from solution: Further manifestations of the inversion of crystallization rates with temperature. *J. Polym. Sci. Polym. Phys. Ed.* **35**, 1775–1791.

Ostwald, W. (1897). Studien über die Bildung und Umwandlung fester Körper. *Zeitschrift für Physikalische Chemie, Stöchiometrie und Verwandschaftslehre* **22**, 289–300.

Pae, K. D., Morrow, D. R., and Sauer, J. A. (1966). Interior morphology of bulk polypropylene. *Nature* **211**, 514–515.

Pardey, R., Zhang, A. Q., Gabori, P. A., Harris, F. W., Cheng, S. Z. D., Adduci, J., Facinnelli, J. V., and Lenz, R. W. (1992). Monotropic liquid crystal behavior in two poly(ester imides) with even and odd flexible spacers. *Macromolecules* **25**, 5060–5608.

Pardey, R., Shen, D. X., Gabori, P. A., Harris, F. W., Cheng, S. Z. D., Adduci, J., Facinelli, J. V., and Lenz, R. W. (1993). Ordered structures in a series of liquid-crystalline poly(ester imide)s. *Macromolecules* **26**, 3687–3697.

Pardey, R., Wu, S. S., Chen, J., Harris, F. W., Cheng, S. Z. D., Keller, A., Adduci, J., Facinelli, J. V., and Lenz, R. W. (1994). Liquid crystal transition and crystallization kinetics in poly(ester imide)s. *Macromolecules* **27**, 5794–5802.

Passaglia, E., and Khoury, F. (1984). Crystal-growth kinetics and the lateral habits of polyethylene crystals. *Polymer* **25**, 621–644.

Patel, D., and Bassett, D. C. (1994). On spherulitic crystallization and the morphology of melt-crystallized poly(4-methylpehtene-1). *Proc. Math. Phys. Sci.* **445**, 577–595.

Pennings, A. J., and Kiel, A. M. (1965). Fractionation of polymers by crystallization from solution, III. On the morphology of fibrillar polyethylene crystals grown in solution. *Kolloid-Zeitschrift & Zeitschrift für Polymere* **205**, 160–162.

Pennings, A. J., van der Mark, J. M. A. A., and Booij, H. C. (1970). Hydrodynamically induced crystallization of polymers from solution. II. The effect of secondary flow. *Kolloid-Zeitschrift & Zeitschrift für Polymere* **236**, 99–111.

Percec, V., and Keller, A. (1990). A thermodynamic interpretation of polymer molecular weight effect on the phase transitions of main-chain and side-chain liquid-crystal polymers. *Macromolecules* **23**, 4347–4350.

Percec, V., and Yourd, R. (1989a). Liquid crystalline polyethers based on conformational isomerism. 2. Thermotropic polyethers and copolyethers based on 1-(4-hydroxyphenyl)-2-(2-methyl-4- hydroxyphenyl)ethane and flexible spacers containing an odd number of methylene units. *Macromolecules* **22**, 524–537.

Percec, V., and Yourd, R. (1989b). Liquid crystalline polyethers and copolyethers based on conformational isomerism. 3. The influence of thermal history on the phase transitions of thermotropic polyethers and copolyethers based on 1-(4-hydroxyphenyl)-2-(2-methyl-4- hydroxyphenyl)ethane and flexible spacers containing an odd number of methylene units. *Macromolecules* **22**, 3229–3242.

Petraccone, V., Pirozzi, B., Frasci, A., and Corradini, P. (1976). Polymorphism of isotactic poly-α-butene. Conformational analysis of the chain and crystalline structure of form 2. *Eur. Polym. J.* **12**, 323–327.

Point, J. J. (1955). Enroulement hélicoïdal dans les sphérolithes der polyéthylène. *Bulletin de la Classe des Sciences, Académie Royale de Belgique* **41**, 982–990.

Pradere, P., and Thomas, E. L. (1990). Antiphase boundaries and ordering defects in syndiotactic polystyrene crystals. *Macromolecules* **23**, 4954–4958.

Prasad, S., Hanne, L., and Dhinojwala, A. (2005). Thermodynamic study of a novel surface ordered phase above the bulk melting temperature in alkyl side chain acrylate polymers. *Macromolecules* **38**, 2541–2543.

Price, F. P. (1959). Extinction patterns of polymer spherulites. *J. Polym. Sci.* **39**, 139–150.

Prime, R. B., and Wunderlich, B. (1969). Extended-chain crystals. III. Size distribution of polyethylene crystals grown under elevated pressure. *J. Polym. Sci. Polym. Phys. Ed.* **7**, 2061–2072.

Putra, E. G. R., and Ungar, G. (2003). In situ solution crystallization study of n-$C_{246}H_{494}$: Self-poisoning and morphology of polymethylene crystals. *Macromolecules* **36**, 5214–5225.

Rastogi, S., and Ungar, G. (1992). Hexagonal columnar phase in 1,4-trans-polybutadiene: Morphology, chain extension, and isothermal phase reversal. *Macromolecules* **25**, 1445–1452.

Rastogi, S., La Camera, D., van der Burgt, F., Terry, A. E., and Cheng, S. Z. D. (2001). Polymorphism in syndiotactic polypropylene: Thermodynamic stable regions for Form I and Form II in pressure-temperature phase diagram. *Macromolecules* **34**, 7730–7736.

Rees, D. V., and Bassett, D. C. (1968). Origin of extended-chain lamellae in polyethylene. *Nature* **219**, 368–370.

Rees, D. V., and Bassett, D. C. (1971). Crystallization of polyethylene at elevated pressures. *J. Polym. Sci. Polym. Phys. Ed.* **9**, 385–406.

Sauer, J. A., Morrow, D. R., and Richardson, G. C. (1965). Morphology of solution-grown polypropylene crystal aggregates. *J. Appl. Phys.* **36**, 3017–3021.

Seto, T., Hara, T., and Tanaka, K. (1968). Phase transformation and deformation processes in oriented polyethylene. *Japn. J. Appl. Phys.* **7**, 31–42.

Shcherbina, M. A., and Ungar, G. (2007). Asymmetric curvature of growth faces of polymer crystals. *Macromolecules* **40**, 402–405.

Sirota, E. B., Pershan, P. S., Sorensen, L. B., and Collett, J. (1987). X-ray and optical studies of the thickness dependence of the phase diagram of liquid-crystal films. *Phys. Rev. A* **36**, 2890–2901.

Somani, R. H., Yang, L., Zhu, L., and Hsiao, B. S. (2005). Flow-induced shish-kebab precursor structures in entangled polymer melts. *Polymer* **46**, 8587–8623.

Song, K., and Krimm, S. (1989). Mixed-integer and fractional-integer chain folding in crystalline lamellae of poly(ethylene oxide): A Raman longitudinal acoustic mode study. *Macromolecules* **22**, 1504–1505.

Song, K., and Krimm, S. (1990a). Raman longitudinal acoustic mode (LAM) studies of folded-chain morphology in poly(ethylene oxide) (PEO). I. Normal mode analysis of LAM of a helical-chain oligomer of PEO. *J. Polym. Sci. Polym. Phys. Ed.* **28**, 35–50.

Song, K., and Krimm, S. (1990b). Raman longitudinal acoustic mode (LAM) studies of folded-chain morphology in poly(ethylene oxide) (PEO). 3. Chain folding in PEO as a function of molecular weight. *Macromolecules* **23**, 1946–1957.

Starkweather, H. W. Jr. (1979). A comparison of the rheological properties of polytetrafluoroethylene below its melting point with certain low-molecular weight smectic states. *J. Polym. Sci. Polym. Phys. Ed.* **17** 73–79.

Starkweather, H. W. Jr., Zoller, P., Jones, G. A., and Vega, A. J. (1982). The heat of fusion of poly (tetrafluoroethylene). *J. Polym. Sci. Polym. Phys. Ed.* **20**, 751–761.

Stocker, W., Schumacher, M., Graff, S., Thierry, A., Wittmann, J.-C., and Lotz, B. (1998). Epitaxial crystallization and AFM investigation of a frustrated polymer structure: Isotactic poly(propylene), β phase. *Macromolecules* **31**, 807–814.

Storks, K. H. (1938). An electron-diffraction examination of some linear high polymers. *J. Am. Chem. Soc.* **60**, 1753–1761.

Suehiro, K., and Takayanagi, M. (1970). Structural studies of the high temperature from of trans-1,4-polybutadiene crystal. *J. Macromol. Sci. Part B: Phys.* **4**, 39–46.

Sutton, S. J., Vaughan, A. S., and Bassett, D. C. (1996). On the morphology and crystallization kinetics of monodisperse polyethylene oligomers crystallized from the melt. *Polymer* **37**, 5735–5738.

Tadokoro, H. (1979). "Structure of Crystalline Polymers." Wiley-Interscience, New York.

Tanda, Y., Kawasaki, N., Imada, K., and Takayanagi, M. (1966). New crystal modifications of isotactic poly-4-methyl-pentene-1. *Reports on Progress in Polymer Physics in Japan* **IX**, 165–168.

Till, P. H. Jr. (1957). The growth of single crystals of linear polyethylene. *J. Polym. Sci.* **24** 301–306.

Toda, A. (1992). Growth of polyethylene single crystals from the melt: Change in lateral habit and regime I-II transition. *Colloid Polym. Sci.* **270**, 667–681.

Toda, A. (1993). Growth mode and curved lateral habits of polyethylene single crystals. *Faraday Discuss.* **95**, 129–143.

Toda, A., and Keller, A. (1993). Growth of polyethylene single crystals from the melt: Morphology. *Colloid Polym. Sci.* **271**, 328–342.

Toda, A., Miyaji, H., and Kiho, H. (1986). Regime II growth of polyethylene single crystals from dilute solution in n-octane. *Polymer* **27**, 1505–1508.

Toda, A., Arita, T., and Hikosaka, M. (2001). Three-dimensional morphology of PVDF single crystals forming banded spherulites. *Polymer* **42**, 2223–2233.

Tonelli, A. E. (1989). "NMR Spectroscopy and Polymer Microstructure: The Conformational Connection." VCH, New York.

Tosaka, M., Tsuji, M., Kohjiya, S., Cartier, L., and Lotz, B. (1999). Crystallization of syndiotactic polystyrene in β-form. 4. Crystal structure of melt-grown modification. *Macromolecules* **32**, 4905–4911.

Tosaka, M., Kamijo, T., Tsuji, M., Kohjiya, S., Ogawa, T., Isoda, S., and Kobayashi, T. (2000). High-resolution transmission electron microscopy of crystal transformation in solution-grown lamellae of isotactic polybutene-1. *Macromolecules* **33**, 9666–9672.

Tracz, A., and Ungar, G. (2005). AFM study of lamellar structure of melt-crystallized n-alkane $C_{390}H_{782}$. *Macromolecules* **38**, 4962–4965.

Turner-Jones, A. (1971). Development of the γ-crystal form in random copolymers of propylene and their analysis by differential scanning calorimetry and X-ray methods. *Polymer* **12**, 487–508.

Turner-Jones, A., Aizlewood, J. M., and Beckett, D. R. (1964). Crystalline forms of isotactic polypropylene. *Die Makromolekulare Chemie* **75**, 134–158.

Tweet, D. J., Holyst, R., Swanson, B. D., Stragier, H., and Sorensen, L. B. (1990). X-ray determination of the molecular tilt and layer fluctuation profiles of freely suspended liquid-crystal films. *Phys. Rev. Lett.* **65**, 2157–2160.

Ungar, G. (1993). Thermotropic hexagonal phases in polymers: Common features and classification. *Polymer* **34**, 2050–2059.

Ungar, G., and Keller, A. (1986). Time-resolved synchrotron X-ray study of chain-folded crystallization of long paraffins. *Polymer* **27**, 1835–1844.

Ungar, G., and Keller, A. (1987). Inversion of the temperature dependence of crystallization rates due to onset of chain folding. *Polymer* **28**, 1899–1907.

Ungar, G., and Zeng, X.-B. (2001). Learning polymer crystallization with the aid of linear, branched and cyclic model compounds. *Chem. Rev.* **101**, 4157–4188.

Ungar, G., Stejny, J., Keller, A., Bidd, I., and Whiting, M. C. (1985). The crystallization of ultralong normal paraffins. The onset of chain folding. *Science* **229**, 386–389.

Ungar, G., Feijoo, J. L., Keller, A., Yourd, R., and Percec, V. (1990). Simultaneous X-ray/DSC study of mesomorphism in polymers with a semiflexible mesogen. *Macromolecules* **23**, 3411–3416.

Ungar, G., Zeng, X. B., Brooke, G. M., and Mohammed, S. (1998). Structure and formation of noninteger and integer folded-chain crystals of linear and branched monodisperse ethylene oligomers. *Macromolecules* **31**, 1875–1879.

Ungar, G., Mandal, P. K., Higgs, P. G., de Silva, D. S. M., Boda, E., and Chen, C. M. (2000). Dilution wave and negative-order crystallization kinetics of chain molecules. *Phys. Rev. Lett.* **85**, 4397–4400.

Ungar, G., Putra, E. G. R., de Silva, D. S. M., Shcherbina, M. A., and Waddon, A. J. (2005). The effect of self-poisoning on crystal morphology and growth rates. *Adv. Polym. Sci.* **180**, 45–87.

Varga, J. (1989). β-Modification of polypropylene and its two-component systems. *J. Therm. Anal.* **35**, 1891–1912.

Vaughan, A. S. (1993). Etching and morphology of poly(vinylidene fluoride) have been examined following the development of an etching technique which allows the study of representative morphologies in this polymer. *J. Mater. Sci.* **28**, 1805–1813.

Waddon, A. J., and Keller, A. (1990). A temperature window of extrudability and reduced flow resistance in high-molecular weight polyethylene; Interpretation in terms of flow induced mobile hexagonal phase. *J. Polym. Sci. Polym. Phys. Ed.* **28**, 1063–1073.

Waddon, A. J., and Keller, A. (1992). The temperature window of minimum flow resistance in melt flow of polyethylene: Further studies on the effects of strain rate and branching. *J. Polym. Sci. Polym. Phys. Ed.* **30**, 923–929.

Weinhold, S., Litt, M. H., and Lando, J. B. (1979). Oriented phase III poly(vinylidene fluoride). *J. Polym. Sci. Polymer Letter Ed.* **17**, 585–589.

Weinhold, S., Litt, M. H., and Lando, J. B. (1980). The crystal structure of the γ phase of poly (vinylidene fluoride). *Macromolecules* **13**, 1178–1183.

Weng, X., Li, C. Y., Jin, S., Zhang, J. Z., Zhang, D., Harris, F. W., Cheng, S. Z. D., and Lotz, B. (2002). Helical senses, liquid crystalline behavior, crystal rotational twinning in a main-chain polyester with molecular asymmetry and odd-even effects on its analogs. *Macromolecules* **35**, 9678–9686.

Wittmann, J. C., and Lotz, B. (1982). Crystallization of paraffins and polyethylene from the "vapor phase." A new surface decoration technique for polymer crystals. *Die Makromolekulare Chemie, Rapid Commun.* **3**, 733–738.

Wittmann, J. C., and Lotz, B. (1985). Polymer decoration: The orientation of polymer folds as revealed by the crystallization of polymer vapors. *J. Polym. Sci. Polym. Phys. Ed.* **23**, 205–226.

Wittmann, J. C., and Lotz, B. (1989). Epitaxial crystallization of monoclinic and orthorhombic polyethylene phases. *Polymer* **30**, 27–34.

Wittmann, J. C., and Lotz, B. (1990). Epitaxial crystallization of polymers on organic and polymeric substrates. *Prog. Polym. Sci.* **15**, 909–948.

Woo, E. M., Sun, Y. S., and Yang, C.-P. (2001). Polymorphism, thermal behavior, and crystal stability in syndiotactic polystyrene vs. its miscible blends. *Prog. Polym. Sci.* **26**, 945–983.

Woodward, A. E., and Morrow, D. R. (1968). Annealing of poly-1-butene single crystals. *J. Polym. Sci. Polym. Phys. Ed.* **6**, 1987–1997.

Wu, X. Z., Ocko, B. M., Sirota, E. B., Sinha, S. K., Deutsch, M., Cao, B. H., and Kim, M. W. (1993). Surface tension measurements of surface freezing in liquid normal alkanes. *Science* **261**, 1018–1021.

Wunderlich, B. (1973). "Macromolecular Physics. Volume I. Crystal Structure, Morphology, Defects." Academy Press, New York.

Wunderlich, B. (1980). "Macromolecular Physics. Volume III. Crystal Melting." Academic Press, New York.

Wunderlich, B., and Arakawa, T. (1964). Polyethylene crystallized from the melt under elevated pressure. *J. Polym. Sci. Polym. Phys. Ed.* **2**, 6397–6706.

Wunderlich, B., and Davidson, T. (1969). Extended-chain crystals. I. General crystallization conditions and review of pressure crystallization of polyethylene. *J. Polym. Sci. Polym. Phys. Ed.* **7**, 2043–2050.

Wunderlich, B., and Melillo, L. (1968). Morphology and growth of extended chain crystals of polyethylene. *Die Makromolekulare Chemie* **118**, 250–264.

Wunderlich, B., and Sullivan, P. (1962). Solution-grown polyethylene dendrites. *J. Polym. Sci.* **61**, 195–221.

Wunderlich, B., James, E. A., and Shu, T.-W. (1964). Crystallization of polyethylene from o-xylene. *J. Polym. Sci. Polym. Phys. Ed.* **2**, 2759–2769.

Wunderlich, B., Möller, M., Grebowicz, J., and Baur, H. (1988). Conformational motion and disorder in low and high molecular mass crystals. *Adv. Polym. Sci.* **87**, 1–137.

Xie, B., Shi, H., Jiang, S., Zhao, Y., Han, C. C., Xu, D., and Wang, D. (2006). Crystallization behaviors of n-nonadecane in confined space: Observation of metastable phase induced by surface freezing. *J. Phys. Chem. B* **110**, 14279–14282.

Xue, C., Jin, S., Weng, X., Ge, J. J., Shen, Z., Shen, H., Graham, M. J., Jeong, K.-U., Huang, H., Zhang, D., Guo, M., Harris, F. W., *et al.* (2004). Self-assembled "supra-molecular" structures via hydrogen bonding and aromatic/aliphatic microphase separation on different length scales in a symmetric-tapered bisamides. *Chem. Mater.* **16**, 1014–1025.

Yamamoto, T., Miyaji, H., and Asai, K. (1977). Structure and properties of a high pressure phase of polyethylene. *Japn. J. Appl. Phys.* **16**, 1891–1898.

Yamashita, M., Hoshino, A., and Kato, M. (2007). Isotactic poly(butene-1) trigonal crystal growth in the melt. *J. Polym. Sci. Polym. Phys. Ed.* **45**, 684–697.

Yan, S., Katzenberg, F., Petermann, J., Yang, D., Shen, Y., Straupe, C., Wittmann, J. C., and Lotz, B. (2000). A novel epitaxy of isotactic polypropylene (α phase) on PTFE and organic substrates. *Polymer* **41**, 2613–2625.

Yandrasits, M. A., Cheng, S. Z. D., Zhang, A., Cheng, J., Wunderlich, B., and Percec, V. (1992). Mesophase behavior in thermotropic polyethers based on the semiflexible mesogen 1-(4-hydroxyphenyl)-2-(2-methly-4-hydroxyphenyl)ethane. *Macromolecules* **25**, 2112–2121.

Yashioka, A., and Tashiro, K. (2003). Polymer-solvent interactions in crystalline δ form of syndiotactic polystyrene viewed from the solvent from the solvent-exchange process in the δ form and the solvent evaporation phenomenon in the thermally induced δ-γ phase transition. *Macromolecules* **36**, 3593–3600.

Yasuniwa, M., Nakafuku, C., and Takemura, T. (1973). Melting and crystallization process of polyethylene under high pressure. *Polym. J. (Japan)* **4**, 526–533.

Yasuniwa, M., Enoshita, R., and Takemura, T. (1976). X-ray studies of polyethylene under high pressure. *Japn. J. Appl. Phys.* **15**, 1421–1428.

Yokouchi, M., Sakakibara, Y., Chatani, Y., Tadokoro, H., Tanaka, T., and Yoda, K. (1976). Structures of two crystalline forms of poly(butylene terephthalate) and reversible transition between them by mechanical deformation. *Macromolecules* **9**, 266–273.

Yoon, Y., Zhang, A. Q., Ho, R.-M., Cheng, S. Z. D., Percec, V., and Chu, P. (1996a). Phase identification in a series of liquid crystalline TPP polyethers and copolyethers having highly ordered mesophase structures. 1. Phase diagrams of odd-numbered polyethers. *Macromolecules* **29**, 294–305.

Yoon, Y., Ho, R.-M., Moon, B.-S., Kim, D., McCreight, K. W., Li, F., Harris, F. W., Cheng, S. Z. D., Perce, V., and Chu, P. (1996b). Mesophase identifications in a series of liquid crystalline biphenylylphenylpropane polyethers and copolyethers having highly ordered mesophase structures 2. Phase diagrams of even-numbered polyethers. *Macromolecules* **29**, 3421–3431.

Zeng, X., and Ungar, G. (1998). Lamellar structure of non-integer folded and extended long-chain n-alkanes by small-angle X-ray diffraction. *Polymer* **39**, 4523–4533.

Zeng, X. B., Ungar, G., Spells, S. J., and King, S. M. (2005). Real-time neutron scattering study of transient phases in polymer crystallization. *Macromolecules* **38**, 7201–7204.

Zhang, J., Yang, D., Thierry, A., Wittmann, J. C., and Lotz, B. (2001). Isochiral form II of syndiotactic polypropylene produced by epitaxial crystallization. *Macromolecules* **34**, 6261–6267.

Zhou, W. S., Cheng, S. Z. D., Putthanarat, S., Eby, R. K., Reneker, D. H., Lotz, B., Magonov, S., Hsieh, E. T., Geerts, R. G., Palackal, S. J., Hawley, G. R., and Welch, M. B. (2000).

Crystallization, melting, and morphology of syndiotactic polypropylene fractions. 4. *In-situ* lamellar single crystal growth and melting in different sectors. *Macromolecules* **33**, 6861–6868.
Zhou, W., Weng, X., Jin, S., Rastogi, S., Lovinger, A. J., Lotz, B., and Cheng, S. Z. D. (2003). Chain orientation and defects in lamellar single crystals of syndiotactic polypropylene fractions. *Macromolecules* **36**, 9485–9491.
Zimmermann, H. J. (1993). Structural analysis of random propylene-ethylene copolymers. *J. Macromol. Sci. Part B: Phys.* **32**, 141–161.
Zwiers, R. J. M., Gogolewski, S., and Pennings, A. J. (1983). General crystallization behavior of poly (L-lactic acid) PLLA: 2. Eutectic crystallization of PLLA. *Polymer* **24**, 167–174.

CHAPTER SIX

Interdependence of Metastable States on Different Length Scales

Contents

The major focus of this chapter is the correlation of metastable states on different length scales. In particular, the interdependence of metastable states on different length scales needs to be understood in order to explain complicated experimental observations. This chapter starts with the topic of two polymorphs each with its own phase-stability (melting temperature) dependence on phase-size (lamellar thickness). The phase-stability inversion between two polymorphs can be caused by the phase size, and in polymer crystals by the lamellar thickness changes. The next topic is liquid–liquid phase separation coupled with vitrification in polymer–solvent mixtures. As soon as the temperature is cooled to the cross-over point between the binodal curve and the glass transition depression line of this two-component system, the polymer-rich phase vitrifies and becomes compositionally invariant. Another interesting class of examples is the liquid–liquid phase separation of a semi-crystalline polymer–solvent mixture or a polymer blend coupled with crystallization of one of the polymeric components. Depending upon the relative positions of the binodal curve and melting temperature depression line, they could intersect, or liquid–liquid phase separation could be underneath the melting temperature depression line, causing this phase separation to become entirely metastable. More complicated examples combine three different processes, such as liquid–liquid phase separation, gelation, and crystallization. Within each of these three processes, metastable states exist with their own physical origins. Yet, on larger scales, these processes can either intersect or override each other. Complicated crystalline and phase morphologies, as well as some materials' properties, can be observed when the systems are developing toward ultimate stability via different metastable states.

1. Combining Phase Size Effects with Polymorphs

1.1. Phase-stability changes in polymorphs based on phase sizes

In most of the common semi-crystalline polymer systems studied, metastable states on different length scales are often interdependent. On the smallest length scale discussed here, the crystal symmetry and structure of polymer lamellar crystals can be defined by following crystallographic rules. For a given polymer, a range of polymorphs in the conventional sense is possible. For example, polyethylene has three polymorphs, which possess orthorhombic, monoclinic, and hexagonal crystal structures. As described in Section 3 of Chapter 3, all but one polymorph is metastable at a fixed temperature and pressure. However, each crystal structure can also be associated with different lamellar thicknesses on a larger length scale, as determined by supercooling (see the discussion in Section 2.3 of Chapter 4). We can thus have a hierarchy of metastabilities, based on the interdependence of one class of metastability related to the crystal lattice and another related to the lamellar thickness. With respect to the latter point, the folded chain lamellar crystals of polymers and uniform oligomers in particular, straddle the boundary of these two categories of metastability.

This discussion implies that each crystallographic polymorph, with its lattice symmetry, can form small crystals. Therefore, each lamellar polymorph has its own size dependence according to the Thomson–Gibbs Eq. (4.6). This size dependence may be different for each polymorph, as determined by the fold surface free energy (γ_e) and heat of fusion (Δh_f). As shown in Fig. 4.18, the melting temperature is linearly related to the reciprocal of lamellar thickness. This linear relationship serves as the phase-stability boundary. The equilibrium melting temperature can be identified by extrapolating the size dependence of the melting temperature to zero reciprocal lamellar thickness (thickness approaches infinity), with the slope being proportional to the ratio between the fold surface free energy and the heat of fusion ($2T_m^0\gamma_e/\Delta h_f$). If we have a semi-crystalline homopolymer that exhibits two polymorphs, the equilibrium melting temperature of the metastable phase is lower than that of the stable phase, $(T_m^0)_{meta} < (T_m^0)_{st}$. Each of these two phases possess their own linear relationship between the melting temperature and reciprocal lamellar thickness, resulting in different phase-stability boundaries and giving rise to two slopes: $2(T_m^0)_{meta}(\gamma_e/\Delta h)_{meta}$ and $2(T_m^0)_{st}(\gamma_e/\Delta h)_{st}$. There are two possibilities for these two phase-stability boundaries. They can either intersect, resulting in an inversion of phase stability at smaller lamellar thicknesses, or not, indicating no phase-stability inversion as a function of phase size. A cross-over between these two phase-stability lines requires that the absolute value of the slope of metastable phase must be smaller than that of the stable phase, such that $\left|(T_{\rm m}^0)_{\rm meta}(\gamma_{\rm e}/\Delta h)_{\rm meta}\right| < \left|(T_{\rm m}^0)_{\rm st}(\gamma_{\rm e}/\Delta h)_{\rm st}\right|$. The cross-over point of these two lines defines a critical lamellar thickness, which suggests that a structurally metastable phase, as defined in Section 2.3 of Chapter 4, can become the stable one when this phase dimension is smaller than the critical lamellar thickness. Conversely, the stable phase can become metastable when it is smaller than the critical lamellar thickness. In other words, one can have a situation in which the stabilities invert with phase size. This description can be understood from the schematic illustration in Fig. 6.1. The possibility of stability inversion with size may have potential consequences in polymer crystallization (Cheng *et al.*, 1999; Keller and Cheng, 1998; Keller *et al.*, 1994, 1996).

The question to ask is: What is the thermodynamic basis of this phase size-stability inversion? Let us first investigate the relationship between free energy and temperature at constant pressure for a metastable phase, a stable ordered phase, and an isotropic melt in a one-component system as shown in Fig. 3.7A, which have been schematically redrawn in Fig. 6.2A. Note that in this figure, these three phases are of infinite size; thus, in thermodynamic equilibrium, only phase structure metastability is taken into account. At no temperature does the metastable phase possess the lowest free energy across the whole temperature range in Fig. 6.2A. In this figure, there are three transition temperatures. The transition between the stable ordered phase and the isotropic melt is the melting temperature, $(T_{\rm m}^0)_{\rm st}$, whereas the transition temperature between the metastable phase and the melt is the melting temperature, $(T_{\rm m}^0)_{\rm meta}$. The transition between the stable and metastable phases is the disordering temperature, $T_{\rm d}^0$; however, the disordering temperature is not observable and thus, is a virtual transition temperature.

There are two possible ways to make the metastable phase stable. One can either move the free energy line of the melt up to make the melt state less stable, as shown

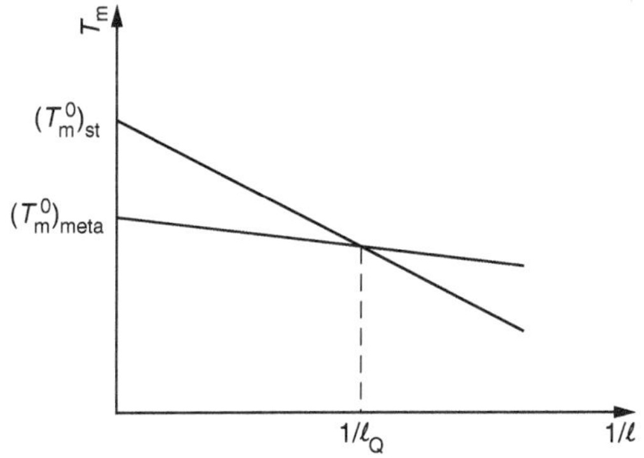

Figure 6.1 Schematic illustration of the relationship between the melting temperatures and reciprocal lamellar thicknesses for both the stable and metastable phases. The stable phase possesses a slope that is larger than that of the metastable phase. Therefore, the phase stability inverts between these two phases when the crystal thickness is smaller than $\ell_{\rm Q}$ [replotted from Cheng *et al.* (1999), with permission].

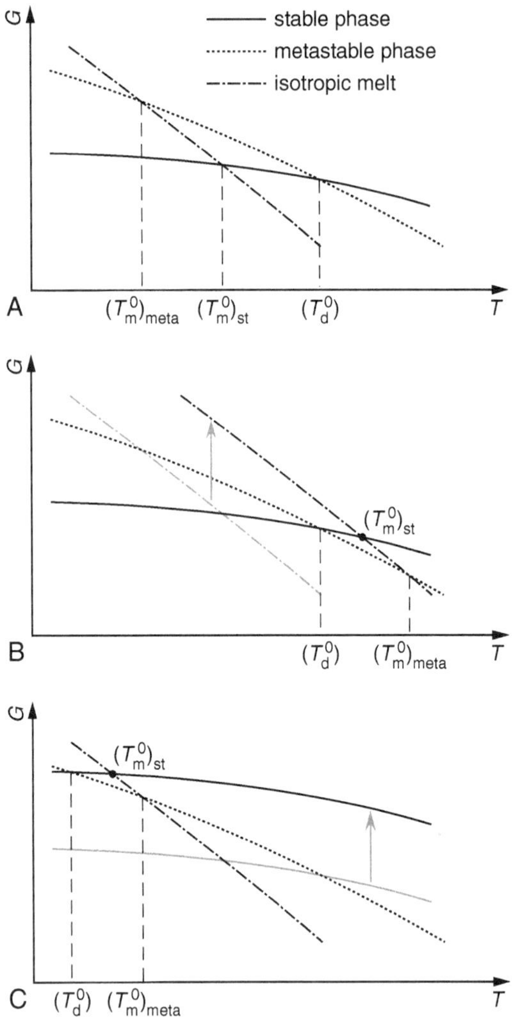

Figure 6.2 Schematic illustration of the free energy versus temperature at constant pressure: (A) a metastable phase which does not have the lowest free energy for the entire temperature region studied; (B) when the free energy line of the melt moves up, the metastable phase becomes stable in the temperature region between the disordering temperature, T_d^0; and the melting temperature $(T_m^0)_{meta}$; (C) when the free energy line of the stable ordered phase moves up, the metastable phase becomes stable in a temperature region between the disordering temperature, T_d^0, and the melting temperature, $(T_m^0)_{meta}$ [replotted from Keller *et al.* (1994), with permission].

in Fig. 6.2B, or move the free energy line of the stable phase up to make the stable phase less stable, as shown in Fig. 6.2C. In both cases, the originally metastable phase now becomes stable between the disordering temperature, T_d^0, and the melting temperature of the metastable phase, $(T_m^0)_{meta}$, where it can be, experimentally observable. Figure 6.2B implies that either the enthalpy increases or the entropy decreases in the melt state or a combination of both. This can be realized in most cases by applying high pressures to the system. The melting temperature increases as described by the Clausius–Clapeyron Eq. (3.1) (as long as the crystalline density is higher than the corresponding amorphous density). If the metastable phase is less affected by the pressure change, this phase may become thermodynamically stable in the temperature region between the disordering temperature, T_d^0, and the melting temperature of the metastable phase, $(T_m^0)_{meta}$, as shown in Fig. 6.2B. Other ways also exist, such as orienting the melt state or preventing the disorientation of an oriented solid state upon melting with external fields. An increase in the free energy line of the melt state can also be realized by tuning the chemical structures. For example, one can increase the rigidity of the chain molecules to reduce the entropy and/or increase molecular interactions, thus increasing the packing density and enthalpy (Keller *et al.*, 1994).

In order to move the free energy line of the stable ordered phase up, we can introduce imperfections into the ordered phase in one of the two ways: by introducing defects within the crystals and/or by decreasing the phase size (Cheng *et al.*, 1999; Keller and Cheng, 1998; Keller *et al.*, 1994). Experimental observations have shown that decreasing the phase size is much more effective at destabilizing a crystal phase due to the drastic increase in the surface free energy to bulk free energy ratio. As a consequence, the metastable phase becomes stable in a temperature region between the disordering temperature, T_d^0, and the melting temperature of the metastable phase, $(T_m^0)_{meta}$, as shown in Fig. 6.2C.

Now, our focus will be on lamellar crystals (which are the structurally stable phase) with different lamellar thicknesses. Assuming that no structural changes occur within the crystal or on the crystal surface, then, the differences in lamellar thicknesses represent and govern the pure size dependence of phase stability. Each lamellar thickness possesses its own free energy line, as shown in Fig. 6.3. With

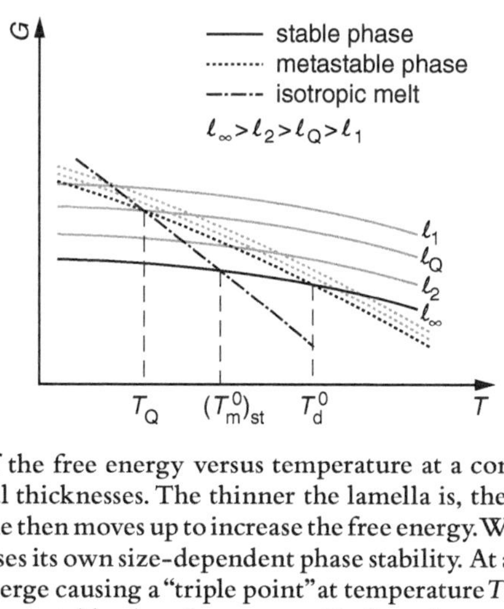

Figure 6.3 Illustration of the free energy versus temperature at a constant pressure for several metastable lamellar crystal thicknesses. The thinner the lamella is, the lower the stability of the crystal. The free energy line then moves up to increase the free energy. When we introduce another metastable phase, it possesses its own size-dependent phase stability. At a thickness of ℓ_Q, the three transition temperatures merge causing a "triple point" at temperature T_Q. For any lamellar thickness smaller than ℓ_Q, the metastable phase becomes stable for at least a finite temperature region [replotted from Keller *et al.* (1994), with permission].

decreasing lamellar thickness, the free energy line moves up, thus representing decreased crystal stability. Therefore, each crystal thickness has a corresponding melting temperature. In other words, the melting temperature represents the stability of the crystals. Correctly determining how the metastable crystal melting temperature depends on lamellar thickness requires that lamellar thickening does not take place. Thickening results in increasing crystal perfection, which changes the system's entropy. The melting temperature without thickening or perfecting is called the "zero-entropy production melting temperature," as described in Section 5 of Chapter 3 (Wunderlich, 1980). These temperatures are used as the melting temperatures in the Thomson–Gibbs Eq. (4.6).

We can then add the morphologically metastable phase into Fig. 6.3 on top of the structurally metastable phase. With decreasing phase size, such as lamellar thickness, the stability of this metastable phase decreases (the free energy line moves up in the figure). However, the size dependence of the metastable phase must be different from that of the stable phase, and in the case described in Fig. 6.1, we consider that the size dependence of the metastable phase is weaker than that of the stable phase. In order to simplify the discussion, we will temporarily exclude the size dependence of the metastable phase in Fig. 6.3. In terms of the melting temperatures, this difference in slopes as shown in Fig. 6.1 means that the melting temperature of the stable phase decreases faster than that of the metastable phase. In Fig. 6.3, decreasing the lamellar thickness to a value of ℓ_Q causes the melting temperature of the stable phase, $(T_m)_{st}$, the metastable phase, $(T_m)_{meta}$, and the disordering temperature, T_d, to be identical. This is the "triple point" shown in Fig. 6.1. When the lamellar thickness is thinner than the triple-point thickness, ℓ_Q, the melting temperature of the stable phase becomes lower than that of the metastable phase. Here the previously metastable phase becomes stable. A similar discussion can be carried out when the weak size dependence of the metastable phase is added, as shown in Fig. 6.3.

If now we plot the melting temperatures, $(T_m)_{st}$ and $(T_m)_{meta}$, and the disordering temperature, T_d, obtained in Fig. 6.3, against the reciprocal lamellar thicknesses, $1/\ell$, we obtain Fig. 6.4. This figure depicts the phase-stability dependence on size, as shown in Fig. 6.1; thus, it is a phase-stability diagram rather than a phase diagram which describes the phase stability at infinite size. There are now three transition temperature lines with respect to the reciprocal lamellar thickness that construct the phase-stability diagram of these phases with respect to the phase size. The various hatched areas represent the areas where the ordered stable or metastable phases are stable or metastable due to phase size changes.

In detail, below the reciprocal thickness of the "triple point," $1/\ell_Q$, is the stable phase-melt, $(T_m)_{st}$, transition line which is solid. With respect to the reciprocal thickness, it is the stable phase-stability boundary. Above this boundary is the dotted line formed by the transition temperature of the stable and metastable phases, T_d, with respect to the reciprocal thickness. This line is not experimentally observable except at the "triple point," so we call this line a virtual phase-stability boundary. Below the stable phase-stability boundary, the dashed line is the transition temperature between the metastable phase and the melt, $(T_m)_{meta}$, and, thus, the metastable phase-stability boundary. The location of these three phase-stability boundaries reverses sequence with respect to transition temperature in the region above the

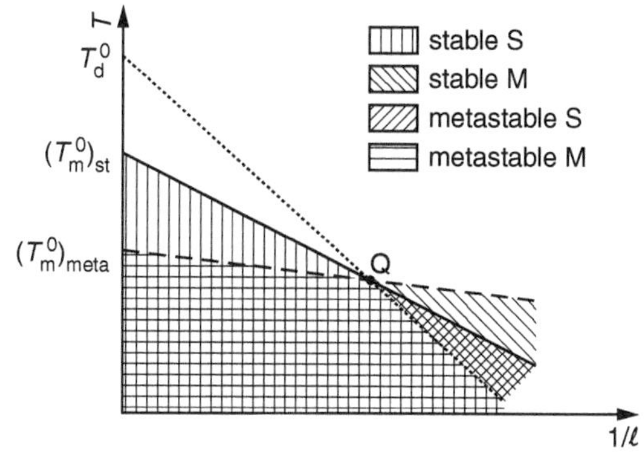

Figure 6.4 Size-dependent phase-stability relationships between transition temperatures and reciprocal thicknesses for three transitions: the structurally stable phase and the melt (solid line), the metastable phase and the melt (dashed line), and the stable and the metastable phases (dotted line). All these transition temperatures can be obtained based on Fig. 6.3. The "triple point," Q, represents a co-existence of three phases. The various hatched areas represent the phase regimes where the S and M phases can exist as a stable or metastable phase. The words stable and metastable refer to the relative phase stability at infinite size, while S and M refer to the relative phase stability at finite sizes [replotted from Keller *et al.* (1994), with permission].

"triple point" reciprocal thickness, $1/\ell_Q$. Namely, the stable phase-stability boundary is still in the middle, while the virtual phase-stability boundary is on the bottom with the metastable phase-stability boundary at the top. This phase-stability inversion originates from the slope difference, as shown in Figs. 6.1 and 6.4. The slopes for both the stable and metastable phases are $2(T_m^0)_{meta}(\gamma_e/\Delta h)_{meta}$ and $2(T_m^0)_{st}(\gamma_e/\Delta h)_{st}$; they can be rewritten in the form, $2(\gamma_e/\Delta s)_{meta}$ and $2(\gamma_e/\Delta s)_{st}$. Hence, the ratio of the fold surface energy and entropy change dominates the phase stability (Ungar, 1986).

Furthermore, these boundaries also provide information about stable and metastable phase regions in the different hatched areas shown in Fig. 6.4. Below the metastable phase-stability boundary at $1/\ell$ values below $1/\ell_Q$ and below the virtual phase-stability boundary at $1/\ell$ values above $1/\ell_Q$, the ordered stable phase is stable, and the metastable phase is metastable. In between the stable phase-stability and the metastable phase-stability boundaries below $1/\ell_Q$, only the stable phase exists. Above $1/\ell_Q$ in the area between the virtual phase and stable phase-stability boundaries, the stable phase is metastable, and the metastable phase is stable. Finally, in the area between the stable phase-stability boundary and the metastable phase-stability boundary above $1/\ell_Q$, only the stable metastable phase exists. This is a phase-stability inversion based on phase size.

Figure 6.1 may provide further information on the connection between phase stability and transformation rates that emerges from the size dependence of stability. The linear phase-stability boundaries plotted between the melting temperature, T_m, and reciprocal lamellar thickness, $1/\ell$, in this figure represent the minimum phase sizes that are stable at each temperature for both the stable and metastable phases. This boundary implies that it is the polymorph that is stable at the smallest size at a specific temperature that will appear first in isothermal phase growth. From the perspective of kinetics, the smallest stable size is also the smallest critical nucleus size. In turn, this critical nucleus size represents an aspect of the barrier for phase growth. The smaller the barrier height is, the faster the phase develops. As we know, existing treatments of metastability, including attempts to explain Ostwald's stage rule (1897), invoke the concept that smaller barrier heights leading to faster transformation rates as the reason for the prominence of metastable phase variants in phase transformations. With the concept of phase size causing stability inversion, one can see that the higher stability of the structurally metastable phase (defined at infinite size) at sufficiently small sizes and the faster rate of formation of this phase are interdependent. In fact, the two concepts are equivalent; they merely approach the issue from two different viewpoints. Accordingly, a phase evolves preferentially in its metastable form, not due to some inherent preference for metastability, but because of the small size of the phase at the beginning of the phase transformation. It is a further matter whether such a phase will retain the same structure throughout phase growth. If so, Ostwald's stage rule will be followed. Alternatively, it may transform into the state of ultimate stability, in which case the memory of the transient initial phase will be lost.

The size dependence of phase stability also significantly changes phase behaviors, which are described in phase-stability diagrams. Polyethylene will be used as an example. The polyethylene temperature–pressure phase diagram is shown in Fig. 5.28. In this figure, a "triple point" can be seen where the isotropic melt, the orthorhombic phase, and the hexagonal phase meet. Now, considering the phase-stability size dependence, this "triple point" can become phase size-dependent at sufficiently small sizes. The "triple point" can then be extended into a "triple line," as first treated by Dafay *et al.* in the case of a vapor–liquid–solid equilibrium (1966). In the case of polyethylene, it is expected that the phase-stability boundaries also move with decreased phase size, as shown in Fig. 6.5. The dashed line in this figure represents the "triple line" (Rastogi and Kurelec, 2000). Studies on this system have revealed an unforeseen singularity in the resulting expression for the size dependence of the "triple point," indicating that the inclusion of size as a variable

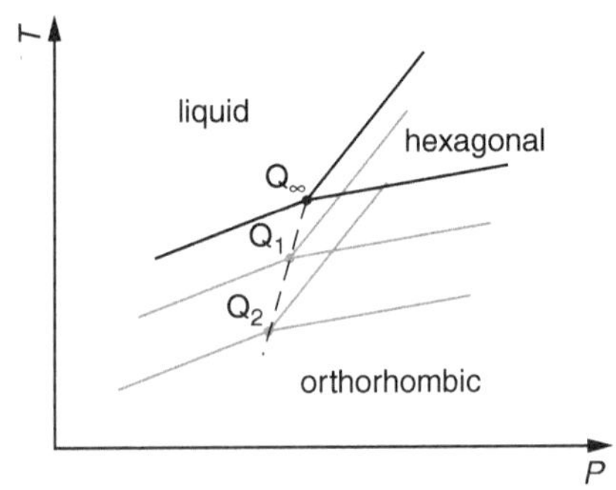

Figure 6.5 Size-dependent phase-stability diagrams for polyethylene at different temperatures and pressures. The dashed line connecting all the "triple points" represents the "triple line." It indicates that the "triple points" are now size dependent [replotted from Rastogi and Kurelec (2000), with permission].

can have a more profound effect than expected. This effect is potentially important for general phase behavior. However, the physical meaning of this singularity has not yet been determined.

Finally, small phase sizes can also be caused by external constraints. The combination of reduced size and altered surface conditions may lead to a shift in phase stability that could result in the development of a metastable phase not seen in a conventional unconstrained system, which is a case of a purely size-induced stability shift. The metastable phase becomes stable under external constraints, with the important difference that the constraints can usually be expressed in terms of thermodynamic variables and, thus, can be described with true phase diagrams (Evans, 1990). The constraints can be small cracks, gaps, or cavities in solids, such as the classical case of capillary condensation (Evans, 1990). In polymer systems, they can be the boundaries of nano-phases that arise from the localized liquid–liquid phase separation of block copolymers where the material constituting the nano-phase is capable of further transformations, such as crystallization or mesophase formation, as described in Section 3.4 of Chapter 4. These processes then must take place within the preexisting confined nano-phases. For example, it was reported that in radiation cross-linked polyethylene at elevated pressure, the "triple point" moves with radiation dose–pressure superposition (Vaughan *et al.*, 1985). Another interesting example of constraints arises with polymer–clay hybrid composites, whose constraint size is less precisely controlled but technologically important. Polymers are intercalated between thin flakes of clay in which the spacing between these layers can be on the range of a few nanometers. There is an early report on nylon 6–clay composites where the nylon 6 crystals are in the metastable γ phase (Kojima *et al.*, 1994). While the authors of this report have not commented on it from the point of view of thermodynamic stability, such observations tend to be interpreted as being a consequence of a constrained-size-induced shift in thermodynamic stability. In fact, because the clay-layer separation is controllable, such systems should lend themselves to a systematic exploration of the size-induced stability inversion and of the controlled production of metastable crystal polymorphs.

Based on the above description, it can be seen that metastability is much more complex and by the same token, much richer in variety than it may appear at first sight. In the absence of the appropriate vocabulary to adequately describe the phenomenon involved on different length scales, a new nomenclature and the use of presently unavoidable, self-contradictory terms, such as "stable metastable" and "metastable stable," have to be utilized. In this way, hopefully the subtle diversity of conceptual issues expressed by this set of nomenclature and definitions can then be translated by the readers into the corresponding effects and structures which arise in actual systems.

1.2. Examples of phase inversion by crossing over the phase-stability boundaries

Let us first carefully understand phase inversion by crossing over the crystal stability boundaries with respect to lamellar thickness in polyethylene. As described in Section 2.2 of Chapter 5, polyethylene has two crystal polymorphs: the orthorhombic and the hexagonal phases (ignoring, for present purposes, the mechanically induced monoclinic polymorph). Originally, the extended chain crystals of polyethylene were observed in experiments at elevated pressures (Geil *et al.*, 1964; Prime and Wunderlich, 1969; Rees and Bassett, 1968, 1971; Wunderlich and Arakawa, 1964; Wunderlich and Melillo, 1968; Wunderlich and Davidson, 1969). It was then recognized that the characteristic extended chain crystal morphology developed upon crystallization is the hexagonal phase (Bassett and Turner, 1972; Yasuniwa *et al.*, 1973, 1976). Subsequently, it was observed that the crystals can start growing in the hexagonal phase below the "triple point" in the region where the orthorhombic phase is stable and the hexagonal phase is metastable (Bassett and Turner, 1974a,b). The resulting temperature–pressure phase diagram is shown in Fig. 5.28. We can plot a schematic temperature–pressure equilibrium phase diagram, as shown in Fig. 6.6 (Hikosaka *et al.*, 1992).

In this figure, three transition temperatures at different pressures are included both above and below the "triple point." They are the transition temperatures between the orthorhombic and the hexagonal phases, the orthorhombic phase and the melt, and the hexagonal phase and the melt. Above the "triple point," the region is bounded by two transition temperature lines. The lower limit of this region is bounded by the transition temperatures from the orthorhombic phase to the hexagonal phase, and the upper limit is the transition temperature from the hexagonal phase to the isotropic melt. Within this region, the hexagonal phase is the only stable phase. Below the "triple point," we can find another region which is bounded by both the melting temperatures of the hexagonal phase (the lower limit) and the orthorhombic phase (the upper limit) to the isotropic melt. The orthorhombic phase is the only stable phase in this region. Therefore, the formation of the hexagonal metastable phase below the "triple point" is possible only underneath

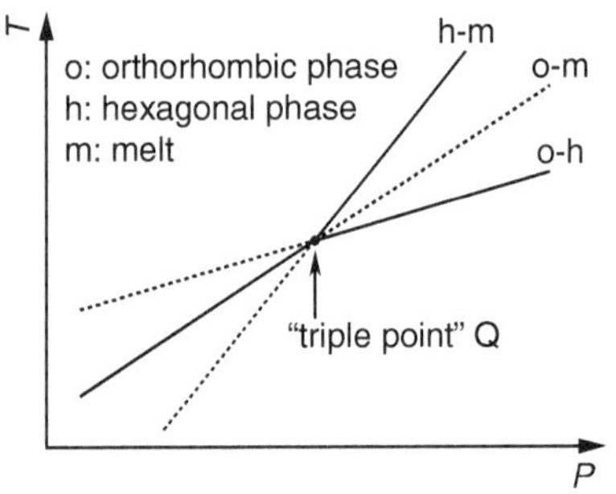

Figure 6.6 Schematic temperature–pressure phase diagram of polyethylene at thermodynamic equilibrium. Above the "triple point" at high temperatures and pressures, there is a region in which the hexagonal phase is stable. This region is demarcated by two transition temperature lines. The lower limit of this region is bounded by the transition temperatures from the orthorhombic phase to the hexagonal phase, and the upper limit is the transition temperature from the hexagonal phase to the isotropic melt. Below the "triple point," we can also find a region which is bounded by the melting temperatures of the hexagonal phase (the lower limit) and the orthorhombic phase (the upper limit) to the isotropic melt [replotted from Hikosaka *et al.* (1992), with permission].

the boundary constructed by the melting temperature of the hexagonal phase to the isotropic melt.

These two ordered structural forms differ not only in terms of symmetry and atomic position, but are also two different states of matter. In the orthorhombic structure, the chains form a regular crystalline lattice, while the hexagonal structure is a mesophase with large molecular mobility along the chain direction. As a consequence of this difference, the orthorhombic phase is the representative structure of folded chain crystals that grow only laterally with a fixed lamellar thickness determined by the supercooling. On the other hand, the hexagonal phase has extended chain crystals, and the crystals also keep growing in the thickness direction. The growth is terminated only by crystal impingement (Hikosaka *et al.*, 1992).

Later studies revealed that slightly below the "triple point" and the hexagonal-isotropic melting temperature boundary, crystallization always starts in the metastable hexagonal phase and proceeds in both the lateral and the thickness directions until a hexagonal-to-orthorhombic phase transformation takes place. In other words, a solid–solid transformation occurs from the metastable hexagonal phase into the ultimately stable orthorhombic phase (see Fig. 6.6). At the same time, a remarkable observation was made that the crystal growth along both the thickness and the lateral directions stops after the hexagonal-to-orthorhombic phase transformation occurs (Hikosaka *et al.*, 1992; Rastogi *et al.*, 1991).

Figure 6.7 shows a set of real-time polarized optical microscopy images to illustrate the assertions made above. At this level of magnification, only the lateral dimensions can be quantitatively measured, while the identification of the orthorhombic and hexagonal phases have been based on optical criteria combined with real-time wide angle x-ray diffraction results (Hikosaka *et al.*, 1992; Rastogi *et al.*, 1991). The original, untransformed hexagonal phase, which is seen growing, begins to melt when the temperature is raised to above the initial crystallization temperature, while the transformed orthorhombic phase, which has stopped growing, remains un-melted. Therefore, even without any reference to particular structures, this is evidence that the phase which starts growing is metastable compared to the phase that has ceased to grow due to the transformation from the hexagonal to the orthorhombic phase. In other words, crystallization only occurs in the metastable state on the time scale of the observation. It must be noted that when the crystallization is carried out above the "triple point" pressure and temperature where the hexagonal phase is stable, the melting temperature of the hexagonal phase is higher than that of the orthorhombic phase, as shown in Fig. 6.6. However, when the crystallization occurs below the "triple point" pressure and temperature underneath the hexagonal-isotropic melting temperature boundary, as shown in Fig. 6.7, the melting temperature of the hexagonal phase is lower than that of the orthorhombic phase. This is another piece of evidence for the phase-size driven stability inversion.

These assessments are clearly shown in Fig. 6.7. Figure 6.7A–D corresponds to different crystallization times (62 min in Fig. 6.7A to 101 min in Fig. 6.7D) for a polyethylene sample crystallized at a constant temperature, corresponding to a supercooling of 7 °C and at a pressure of 2.82 kbar. In these images, our attention is focused on the four crystals labeled A, B, C, and D. With increasing time, both

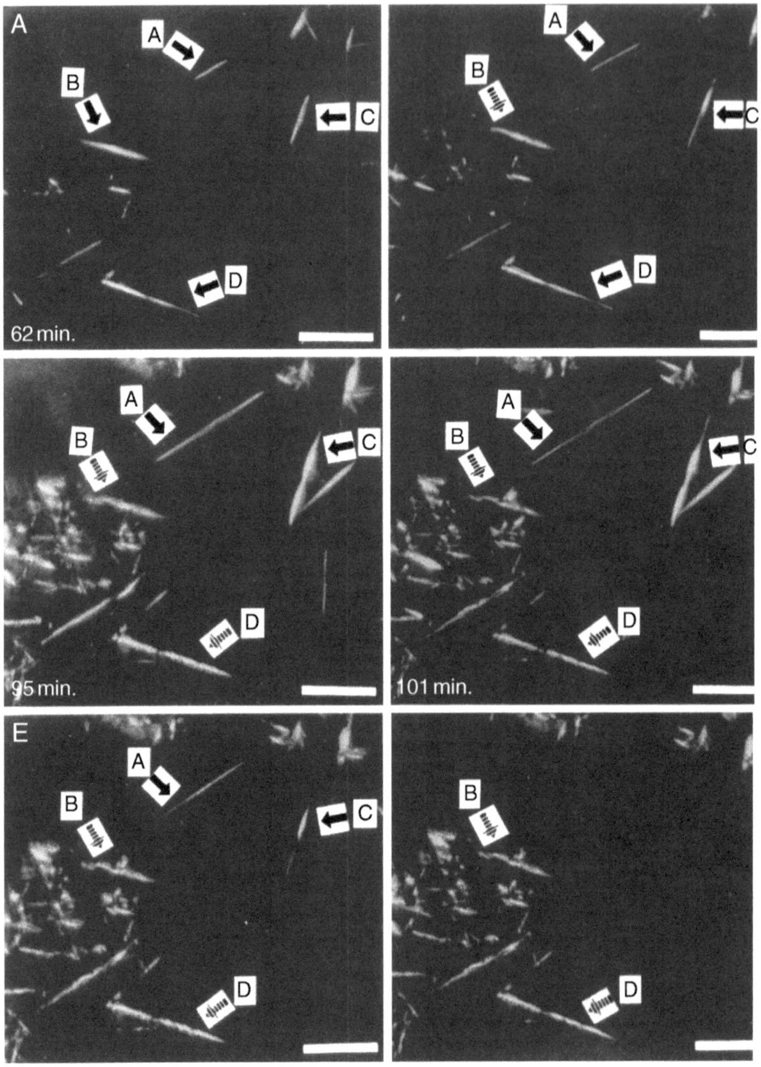

Figure 6.7 Real-time polarized optical microscopic images of a polyethylene sample crystallized at an elevated pressure of 2.82 kbar. (A–D) are the images taken when the sample was crystallized at a constant temperature that corresponds to a supercooling of 7 °C at different times: (A) 62 min, (B) 70 min, (C) 95 min, and (D) 101 min. Crystals A and C grow faster than crystals B and D. (E) and (F) are two images after the temperature of the sample was changed to correspond to a supercooling of 2 °C. The isothermal time of (F) is longer than that of (E). At this temperature, crystals A and C gradually melt, while crystals B and D are retained. The scale bars represent 50 μm [reprinted from Hikosaka *et al.* (1992); Rastogi *et al.* (1991), with permission].

crystals A and C grow faster than crystals B and D. The sample was then heated up to a temperature that corresponds to a supercooling of 2 °C, as shown in Fig. 6.7E and F. Figure 6.7F represents a longer residence time at this supercooling compared to Fig. 6.7E. Both A and C gradually disappear, while crystals B and D remain

(Hikosaka *et al.*, 1992; Rastogi *et al.*, 1991), revealing that A and C are metastable (hexagonal phase) with respect to B and D (orthorhombic phase).

Since Fig. 6.7 is a real-time experimental record, we can measure the size-change of these crystals to determine the growth rates of both the hexagonal and orthorhombic phases. Figure 6.8 illustrates these kinetic results. This figure records the growth of two crystals (1 and 2). At a temperature that corresponds to a supercooling of 5.2 °C, these two crystals at 3.2 kbar have the same growth rate (identical slopes in this figure). The isothermal temperature was then quickly elevated to a temperature that corresponds to a 2.5 °C supercooling. The growth rate of crystal 1 suddenly decreases to zero (no size change) after isothermally crystallizing for 26 min. Even before the temperature increase, the growth of crystal 1 had stopped. On the other hand, the growth rate of crystal 2 responds to the temperature change. As soon as the temperature is changed, the growth rate becomes negative, indicating that the crystal is melting for 28 min, as shown in Fig. 6.8. The isothermal temperature is then rapidly decreased back to a 5.9 °C supercooling. Crystal 1 still maintains its size, while the growth rate of crystal 2 increases again (Hikosaka *et al.*, 1992; Rastogi *et al.*, 1991). In fact, this figure provides quantitative evidence that compared with the ultimately stable orthorhombic phase, the ultimately metastable hexagonal phase possesses higher stability at small sizes, and this phase also grows at a faster rate, which is the kinetic counterpart of higher stability. This process accounts for Ostwald's stage rule (1897), and above all, encompasses the phase-stability inversion effect itself.

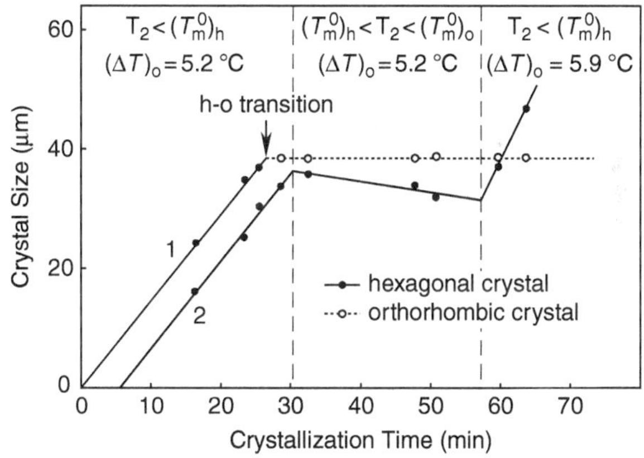

Figure 6.8 Crystal size changes with crystallization temperature and time at an elevated pressure of 3.2 kbar. The data were from real-time polarized optical microscopy measurements. In the first 31 min, the isothermal crystallization was conducted at a supercooling of 5.2 °C. Both crystals 1 and 2 grow at the same rates. However, at a time of 26 min, crystal 1 abruptly changed its growth rate, indicating a transition from a hexagonal to an orthorhombic phase. In the following 28 min, the supercooling was decreased to 2.5 °C. Crystal 1 does not grow, while the rate of crystal 2 becomes negative indicating a melting process. Only after the supercooling was increased to 5.9 °C again does crystal 2 resume growing, while the crystal 1 maintains a constant size [replotted from Hikosaka *et al.* (1992); Rastogi *et al.* (1991), with permission].

The important question is: Why does lateral growth stop after the hexagonal-to-orthorhombic phase transition? It appears that the conditions are still perfect for the nucleation and growth of thinner hexagonal lamellae on the thicker orthorhombic substrates. This may be one of the best proofs for "self-poisoning," as discussed in Section 1.5 of Chapter 4 (Organ *et al.*, 1989; Putra and Ungar, 2003; Ungar and Keller, 1986, 1987; Ungar and Zeng, 2001; Ungar *et al.*, 2005). In the hexagonal phase, sliding diffusion along the thickness direction causes the growth front to be covered with crystallographically irregular chains. Once in the orthorhombic phase, incorrectly folded chains are pinned hard to the substrate surface and cannot refold. There cannot be a sizable hexagonal layer on an orthorhombic substrate, since the hexagonal-to-orthorhombic transition front sweeps through right to the substrate surface.

All these results are unambiguous for the phase growth along the lateral direction. This condition equally applies to the growth in the thickness direction, as evidenced by quantitative measurements with transmission electron microscopy (Hikosaka *et al.*, 1997, 2000), even though the polarized optical microscopy images already have made this qualitatively apparent (Hikosaka *et al.*, 1992; Rastogi *et al.*, 1991). The explanation of this thickening growth requires introducing size (ℓ) as the stability controlling factor. We shall now apply this principle to the special case of polyethylene.

By a number of indicators, lateral growth, as observed in Figs. 6.7 and 6.8, is interrelated to thickening. There are two possible reasons for the interdependence of the lateral growth and thickening in an isolated lamellar single crystal. First, lateral growth and thickening simultaneously cease immediately after the phase transformation from the hexagonal to the orthorhombic phase. Second, lateral growth continues at a constant thickness (Gutzow and Toschev, 1968). It has been observed that in a temperature and pressure region which is somewhat lower than that pertaining to Figs. 6.7 and 6.8, following the hexagonal-to-orthorhombic transformation, lateral growth could continue after cessation of the thickening through branching of lamellae (Hikosaka *et al.*, 1995; Rastogi and Kurelec, 2000).

The concept of stability inversion, which is expressed in Figs. 6.1 and 6.4, becomes readily applicable to polyethylene if the phases, there referred to as stable and metastable, are taken as the orthorhombic and hexagonal phases, respectively, and the stability determining dimension, ℓ, is taken as the lamellar thickness. Figure 6.9 illustrates isothermal crystal growth in terms of a phase-stability diagram with notation appropriate to polyethylene. First, one should note the fact that stability inversion with size applies only to temperatures that are lower than the "triple point" temperature, T_Q. Above this temperature, the ultimately stable orthorhombic phase directly grows isothermally from the isotropic melt in the conventional way, as shown by the top horizontal arrows in regime A of Fig. 6.9. However, below the "triple point" temperature T_Q, the phase that evolves first in polyethylene isothermal experiments is the stable hexagonal phase due to its small size (note that the hexagonal phase is metastable at infinite size). This small size is the kinetically defined critical nucleus (see e.g., Section 1.5 of Chapter 4). As the hexagonal phase thickens to larger than size, ℓ_d, it transforms into the orthorhombic phase, which is ultimately stable. This transformation phase is shown by the bottom

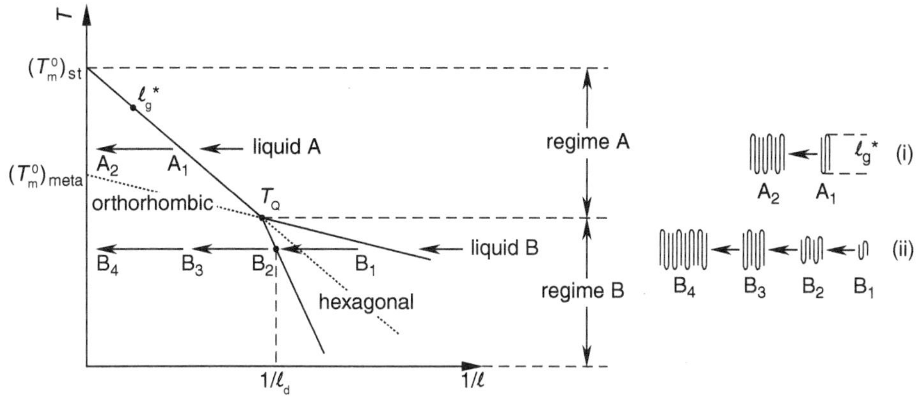

Figure 6.9 Crystal growth at an isothermal temperature versus the reciprocal lamellar thickness. The phase-stability diagram is identical to Fig. 6.4. The solid lines represent the stable phase demarcation lines and the dashed lines are metastable phase demarcation lines. The two sets of horizontal arrows toward the value of $1/\ell = 0$ are two isothermal crystallization processes at different temperatures. One is above the temperature of the "triple point" in regime A, and another is below in regime B. The horizontal move indicates thickness growth. The critical nucleus thickness, ℓ^*, determines the size of limiting stability (critical nuclei) of the respective phases. The molecular schematic of (i) illustrates a traditionally envisioned mode of growth, which is exclusively lateral at a fixed, kinetically determined thickness ℓ_g^*, where $\ell_g^* > \ell_A^*$ in regime A, and (ii) shows simultaneous growth in both the lateral and thickness directions. The thickness growth is terminated by the hexagonal-to-orthorhombic transformation somewhere in the orthorhombic phase stable region along the arrow in regime B [replotted from Keller *et al.* (1994), with permission].

set of horizontal arrows in regime B of Fig. 6.9. It is dependent upon the kinetics of the solid-state transformation which leaves two alternatives. First, there is no solid-state transformation on the experimental time scale, and the end result is a macroscopic metastable crystal. Although this is a possible explanation for the existence of metastable polymorphs in most simple substances, it certainly does not happen in polyethylene. Experience has shown that the hexagonal phase cannot be "quenched in," so it always transforms into the orthorhombic phase. Second, when growing past the lamellar thickness of ℓ_d, the solid-state transformation leading to the phase of ultimate stability sets in. Therefore, one expects all traces of the growth history to be erased, leaving no memory of the initial hexagonal phase. This last statement, while true for most crystals, does not apply to polymers such as polyethylene. Specifically, in the hexagonal-to-orthorhombic phase transformation, thickening growth stops, and the thickness prevailing at the transformation becomes a permanent feature of the morphology.

The above discussion, when it applies, has two wide-ranging consequences. The first is that we have a morphological indicator marking the size where the structure changes, providing evidence that it started growing with a different structure than the initial one. The second consequence of particular importance for polymers is that it provides an explanation for the limited finite lamellar thickness of folded chain polymer crystals. The reason for this is clear from the sketches under regime B in

Fig. 6.9. We emphasize that this need not be the only explanation and need not exclude the presently held views embodied by the kinetic theories of chain folding indicated by the sketches under regime A in Fig. 6.9. In fact, these theories would apply, as before, at crystallization temperatures above the temperature of the "triple point," T_Q, in regime A. The new picture is generalized from the hexagonal–orthorhombic phase transformation in polyethylene arresting thickening at crystallization temperatures below the temperature of the "triple point," T_Q, in regime B, if this growth does exist. It can be shown that the well-documented reciprocal supercooling dependence, $1/\Delta T$, of the lamellar thickness, ℓ, follows from both mechanisms, and the two cannot be distinguished or tested on that basis alone (Hikosaka *et al.*, 1995).

As we now know, folded chain crystal lamellae of limited thickness are metastable because of their small thickness, irrespective of the stability or metastability of the crystal structure. We draw attention again to the hierarchy of metastability concept we are developing in this chapter. The present example of polyethylene is an illustration to show the intricate interplay between stability and metastability on different length scales in a way that is relevant to polyethylene and similar polymers. Namely, the interdependence between these two different length scales (crystal structure and lamellar thickness) takes a unique form. For thicknesses when the hexagonal structure is stable, the crystal thickening growth will counteract the effect of size-determined stability until such growth ceases, as a result of the hexagonal–orthorhombic transformation. Therefore, the thickening growth, possible only in the hexagonal phase, is self-terminating. As a consequence, this phase structure determines the lamellar thickness on the larger length scale.

At this stage, a question needs to be raised as to the relationship in Fig. 6.4 (and Fig. 6.9) with respect to the equilibrium temperature–pressure phase diagram shown in Figs. 5.28 and 6.6. In fact, these two figures represent a section in a three-dimensional pressure–temperature–reciprocal thickness phase-stability diagram when the reciprocal thickness is zero ($1/\ell = 0$): namely, the lamellar thickness is infinite ($\ell \rightarrow \infty$). A speculative general pressure–temperature–reciprocal thickness phase-stability diagram can be shown in Fig. 6.10, since at this stage we only know phase stabilities at elevated pressures. However, it is certain that the alternative temperature versus reciprocal thickness phase-stability diagrams, such as in Figs. 6.4 and 6.9, are sections of Fig. 6.10 at a constant pressure, where the pressure needs to be lower than the pressure of the "triple point," P_Q ($P < P_Q$), as shown in the equilibrium pressure–temperature phase diagram of Figs. 5.28 and 6.6. Figure 6.10 is, in fact, the complete representation of phase stability with the size (ℓ) [under the condition that the inequality $\left|(T_m^0)_{meta}(\gamma_e/\Delta h)_{meta}\right| < \left|(T_m^0)_{st}(\gamma_e/\Delta h)_{st}\right|$ is true]. We also see that the conventional "triple point" extends into a "triple line" that marks the singularities mentioned in our previous discussion.

A somewhat different, yet related, approach, aimed at providing support for the picture we discussed for polyethylene, is to establish the existence of a stable hexagonal phase region in a two-dimensional temperature–reciprocal thickness section of a three-dimensional pressure–temperature–reciprocal thickness phase-stability diagram, such as in Fig. 6.10. This approach is best accomplished by heating

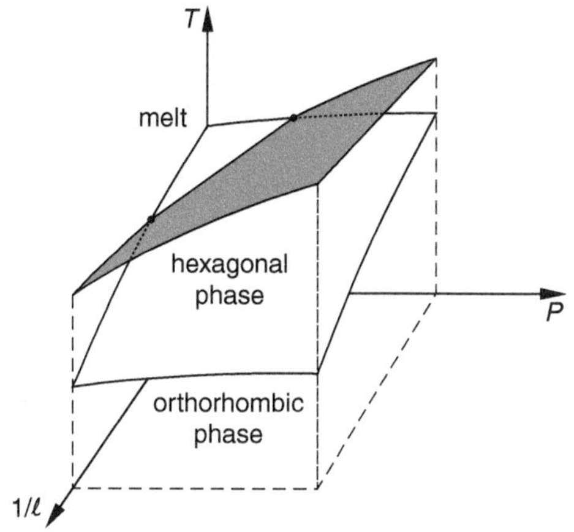

Figure 6.10 A three-dimensional temperature–pressure–reciprocal lamellar thickness phase-stability diagram for the polyethylene. The continuous shaded region (a volume in this three-dimensional space) is where the metastable hexagonal phase is stable. The section at which the pressure is zero yields the phase-stability diagrams of Figs. 6.1 and 6.4. The section where the reciprocal thickness equals to zero yields the phase diagram of Fig. 5.27. The "triple point" in all those figures now becomes the new "triple line," which defines the boundary below which the hexagonal phase can (or cannot) exist [replotted from Keller *et al.* (1994), with permission].

pre-existing crystals of specified thickness isobarically and monitoring their change as a function of temperature, using real-time wide angle x-ray diffraction experiments. It will be evident that for sufficiently small lamellar thicknesses, much smaller than the "triple point" thicknesses (see Fig. 6.4), the initial orthorhombic phase during heating should transform into a hexagonal phase before this small orthorhombic crystal melts. On the other hand, when the thickness is much thicker than that defined by the "triple point," the orthorhombic phase should pass directly into the melt without any intervening hexagonal phase.

The experimental results in Figs. 6.7 and 6.8 are all from elevated pressures which are slightly lower than the pressure of the "triple point." Now one of the remaining questions is: Can we perform such experiments at or close to atmospheric pressure? In other words, can we establish a phase-stability relationship between the pressures near both the "triple point" and atmospheric pressure? This question is important since we would like to prove that Fig. 6.10 is the real phase-stability diagram (with Figs. 6.4 and 6.9) and, thus, demonstrate the existence of a size-induced phase inversion. Efforts to extend experiments to lower pressures and temperatures even farther away from the equilibrium "triple point" were made by Rastogi and Kurelec (2000). Figure 6.11 shows an experimental temperature–pressure phase diagram for a polyethylene fraction with a weight average molecular weight of 120 kg/mol and a polydispersity of 1.2. Two sets of transition temperatures are observable. The higher temperature set is for the melting temperatures attributed to the transition from the orthorhombic phase to the isotropic melt, and the lower temperature set is for the melting temperatures attributed to the hexagonal phase-isotropic melt transition. These two sets of data construct the phase boundaries shown in Fig. 6.6 below the "triple point" (Rastogi and Kurelec, 2000). Since all of these melting

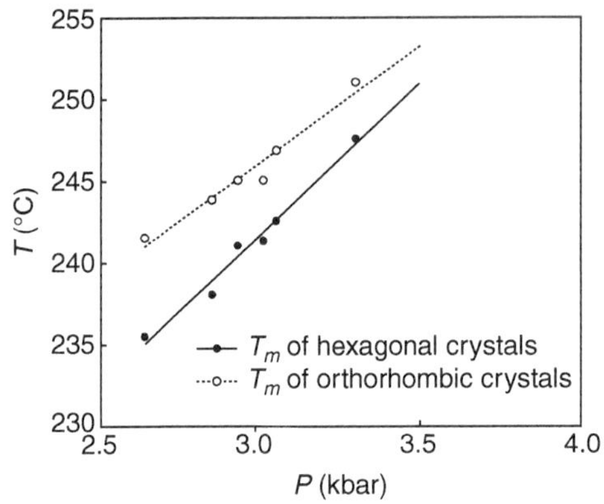

Figure 6.11 Temperature–pressure phase diagram for one polyethylene fraction with a weight average molecular weight of 120 kg/mol and polydispersity of 1.2. The melting temperatures were obtained utilizing the real-time isobaric melting of the single crystals [replotted from Rastogi and Kurelec (2000), with permission].

temperatures are for extended chain crystals, they are in thermodynamic equilibrium. The next step will be to obtain the temperature–pressure diagram for folded chain crystals, at least in the case of the orthorhombic phase. Therefore, one could systematically observe the size dependence at the "triple point" to form a "triple line" that has been deduced and analyzed here.

Obtaining experimental evidence to systematically understand the size dependence of the "triple point" is difficult. One of the practical problems which complicates the experiment, but at the same time provides further support for the whole picture, is that the thickness does not stay constant during the course of the heating experiments. The thickness has been found to increase; in fact, such an increase starts, or at least is greatly accelerated, upon entering the hexagonal region (Rastogi and Kurelec, 2000). On the one hand, this discovery has been fully consistent with the expected high chain mobility in the hexagonal phase, which induces or promotes lamellar thickening through chain refolding via sliding diffusion (Hikosaka, 1987, 1990). Note that this process corresponds to a secondary crystallization process, which is the perfection of an already formed crystal, in contrast to the previously considered thickening which involves the addition of new material to the growing crystal. Such crystal thickening can move the experimental pathway from the initially chosen thickness, which is much smaller than that at the "triple point," toward infinite thickness. In the case of thickening, we are changing both the lamellar thickness as well as the temperature, and therefore moving to smaller reciprocal lamellar thickness values, while the temperature moves toward higher values in Fig. 6.9. In work using synchrotron real-time x-ray diffraction, it is possible to control and follow the lamellar thickening with increasing temperatures, to a certain extent. In fact, it can be shown that as long as the lamellar thickness remained thinner than that at the "triple point," the hexagonal region could be exited by increasing the thickness, creating a hexagonal-to-orthorhombic phase transformation when moving horizontally along the reciprocal lamellar thickness axis in Fig. 6.9, or a hexagonal phase reentry, by moving vertically along the temperature axis in Fig. 6.9 with continued heating.

Does this polyethylene example represent some universal principle for polymer crystallization with polymorphs? There are two other examples with similar crystallization behavior. The first example is syndiotactic polystyrene in the polymorphic α-(trigonal) and β-(orthorhombic) phases crystallized from the melt (Cartier *et al.*, 1998; De Rosa, 1996; De Rosa *et al.*, 1992; Guerra *et al.*, 1990, 1991; Pradère and Thomas, 1990; Tosaka *et al.*, 1999). A detailed description of this polymorphism can be found in Section 2.1 of Chapter 5. Depending upon the crystallization conditions, such as crystallization temperature and residence temperature in the melt, pure α-or pure β-phases can grow. Thermodynamically, it was found that the β-phase is more stable than the α-phase at infinite sizes. Namely, the equilibrium melting temperature of the β-phase is higher than that of the α-phase. However, in metastable thin lamellar crystals, the melting temperature of the α-phase is experimentally higher than that of the β-phase, and the α-phase grows faster. With increased crystallization/annealing time or heating the sample to higher temperatures, the β-phase gradually develops, which may be an example of a phase-stability inversion with crystal lamellar thickening.

A metastability analysis similar to the case shown in Fig. 6.9 was conducted on this syndiotactic polystyrene system (Ho *et al.*, 2000). The α- to β-phase transformation was observed using wide angle x-ray and electron diffraction techniques. Both crystal phases have the identical *c* axis orientation, with three different crystallographic correlations between these two phase structures in the *ab* plane (Ho *et al.*, 2001). The issues that remain are determining how the α- to β-phase transformation takes place and locating the "triple point" of the phase-stability diagram. Does the frustrated trigonal α-phase have large mobility along the *c* axis, similar to the hexagonal phase in polyethylene? Molecular dynamics in the α-phase need to be investigated by utilizing solid state carbon-13 nuclear magnetic resonance and other experimental techniques. In addition, a quantitative measurement of the lamellar thickness changes coupled with the phase transformation is necessary to verify the phase-size-induced stability inversion.

Another example is the competition between the α-(monoclinic) and γ-(orthorhombic) crystal phases in isotactic polypropylene (Brückner and Meille, 1989; Lotz *et al.*, 1996; Turner-Jones *et al.*, 1964). It is known that isotactic polypropylene has many polymorphic behaviors, and the α- and γ-phases are two of the four modifications. In homo-polypropylene with very high isotacticity, it is difficult to form the γ-phase. Detailed formation conditions can be found in Section 2.1 of Chapter 5.

Since the equilibrium melting temperature of the γ-phase of isotactic polypropylene is slightly higher that that of the α-phase, and their heats of fusion and fold surface free energies are rather similar, the "triple point" is located at large lamellar thicknesses. This is why the γ-phase is difficult to form at usual crystallization conditions. However, increasing the ethylene comonomer content enhances the γ-phase content, which may be explained by a significant down-shift of the "triple point" (Foresta *et al.*, 2001). This is another example of transforming the "triple point" to a "triple line." Yet, in this case, it is due to the comonomer chemical composition change rather than changing crystallization conditions. Furthermore, there is no possibility of phase transformation from the α- to γ-phase, since the α-phase is a true crystal phase (monoclinic), and the chain mobility in this phase is very limited. Moreover, the γ-phase formation requires a completely different chain orientation in the crystal than the α-phase. This study also has not taken into account the fact that the γ-phase requires epitaxial growth on the α-phase, meaning that the α-phase is always formed before the γ-phase develops.

Finally, we return to the broader perspectives indicated in our discussions. Using our terminology, everything considered in the present section would count as "stable." Nevertheless, rigorous thermodynamic definition requires infinite size, such that the hexagonal phase observed at a pressure and temperature below the equilibrium phase transition temperatures in polyethylene, as shown in Figs. 5.28 and 6.6 (or the two-dimensional section of Fig. 6.10 where the reciprocal thickness approaches zero) would be classified as metastable. Accordingly, all the materials referred to in the present section spotlight the problem of terminology, which arises in the description and even in the definition of metastability. Nevertheless, there are regions within the phase-stability diagrams that are metastable by any criterion. These regions are defined in the legend of Fig. 6.4 (the hatched metastable and stable regions) with corresponding surfaces in the three-dimensional extension in Fig. 6.10 (not drawn). Such phase regions are truly metastable in the sense that they cannot be regarded as "stable" under the classical definition of stable, nor can they be metastable merely because of size considerations or externally imposed constraints. This condition has been applied throughout this section and also in some other sections of this chapter. This distinction between merely shifting stability–metastability boundaries and "true" metastability has, to our knowledge, not been meaningfully recognized in the past.

Of course, we accept that truly metastable phases (which are without external constraints, but are of infinite size) do exist, like all conventional polymorphs and without which Ostwald's stage rule (1897) would never have been formulated. Yet, as we have shown, it is possible to consider such metastable phases as arising through a size-induced shift of stability in the initial stages of their development, as opposed to a system having some intrinsic preference toward metastability. Without the recognition of such possibilities, the role and importance of a truly metastable phase, as unrelated to any shift in stability criteria due to whatever cause and at whatever stage of the phase evolution, cannot be assessed or even adequately discussed. It is hoped that by at least raising these issues, a step in this direction has been made.

1.3. Examples of phase inversion without crossing the phase-stability boundaries

Another possible phase-stability relationship encountered when applying the Thomson–Gibbs Eq. (4.6) to two polymorphs is that these two phases do not apparently invert their stability throughout their entire size range. This situation is illustrated in Fig. 6.12. This phase-stability relationship appears to be less exciting because the two phases possess similar size dependencies.

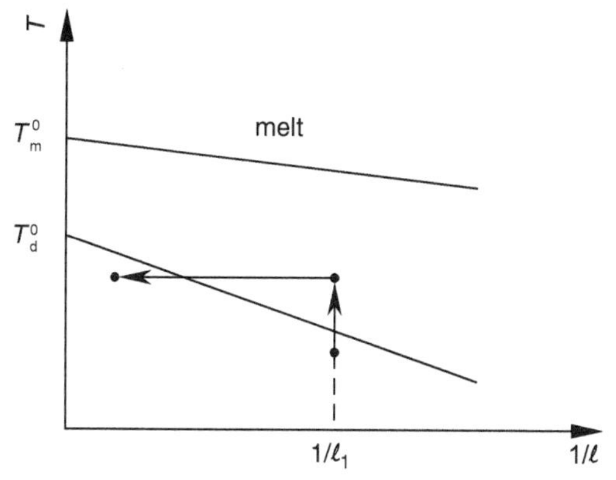

Figure 6.12 Schematic illustrations of two relationships between the melting temperatures and reciprocal lamellar thicknesses for both polymorphs. The absolute values of these two slopes are similar, and therefore no phase-stability inversion apparently occurs with changing phase size. The arrows represent an experimental pathway designed to observe the phase transformation; see text [replotted from Keller *et al.* (1994), with permission].

However, the size effect on phase stability can also be identified in this case. As shown in Fig. 6.12, two important features are: First, the equilibrium melting temperature of the stable phase is always higher than that of the metastable state at thermodynamic equilibrium (here, the lamellar thickness is infinite), such that $(T^0_m)_{st} > (T^0_m)_{meta}$. Second, the absolute value of the slopes of both the phase-stability boundaries are similar, $\left|(T^0_m)_{meta}(\gamma_e/\Delta h)_{meta}\right| \approx \left|(T^0_m)_{st}(\gamma_e/\Delta h)_{st}\right|$. Here, the phase-stability boundaries are more or less parallel to each other, and they never intersect. Therefore, regardless of the lamellar thickness, either small or large, the melting temperature of the metastable phase is always lower than that of the stable phase.

An example of this kind of phase relationship can be found in *trans*-1,4-polybutadiene. This polymer possesses two polymorphs. At low temperatures, a monoclinic crystal phase has been identified (Iwayanagi *et al.*, 1968). This phase transforms at around 70 °C, depending on the lamellar thickness, to a hexagonal phase that possesses considerable conformational disorder (Suehiro and Takayanagi, 1970). The phase transitions can be observed in differential scanning calorimetric experiments, as shown in Fig. 6.13. The lamellar thicknesses of these two phases were obtained utilizing small angle x-ray scattering. The phase-stability diagram of *trans*-1,4-polybutadiene can be illustrated using the relationship between the melting temperature of these two phases and their corresponding reciprocal lamellar thicknesses. Figure 6.14 shows such a phase-stability diagram (Finter and Wegner, 1981).

It should be noted that above 70 °C, the *trans*-1,4-polybutadiene hexagonal phase possesses large chain mobility, which leads to sliding motion along the chain axis. Therefore, a substantial lamellar thickening process in the hexagonal phase can also be obtained, as described in Section 2.2 of Chapter 5. However, in order to observe the transformation between the hexagonal and the monoclinic phases, we need to design a specific experiment. The experimental pathway is shown by solid

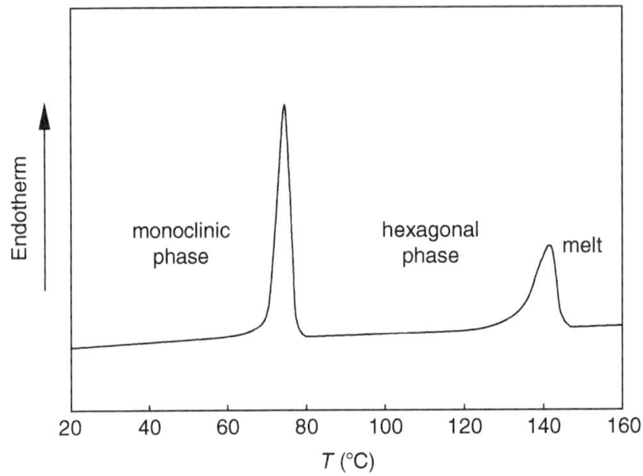

Figure 6.13 Thermo-diagram of *trans*-1,4-polybutadiene in a differential scanning calorimetric experiment during heating. Below 76 °C, the sample is in the monoclinic phase. At 76 °C a phase transition takes place, and above this temperature, the sample enters the hexagonal phase. This hexagonal phase melts at ~140 °C [replotted from Finter and Wegner (1981), with permission].

lines and arrows in Fig. 6.12. First, we need to grow lamellar crystals of *trans*-1,4-polybutadiene with a fixed thickness, ℓ_1, as shown in Fig. 6.12. This was accomplished via isothermal crystallization in solution. The crystals were precipitated and collected into single crystal mats. These samples were then heated up to a temperature that is slightly higher than the transition temperature from the monoclinic to hexagonal phase and isothermally annealed at that temperature. Note that this transition temperature is lamellar thickness dependent, as shown in Fig. 6.14. During this annealing, the lamellae started to thicken, a phenomenon that is equivalent to a horizontal shift from the right toward the left in Fig. 6.12. As soon as the lamellar thickness crosses the stability boundary of the monoclinic phase, there must be a transformation of the hexagonal phase back to the monoclinic phase.

This is indeed what was detected using synchrotron one-dimensional wide angle x-ray diffraction experiments, as shown in Fig. 6.15 (Rastogi and Ungar, 1992). In this figure, the diffraction patterns follow the precisely described heating and annealing processes shown in Fig. 6.12. The strong diffraction peak at $2\theta = 22.4°$ is the characteristic diffraction of the (200) planes of the monoclinic phase at room temperature. This diffraction peak disappears when the sample is heated to 68.5 °C, and a new diffraction peak develops at $2\theta = 20.7°$. This newly developed diffraction belongs to the (100) reflection of the hexagonal phase. However, after the sample was annealed at 68.5 °C for 20 min, the (200) diffraction of the monoclinic phase at $2\theta = 22.4°$ reappears, accompanied by a decrease in the intensity of the (100) diffraction peak of the hexagonal phase. This sample clearly indicates a transformation of the hexagonal phase back to the monoclinic phase with increasing lamellar thickness. Namely, the hexagonal-to-monoclinic phase transformation is size

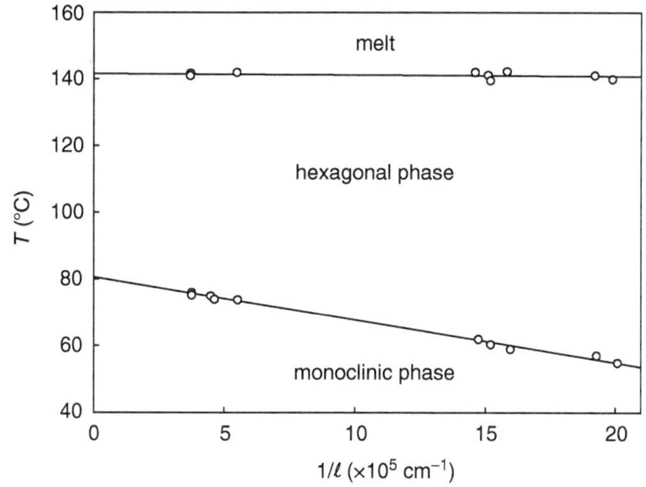

Figure 6.14 Phase-stability diagram for *trans*-1,4-polybutadiene by plotting the melting temperatures of these two phases versus their corresponding crystal thicknesses. Since these two slopes are divergent, no phase-stability inversion occurs by changing the phase size [replotted from Finter and Wegner (1981), with permission].

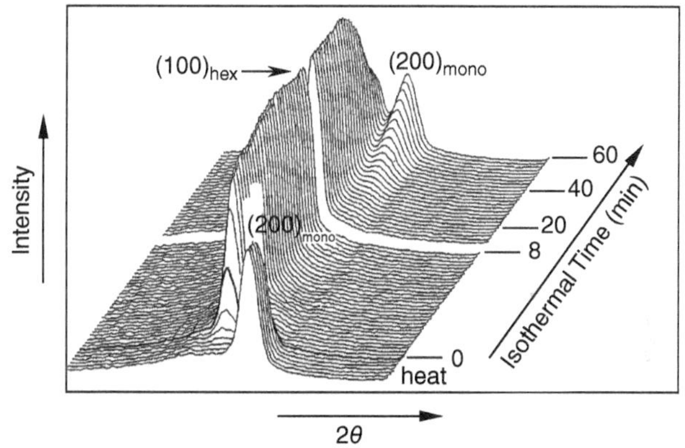

Figure 6.15 Set of wide angle x-ray diffraction patterns from real-time synchrotron x-ray experiments for *trans*-1,4,-polybutadiene samples. The thin lamellar crystal sample was first heated from room temperature to 68.5 °C. The (200) diffraction peak of the monoclinic phase disappears and the (100) diffraction peak of the hexagonal phase develops at that temperature. After the sample was isothermally annealed at 68.5 °C for 20 min or longer, the (200) diffraction of the monoclinic phase reappears accompanied by a decrease in intensity of the (100) diffraction of the hexagonal phase [replotted from Rastogi and Ungar (1992), with permission].

induced. Furthermore, this experimental observation shows that the appearance of the hexagonal phase at 68.5 °C (instead of at 76 °C for thick lamellar crystals, shown in Fig. 6.14) is the result of the existence of thin lamellar crystals, and returning back to the monoclinic phase is a consequence of the thickening. Again, this hexagonal phase possesses large chain mobility which allows the molecules to perform sliding motions along the chain direction, thus allowing unfolding of the chains to enable thickening. The simultaneous small angle x-ray scattering data showed rapid thickening upon the monoclinic-to-hexagonal phase transition (Rastogi and Ungar, 1992). It can be expected that as soon as the hexagonal phase reverts back to the monoclinic phase, this thickening process must stop. This process is identical to the thickening process described in regime B of Fig. 6.9. Although there is no cross-over of the phase-stability boundaries between the monoclinic and hexagonal phases in *trans*-1,4,-polydutadiene, this example reveals that phase-stability inversion due to phase size is also found in this case.

To our knowledge, this is the first and only example, so far, demonstrating phase-stability inversion in polymorphs where the two phase-stability boundaries do not cross. We also note that *trans*-1,4-polybutadiene does not require high pressure to crystallize, due to the fact that the hexagonal phase with high mobility exists in *trans*-1,4-polybutadiene at atmospheric pressure. We can speculate that several other polymers that possess mobile hexagonal phases, such as the fluoro-hydrocarbon polymers described in Section 2.2 of Chapter 5, can also exhibit this phase-stability inversion at ambient pressures.

2. Liquid–Liquid Phase Separation Coupled with Vitrification

Next, we introduce vitrification as a means for creating metastability, not only through the metastable glass, but also by interrupting other phase transformations, such as liquid–liquid phase separation. For illustrative purposes, we restrict ourselves to the liquid–liquid phase separation of a binary polymer–solvent mixture displaying an upper critical solution temperature.

Let us first review the vitrification process in polymer–solvent systems. In a single-component liquid system, vitrification occurs at the glass transition temperature (T_g) during cooling, as described in Section 1.1 of Chapter 4. In a fully miscible binary mixture system, such as a polymer and solvent, the glass transition temperature of the polymer is depressed. This depression is schematically shown in Fig. 6.16. Similar glass transition depression behavior can also be observed in completely miscible polymer blends. To quantitatively describe the glass transition temperature depression phenomena in binary systems, various relationships between the glass transition temperature and concentration have been established (see e.g., Paul and Bucknall, 2000; Utracki, 2002). These relationships show that when the interactions and segmental volume occupied by the solute and solvent are equivalent to the interactions among the individual components, the glass transition temperature changes linearly with the concentration, as shown in Fig. 6.16A. However, when the interactions and segmental volume of the solute and solvent become slightly different, the glass transition temperature changes may exhibit a concave or convex shape with respect to concentration (Fig. 6.16B and C).

Polymer–solvent binary systems with a typical isobaric, upper critical solution temperature phase behavior have been discussed in detail in Section 3.1 of Chapter 4

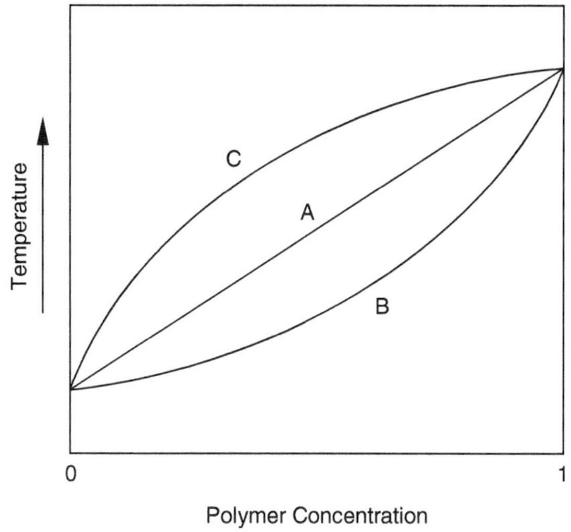

Figure 6.16 Relationship between polymer glass transition temperatures and concentrations of polymers in a polymer–solvent system. The highest glass transition temperature is for the pure polymer, and the lowest glass transition temperature is for the pure solvent. There are three types of glass transition changes with respect to polymer concentration: (A) linear, (B) concave, and (C) convex.

and shown in Fig. 4.24A. As described previously, with decreasing temperature, the system enters the phase separation region bounded by the binodal curve, and liquid–liquid separation takes place. The system divides into two distinct phases with their concentrations defined by the end points of the tie line on the binodal curve at the given temperature. One phase has more solvent (solvent-rich), while the other has more polymer (solute-rich). Moving to lower temperatures, the two phase-separated concentrations become increasingly different. As shown in Fig. 4.24A, we define the binodal and spinodal curves based on thermodynamics. In the region between these two curves, phase separation has to proceed by overcoming a nucleation barrier. On the other hand, spontaneous, barrier-free spinodal decomposition takes place in the area bound by the spinodal curve.

The phase separation morphology corresponding to ultimate stability is a two-layered liquid, as shown in Fig. 4.24B, no matter which phase separation mechanism occurs. Again, the compositions of the layers are determined by the thermodynamic phase boundaries defined by the binodal curve of the phase diagram. However, the kinetic pathway to reach this final morphology can be very different depending on whether the phase separation occurred via nucleation or spinodal decomposition mechanisms. If we freeze and investigate the morphology of a liquid–liquid phase separation during its development, droplets generally indicate a nucleation-limited process, while a bicontinuous network indicates spinodal decomposition. In reality, there is a spectacularly large variety of phase-separated morphologies due to the competition between thermodynamics and phase coarsening via ripening.

Our interest is in binary systems that undergo liquid–liquid phase separation with an upper critical solution temperature, in addition to undergoing a vitrification process. Qualitatively speaking, the influence of vitrification in the early stage of phase separation may cause the equilibrium composition partitioning to be thwarted, creating another class of metastability. In the late stage of phase separation, the vitrification may also freeze the ripening of the phase-separated morphologies. If the glass transition curve intersects the binodal curve, as shown in Fig. 6.17, the intersection is called the Berghmans' point (Arnauts and Berghmans, 1987). Following the vertical arrow in Fig. 6.17 and decreasing the temperature to enter the liquid–liquid phase separation region, two liquids with different concentrations of polymer and solvent develop, as defined by the tie line at each temperature. However, when the temperature tie line passes through the Berghmans' point, the polymer-rich phase vitrifies. Figure 6.17 illustrates several important resulting consequences that will be described.

First, the phase morphology at vitrification will be preserved in a morphologically metastable state. The nature of the morphology itself will depend on the cooling rate and on the initial concentration of the mixture. These parameters will determine which component is dispersed, which component is the matrix, and the mechanism of phase separation. A full range of morphologies can be experimentally observed and systematically produced (Arnauts *et al.*, 1993; Callister *et al.*, 1990; Hikmet *et al.*, 1988). As a second important consequence, at and below the temperature of Berghmans' point not only is the morphological development arrested, but so is the compositional change.

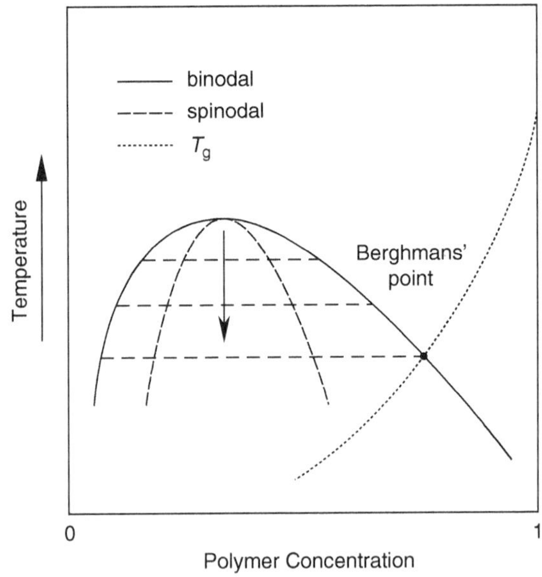

Figure 6.17 Liquid–liquid phase separation coupled with the glass transition temperature depression. The glass transition temperature line crosses over the binodal curve of a polymer–solvent binary mixture. The intersection is Berghmans' point. When temperature decreases to this point, the polymer-rich phase vitrifies [replotted from Keller (1995), with permission].

The most direct result of the glass transition fixing the phase composition is the observation that along the tie line where the Berghmans' point is reached, the glass transition temperature of the polymer-rich phase after phase separation becomes invariant with respect to the initial concentrations of the system (Arnauts and Berghmans, 1987). Figure 6.18 shows this effect in anionically polymerized atactic polystyrene with a weight average molecular weight of 275 kg/mol and a very narrow polydispersity in cyclohexanol at different concentrations. The binodal curve of the liquid–liquid phase separation in this binary system is shown in this figure. This curve was experimentally determined via "cloud point" measurements in solution. The spinodal curve is represented by the dashed line in Fig. 6.18B obtained by theoretical

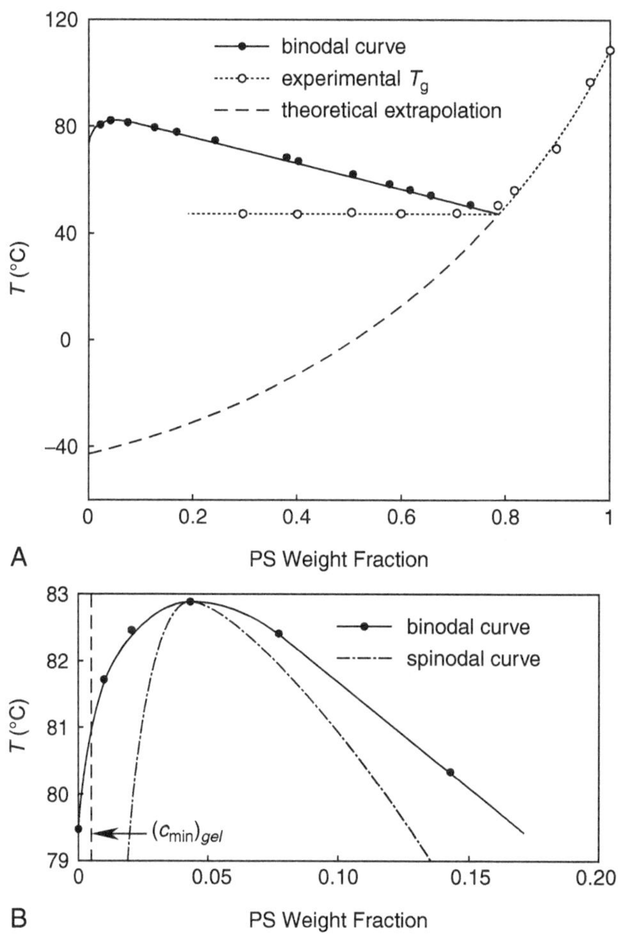

Figure 6.18 Phase diagram of amorphous polystyrene (with a weight average molecular weight of 275 kg/mol) in a cyclohexanol binary mixture depicted in relation to the glass transition line (the broken line is a theoretical extrapolation) (A); enlarged details of (A) about the commensurate point (including the spinodal line presented by dot-dashed lines) (B) [replotted from Hikmet *et al.* (1988), with permission].

calculation. In Fig. 6.18A, the binodal curve intersects the glass transition temperature line. As soon as Berghmans' point is reached, the glass transition temperature with respect to concentration becomes constant. This invariance is a consequence of the fact that not only is the morphology "locked in," but so is the composition. Therefore, in addition to the morphological metastability, there is a compositional metastability. This last issue has been addressed theoretically in diagrammatic form (Frank and Keller, 1988) and experimentally demonstrated through the "compositional aging" effect (Arnauts *et al.*, 1993). However, both are only rather cursory overviews. Further in-depth study of this important subject is needed, mainly to explore the recent opportunity of gaining access to the early and intermediate stages of phase segregation with regard to composition along a given tie line through vitrification.

Although the glass transition temperature of the polymer-rich phase is invariant after this phase is vitrified at Berghmans' point, the phase-separated morphology differs greatly. This difference exists because depending upon the initial concentration of polymer, the polymer-rich phase in the liquid–liquid phase-separated morphology may either be the minority or the majority phase. If the polymer-rich phase is a minority phase, the resulting system morphology will be solid droplets immersed in a liquid matrix. On the other hand, if the polymer-rich phase is a majority phase, the system will be liquid droplets embedded in a solid matrix. In between these two cases, both the vitrified-solid and liquid phases will form bicontinuous phases. The morphological changes based on the initial polymer concentration are illustrated in Fig. 6.19. Needless to say, significant differences in the mechanical behavior of these binary mixtures are expected, ranging from liquid to rubbery to solid.

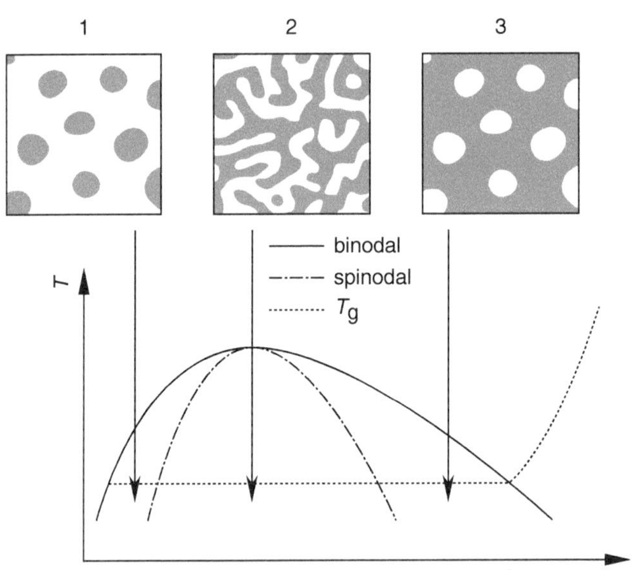

Figure 6.19 Liquid–liquid phase separation coupled with the glass transition temperature line for three different initial polymer concentrations (1, 2, and 3). Their phase separated morphologies are vastly different [replotted from Callister *et al.* (1990), with permission].

Therefore, the question is: What will happen if we quickly quench the binary mixture to a temperature below Berghmans' point? In this case, although liquid–liquid phase separation can occur, the thermodynamic partitioning can no longer be reached since it will be stopped as soon as the concentration touches the glass transition temperature line. One of the separated phases will then become glassy before it reaches the final equilibrium concentration determined by the binodal curve. The vitrified solid thus possesses a concentration which is at the glass transition temperature line, rather than located on the binodal boundary. The corresponding solvent-rich phase will also not reach the concentration defined by the binodal boundary, and the phase-separated system now becomes thermodynamically unstable but lasts a very long time due to vitrification. Two larger issues are: Can we still use equilibrium thermodynamics to provide an explanation for this observation when one of the phases is frozen, and how long can the glassy phase prevent the further liquid–liquid phase separation driven by thermodynamics?

This issue was specifically addressed by Frank and Keller (1988). Under the assumptions that above Berghmans' point the system is completely in thermodynamic equilibrium, the liquid–liquid phase separation driven by thermodynamics occurs spontaneously and rapidly, such that the kinetics of phase separation need not be considered here. The glass transition temperature is sharp and occurs at one temperature, and its concentration dependence in a binary mixture system can be clearly defined; namely, above this glass transition temperature, there is no effect on the liquid–liquid phase separation and below this temperature, the polymer-rich phase freezes. Furthermore, the binodal curve's and the glass transition temperature's dependence on the concentration can be extrapolated to below the point where the phase is frozen. Figure 6.20 shows a temperature–concentration phase diagram

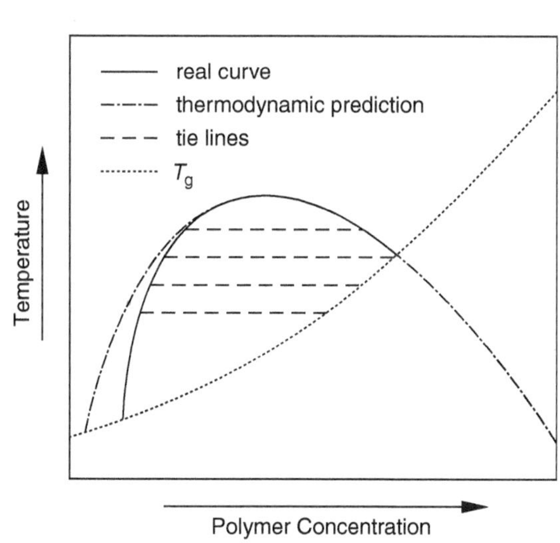

Figure 6.20 Liquid–liquid phase separation coupled with the glass transition temperature line for a binary mixture with an upper critical solution temperature. The dot-dashed line represents the true binodal curve defined by thermodynamics, and the solid line is for the experimentally observable curve [replotted from Frank and Keller (1988), with permission].

for a binary mixture in which the upper critical solution temperature intersects a vitrification process. As pointed out earlier, below Berghmans' point, the concentration of the solvent-rich phase cannot reach the binodal curve, since the polymer-rich phase becomes solid before it reaches its equilibrium concentration, as defined by the binodal curve. In this figure, therefore, the real "binodal" curve constructed by the concentrations of the solvent-rich phase (the solid line in Fig. 6.20) is compared with the true (thermodynamically determined) binodal curve (the dot-dashed line in Fig. 6.20) on the lower polymer concentration side (Frank and Keller, 1988). However, Fig. 6.20 still requires experimental proof. Callister *et al.* reported some results from specifically designed exploratory experiments to show that Fig. 6.20 is representative of the phase separation interrupted by vitrification. The liquid–liquid phase separation is drastically hampered by vitrification, yet it still proceeds in constrained local areas on much smaller scales within the glassy state (Callister *et al.*, 1990).

3. Liquid–Liquid Phase Separation Coupled with Crystallization

3.1. Liquid–liquid phase separation intervened by crystallization in solution

In this section, we will focus on a liquid–liquid phase separation coupled with a crystallization process in polymer-solvent binary mixtures. When a polymer in the system is capable of crystallizing, a crystal with the most stable crystal structure forms, which is the state of ultimate stability at that temperature (disregarding, for the present purpose, the different levels of the morphological metastability introduced by the folded chain crystals with different lamellar thicknesses). Metastable states arise because of the higher supercooling that is often required for this crystal growth to take place at practical rates. In the course of supercooling, the system may enter into the stability regimes of other metastable states. From the kinetics perspective, once the metastable state forms, it will develop faster than the ultimately stable phase and will dominate the phase transition. Therefore, the metastable state is reached first.

Let us first focus our attention on the relationships between the liquid–liquid phase separation and the crystallization processes in binary mixtures. More than 20 years ago, these relationships were summarized by Tanaka and Nishi (1985, 1989). Figure 6.21 schematically illustrates six different cases in terms of the relative positions between the binodal curve of liquid–liquid phase separation and the melting temperature depression curve after one of the components is crystallized. This figure represents thermodynamic phase diagrams which are experimentally accessible but theoretically incomplete based on the phase rule. The top three diagrams of Fig. 6.21 are upper critical solution temperature behavior coupled with melting temperature depression, and the bottom three are lower critical solution temperature behavior coupled with melting temperature depression. In the first case, these two processes cross-over, as shown in Fig. 6.21A and D. In the next case, the two processes are only tangent to each other; namely, the melting

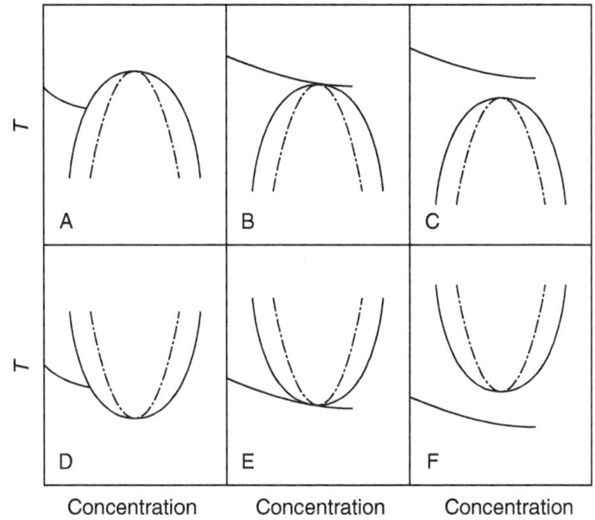

Figure 6.21 Schematic illustration of liquid–liquid phase separation coupled with melting temperature depression. There are six possible cases based on the relative positions of the binodal and the melting temperature depression curves. Among them, the top three are upper critical solution temperature diagrams, and the bottom three are lower critical solution temperature diagrams. These two curves cross over in (A) and (D), are tangential to each other at the critical point in (B) and (E), and do not intersect in (C) and (F) [replotted from Tanaka and Nishi (1989), with permission].

temperature depression curve touches the critical point, as shown in Fig. 6.21B and E. Finally, as shown in Fig. 6.21C and F, these two processes do not intersect at all. In Fig. 6.21C, for binary mixtures with upper critical solution temperature behavior, the melting temperature depression curve is above the binodal curve. In Fig. 6.21F for binary mixtures with lower critical solution temperature behavior, the melting temperature depression line is underneath the binodal curve. Furthermore, if the crystals are in a folded chain metastable state, using our language, in a "stable metastable" phase, the melting temperature depression curve must move down, depending upon the stability of the crystals. However, the binodal curve of the liquid–liquid phase separation remains in thermodynamic equilibrium.

Generally speaking, polymer crystallization from solution, specifically very dilute solution, requires that a crystalline polymer be dissolved into the solvent at a high temperature, indicating that the solvent becomes a relatively good solvent for the polymer. When the temperature is decreased for crystallization, the solvent becomes close to a θ-solvent for the polymer, and the polymer will then crystallize out of the solution since the interaction among the polymer chains becomes stronger than the interaction between the polymer and the solvent. In very dilute solution, single crystals of polymers may grow, as discussed in Section 1.1 of Chapter 5. In concentrated solution, on the other hand, polycrystalline aggregates are generally observed. The morphology depends upon how far the concentration is away from the polymer single-component melt and crystallization conditions.

Figure 6.21 is applicable in both crystalline polymer–solvent systems and polymer blends in which at least one component is crystallizable, since both cases share the same thermodynamic principles, except that in polymer blends the volume fraction is used to describe the thermodynamic behavior. We now focus on crystalline polymer–solvent binary mixtures. When a binary mixture undergoes a liquid–liquid phase separation, as shown in Fig. 6.21A or Fig. 6.21D, two types of morphological textures usually coexist. One corresponds to the polymer-rich phase, and the other is for the solvent-rich phase. Even in the solvent-rich phase, individual single crystals will not usually form; however, stacks of lamellar crystals will. In the polymer-rich phase, denser polycrystalline aggregates are frequently found.

For a long time, people reported observing "abnormal" morphologies in the solution crystallization of polymers. For example, Garber and Geil reported that in solution crystallization of polyethylene, they observed small "globular" particles with a roughly spherical shape (1966). They associated these particles with phase separation of the polyethylene from its solvent. Hay and Keller also reached the same conclusion (1965). Although these observations were the first links established between liquid–liquid phase separation and crystallization, these results have been largely overlooked or forgotten. There is renewed interest in re-examining all of these experimental observations in terms of interdependent metastable states on two length scales.

We shall first describe a detailed study in which liquid–liquid phase separation and crystallization intersect in binary mixtures of polyethylene with different solvents. In essence, the liquid–liquid phase separation and crystallization compete with each other. The most detailed investigation of the phase diagram and final crystalline morphologies of polyethylene in poor solvents, so far, has been carried out by Schaaf *et al.* (1987). We know that there are a number of solvents that can dissolve polyethylene at high temperatures and that phase separate at lower temperatures. The phase diagrams of these mixtures are known, almost all of which are upper critical solution temperature systems. One of the earliest examples of a phase diagram for polyethylene in a poor solvent was reported for a branched polyethylene in amyl acetate and in nitrobenzene by Richards over 60 years ago (1946). Later, Nakajima *et al.* demonstrated similar phase diagrams for linear polyethylene in a number of solvents (1966). As previously pointed out, we are most interested in the case shown in Fig. 6.19C, in which both processes compete. Figure 6.22 is a schematic phase diagram for a binary mixture of a monodisperse polymer and solvent. The polymer can crystallize. In this figure, not only the liquid–liquid phase separation, but also the melting temperature depression of the polymer crystals is assumed to be in equilibrium.

As shown in Fig. 6.22, three regions exist. Region I at high temperatures is where the polymer and solvent are miscible and form an isotropic single-phase liquid. Region II is bounded by the binodal curve within which the liquid–liquid phase separation occurs in an upper critical solution temperature system. Below the equilibrium melting temperature depression line in this figure is region III. There are two areas that need to be discussed. First is the melting temperature depression line in the single liquid phase at higher polymer concentrations before it intersects the binodal curve (denoted AB in the figure). This example exhibits a typical

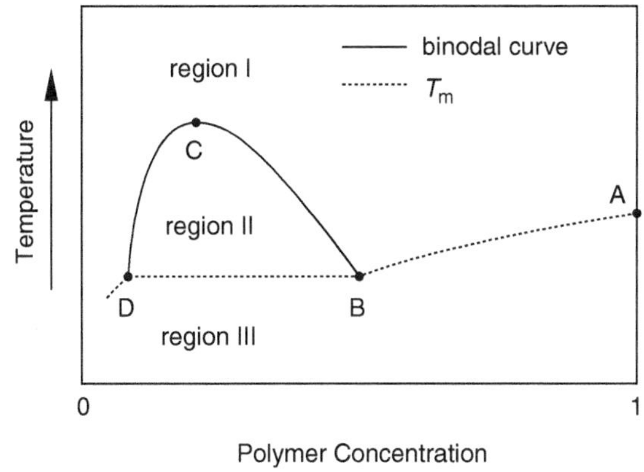

Figure 6.22 Schematic illustration of an equilibrium phase diagram of a binary mixture consisting of a monodisperse polymer and a poor solvent. The three regions of a polymer-poor solvent binary mixture are labeled I, II, and III. In region I at high temperature, a single liquid phase forms. The liquid–liquid phase separation occurs in region II. The melting temperature depression line in between AB indicates that the polymer crystallizes directly from the solution. However, after this line intersects the binodal curve, the melting temperature becomes invariant between BD [replotted from Schaaf *et al.* (1987), with permission].

phenomenon where the polymer crystallizes and separates from the solution as the temperature falls below the melting temperature depression line. With decreasing polymer concentration, the melting temperature is continuously depressed. However, when this melting temperature depression line intersects the binodal curve, the equilibrium melting temperature of the crystals formed within this liquid–liquid phase separation region becomes invariant (denoted BD in this figure). Thus, the question is: Why are these melting temperatures invariant? Or more specifically: Why are they polymer concentration independent? The reason is due to the intervention of the liquid–liquid phase separation to form the polymer-rich and solvent-rich phases. When the binary mixture is quenched to a temperature indicated by the BD line in Fig. 6.22, two phases are formed. When the polymer concentration is low, the polymer-rich phase is always the minority phase, and it usually forms relatively concentrated droplets within the solvent-rich majority phase. Therefore, disregarding the initial polymer concentrations along the BD line, the minority polymer-rich phase always has a fixed composition at B after the liquid–liquid phase separation. After crystallization, therefore, the melting temperatures are always depressed by the existence of the solvent at a constant concentration. It needs to be pointed out again that we are only dealing with the thermodynamic equilibrium situation and ignoring any kinetic factors in polymer crystallization.

However, polymer crystallization is a kinetic process dominated by nucleation and growth (Section 1.5 of Chapter 4), so in order for the polymer to crystallize, supercooling is necessary. The final crystalline morphology is dependent upon the amount of supercooling and the rate of the cooling process. Because of different thermal histories, either heterogeneous or homogeneous nucleation can be dominant. Furthermore, the extent to which we allow liquid–liquid phase separation to develop before polymer crystallization occurs is another important factor affecting the final crystalline morphology (Schaaf *et al.*, 1987) because of the time dependence of the morphological ripening process in the liquid–liquid phase separation, which is described in Section 3.4 of Chapter 2. Therefore, liquid–liquid phase separation and polymer solution crystallization compete with each other. The final crystalline morphologies are determined by many different factors, ranging from the interactions between the polymer and solvent, concentration, as well as both kinetics of crystallization and liquid–liquid phase separation.

The formation of polymer crystalline "globules" or dense crystal aggregates of other shapes has been reported for a number of polymers. They are consistently obtained when relatively poor solvents are used. Many of these observations for the polymer–solvent binary mixtures may be linked with the liquid–liquid phase separation. For example, binary mixtures of polyamides with benzoic acid showed elongated "globules" having an empty cigar-like shape. The anisotropic shape is most likely due to the preferred crystal growth direction in polyamides along the hydrogen bonding direction (Wittmann *et al.*, 1983). Other types of crystalline morphologies that are possibly associated with liquid–liquid phase separation are bowl-shaped crystal aggregates found in poly(D,L-benzyl glutamate), which possesses a hexagonal unit cell, in dioxane (Schaaf *et al.*, 1987) and in "semi-hexagonal" hexafluoroisopropanol (Price *et al.*, 1975, 1979). These observations most likely are due to the fact that all of these polymers possess high rotational symmetry crystal lattices (see Section 1.1 of Chapter 5). The development of "fish-bowl"-like crystal aggregates, for example, may also be associated with the unit cell and single crystal rotational symmetry in the relatively high supercooling region to generate curved crystals. However, a complete understanding of their thermodynamic phase behaviors (phase diagrams shown in Fig. 6.22) is still required to quantitatively describe the liquid–liquid phase separation intersection with crystallization.

The final crystal morphology may also, in turn, provide evidence to judge whether the liquid–liquid phase separation intersects with crystallization. In binary mixtures having the phase separation intersect the crystallization, the crystallized spherical "globules" and other dense aggregates, which are usually attributed to the crystallization of the polymer-rich minority phase, may coexist with another type of crystal morphology in the solvent-rich majority phase. This could be less dense aggregates or even in a few cases, single crystals. Finally, it needs to be emphasized again that all of these crystals are not in ultimate equilibrium, but rather in different metastable states (see Section 2.3 of Chapter 4).

3.2. Sequential liquid–liquid phase separation and crystallization in solution

When the solvent in binary mixtures becomes increasingly better, what will happen? The binodal curve of a liquid–liquid phase separation will be depressed to lower temperatures, and will give rise to the case in Fig. 6.21C. However, when the binodal curve is "buried," but not too far beneath the melting temperature depression line, we still have the possibility to enter the liquid–liquid phase separation region using rapid quenching if the crystallization of the polymer is not too fast.

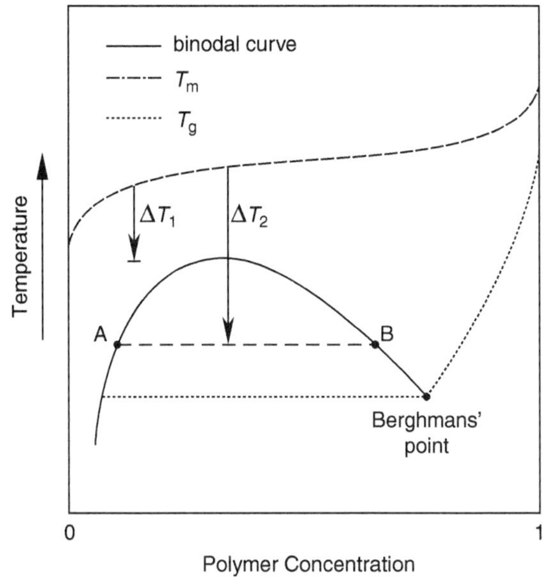

Figure 6.23 Schematic illustration of an equilibrium phase diagram for a fully miscible polymer solution having a metastable liquid–liquid phase separation. This phase separation is beneath the stable liquid-crystal phase line. Depending on the quenching depth (ΔT) and/or cooling rate, we can have crystallization from solution at supercooling ΔT_1, liquid–liquid phase separation at supercooling ΔT_2, or vitrification at a larger supercooling [replotted from Berghmans *et al.* (1995); Keller and Cheng (1998), with permission].

These successive states represent a basic influence on the behaviors of different materials. In this case, the phase separation is entirely metastable compared with the crystal phase. In Fig. 6.23, we also add a vitrification process. Any or all of the corresponding phase transitions can also be interrupted by vitrification.

Figure 6.23 only represents a schematic illustration of the basic principles involved in the case where liquid–liquid phase separation is beneath the crystallization. In this figure, the upper line corresponds to the stable crystal–liquid phase boundary representing crystal melting and crystallization in thermodynamic equilibrium. The binodal curve corresponding to the liquid–liquid phase separation is beneath the crystal–liquid phase boundary. For crystallization to occur, supercooling is required. In this figure, two cases are illustrated with two different supercoolings. For smaller supercoolings, ΔT_1, crystallization is expected to take place from the initial, completely miscible solution, provided sufficient time is allowed. However, at larger supercooling, ΔT_2, the liquid–liquid phase separation will occur first. Depending on the extent of supercooling, we can have either a situation in which crystallization is occurring within the polymer-rich phase (at point B in Fig. 6.23), or a situation where the concentrated phase vitrifies at Berghmans' point (point BP in Fig. 6.23). In the latter case, not only will the liquid–liquid phase separation process become arrested, but also crystallization will be prevented.

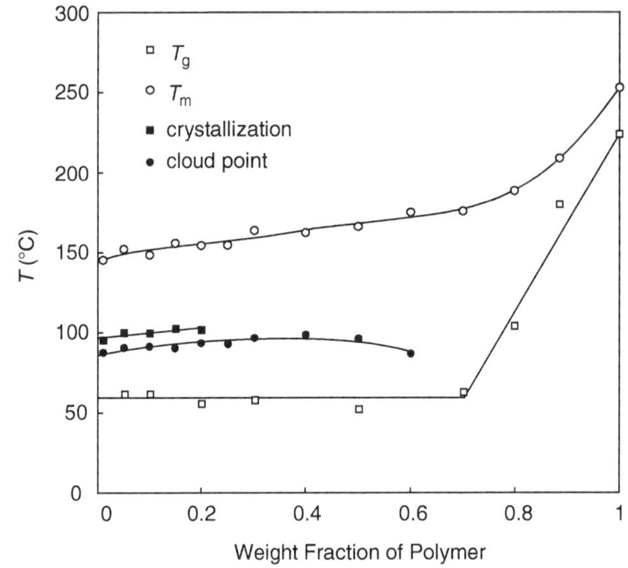

Figure 6.24 Temperature–concentration phase boundary diagram for poly(2,6-dimethyl-1,4-phenylene ether) in cyclohexanol. The polymer has a molecular weight of 30 kg/mol. The open squares are the glass transition temperatures, the open circles are the crystal melting temperatures measured during heating at 10 °C/min, and the dark squares are the crystallization temperatures measured during cooling at 5 °C/min in calorimetric experiments. The dark circles are cloud points during slow cooling [replotted from Berghmans *et al.* (1995), with permission].

Experimental observations of the case in Fig. 6.23 have been reported, such as poly(2,6-dimethyl-1,4-phenylene ether) in cyclohexanol (Berghmans *et al.*, 1995), as shown in Fig. 6.24. Four transition processes can be observed in this figure. On the low temperature side, the first observation is the glass transition temperature change with the concentration of poly(2,6-dimethyl-1,4-phenylene ether) in cyclohexanol. Above the 70% polymer concentration, the glass transition temperature increases quickly with increasing polymer concentration, indicating a miscible mixture in this concentration region. Below this polymer concentration, the glass transition temperature becomes concentration independent, implying that the mixture is now entering a liquid-liquid phase separation region. This phase separation can be verified by the cloud point measurements shown in Fig. 6.24. Therefore, the line constructed by cloud points should represent the binodal curve of this binary mixture. Above the cloud point line is a crystallization process which is exothermic in calorimetric experiments at a cooling rate of 5 °C/min (note that the experiments are under a constant cooling rate, not isothermal). The highest phase transition boundary is attributed to the crystal melting measured at 10 °C/min during heating in calorimetric experiments. Although this figure does not provide a phase diagram under thermodynamic equilibrium, it gives rise to experimental evidence for deducing Fig. 6.23.

A more complicated case is syndiotactic polystyrene in cyclohexanol. This binary mixture shows a similar phase behavior, as in Fig. 6.23. However, since syndiotactic polystyrene possesses polymorphs (Section 2.1 of Chapter 5), it is possible to crystallize the phase separated syndiotactic polystyrene into different crystal forms. De Rudder *et al.* carried out a detailed study of this system, and the phase diagram is schematically illustrated in Fig. 6.25 (1999). First, the liquid–liquid phase separation is deeply buried below the crystal melting temperature. The glass transition temperature dependence on concentration can also be observed, and this temperature becomes constant at 36 °C when the polymer concentration is below 70%, similar to the atactic isomer. When deeply quenching the binary mixture to low temperatures and directly entering the liquid–liquid phase separation region slightly below the glass transition temperature, only amorphous aggregates will be obtained. Heating to above the glass transition temperature, the polymer-rich phase starts to crystallize into the γ-phase, which can only be formed in solution (Section 2.1 of Chapter 5). Further heating leads to a melting and recrystallization of the syndiotactic polystyrene sample. Namely, the γ-phase melts and recrystallizes into the β-phase, with a zigzag chain conformation. Note that the β-phase can only be grown by melt crystallization of syndiotactic polystyrene. What this implies is that the crystals of the γ-phase do not dissolve back into cyclohexanol but remain demixed, even at temperatures which are far above the θ-temperature where the solvent has become relatively good (De Rudder *et al.*, 1999).

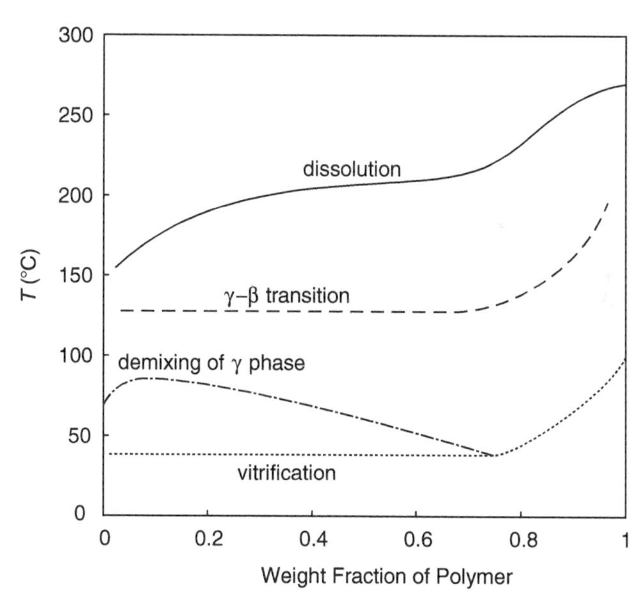

Figure 6.25 Schematic temperature–concentration phase diagram for syndiotactic polystyrene in cyclohexanol. The liquid–liquid phase separation is bounded by the binodal curve. The vitrification process is represented by the glass transition temperature change above a polymer concentration of 70% and invariant below 70%. The dashed line indicates the melting of the γ-phase and recrystallization into the β-phase. The top boundary at the high temperature side corresponds to the β-phase crystal melting [replotted from De Rudder *et al.* (1999), with permission].

Other possible examples of liquid–liquid phase separation beneath crystallization might be found in poly(4-methly-1-pentene) (Khoury and Barnes, 1972), polyoxymethylene (Khoury and Barnes, 1974a), polychloro-trifluoroethylene (Khoury and Barnes, 1974b) in different solvents, and *trans*-1,4-polyisoprene in amyl acetate (Kuo and Woodward, 1984; Xu and Woodward, 1986). The common feature of these examples is that at low supercoolings, all of them grow lamellar single crystals, implying that the crystals are directly formed in solution. With increased supercooling, the crystal morphology changes into hollow "fish-bowl"-like or cup-shaped lamellar crystal aggregates. The appearance of these aggregates may provide strong evidence that the binary mixtures have entered the liquid–liquid phase separation region. The final caution is that a competition between the liquid–liquid phase separation and the crystallization kinetics will ultimately determine the evolution of the final crystal morphology. Again, the morphologies can be arrested if vitrification is involved.

3.3. Liquid–liquid phase separation intersected by crystallization in polymer blends

So far, we have discussed polymer–solvent binary mixtures. If we substitute a polymer for the solvent, we obtain polymer blends. We are interested in the case in which at least one component in the blend is crystallizable. Therefore, all the cases in Fig. 6.21 can also be used to represent the relationships between the liquid–liquid phase separation (either upper or lower critical solution temperature) and crystallization (either one component or both components can crystallize) in polymer blends. There is a large collection of data for crystalline polymer blends available in literature (see e.g., Paul and Bucknall, 2000; Utracki, 2002). However, our focus will only be on the cases in which liquid–liquid phase separation intervenes in crystallization, such that the two different levels of metastable states are interdependent.

Among many examples, we have chosen several systems to illustrate the principles we want to convey. The first system selected is a binary mixture of a poly(ε-caprolactone) with a number average molecular weight of 10.7 kg/mol and a polydispersity of 3 and a low molecular weight polystyrene with a number average molecular weight of 840 g/mol and a polydispersity of 1.1. This is a historical example published by Tanaka and Nishi where perhaps for the first time the interdependence between the liquid–liquid phase separation and crystallization was unequivocally proposed (1985, 1989). Figure 6.26 shows the isobaric temperature–concentration phase diagram of this binary blend that is experimentally accessible. At high temperatures, these two polymers are miscible. With decreasing temperature, the system undergoes a liquid–liquid phase separation as measured by cloud points and a scattering technique indicating an upper critical solution temperature system. In addition, the poly(ε-caprolactone) can crystallize, and the melting temperatures of its crystals, shown in Fig. 6.26, decrease with the concentration of this polymer. Thus, the crystalline solid-melt phase boundary reflects the melting temperature depression in this binary blend. This melting temperature depression line intersects with the binodal curve. Within the binodal curve, the

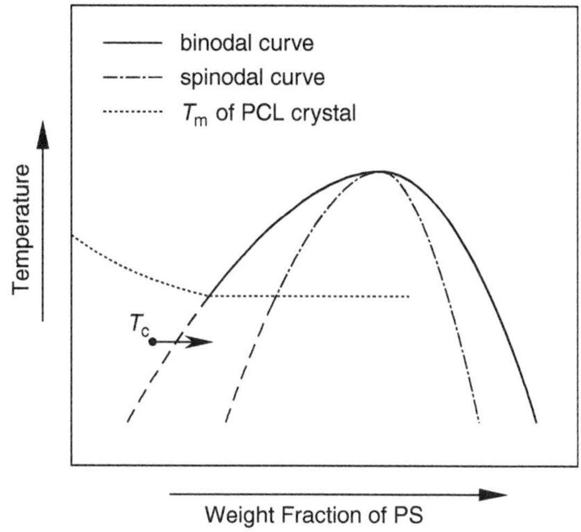

Figure 6.26 Temperature–concentration phase diagram of a binary blend of a poly(ε-caprolactone) having a number average molecular weight of 10.7 kg/mol and a polydispersity of 3 and polystyrene with a number average molecular weight of 840 g/mol and a polydispersity of 1.1. The liquid–liquid phase separation is upper critical solution temperature bounded by the binodal curve. The melting temperatures of the poly(ε-caprolactone) crystals decrease outside of the binodal curve with decreasing concentration. The arrow indicates the concentration changes while carrying out the crystallization process [replotted from Tanaka and Nishi (1989), with permission].

liquid–liquid phase separation takes place, as shown in Fig. 6.26, yet this phase separation may not reach the final equilibrium state due to slow kinetics. Using our language, this is a "stable metastable" phase on the phase morphological scale. On the other hand, crystallization may take place when the temperature is below this melting temperature depression line, depending upon the degree of supercooling. Again, this phase is another "stable metastable" phase. Therefore, these two "stable metastable" phases are on two different (crystalline and phase domain) length scales. In this case where these two processes compete, kinetics determines which one dominates.

By quenching a poly(ε-caprolactone) and polystyrene blend sample with a concentration ratio of 70:30 to a lower temperature, the poly(ε-caprolactone) initially crystallized in the miscible region. The polystyrene molecules needed to be rejected from the growth front and only the poly(ε-caprolactone) molecules were allowed to crystallize. This is a case of "poisoning," as described in Section 1.5 of Chapter 4. Depending upon the diffusion rate of polystyrene, the crystal growth rate of poly(ε-caprolactone) became slower with diffusion-limited (nonlinear) growth rates. However, by continuing the crystallization of poly(ε-caprolactone) in the system, the overall concentration of poly(ε-caprolactone) in the remaining solution would decrease and move horizontally toward the right to pass through the binodal curve and enter the liquid–liquid phase separation region (see Fig. 6.26) where phase separation would begin. The competition between these two processes began with the local phase separation at the growth front of poly(ε-caprolactone) spherulites where the concentration of poly(ε-caprolactone) was the lowest (Tanaka and Nishi, 1989). The low molecular weight polystyrene was used for two reasons. First, the binodal curve would move toward higher temperatures with increasing molecular weight of polystyrene to reduce the possibility of polymer decomposition. Second, with increased molecular weight of polystyrene, the glass transition temperature of the blend could increase. This would significantly affect the diffusion rate of polystyrene, and finally, when the glass transition temperature reached the crystallization temperature, the whole system would be frozen. Phase separation and crystallization in a blend of poly(ε-caprolactone) with a narrower molecular weight distribution and polystyrene oligomer was also reported by Nojima *et al.* (1991), and similar results were achieved.

The second example we have chosen is polyolefin blends, which are technologically important. With the development of single-site catalysts, the ability to precisely control the chemical structures and short chain branches during synthesis of polyolefins has been drastically improved. Commercially successful polyolefin businesses require a variety of binary blends to improve and optimize specifically targeted properties. We focus on the blends in which both polymer components have short chain branches but with different sizes and compositions or where one polymer component is linear without branches. There have been substantial investigations in the binary blends with model polyolefins, as well as with commercial polymers. Research contributions on these binary blends have generated a tremendous amount of discussion regarding liquid–liquid phase separation and crystallization-induced phase separation in these systems. An earlier review summarized the major facts and controversies of those discussions (Crist and Hill, 1997).

Let us look at a series of binary blends consisting of two polyethylenes with hexane and butane branches, respectively (Wang *et al.*, 2002). The samples were specifically designed to have very different branch contents. In the polyethylene with the hexane branches, the hexane branch content was 9 per 1000 carbons with a molecular weight of 112 kg/mol, while in the polyethylene with the butane branches, the branch content was as high as 77 per 1000 carbons with a molecular weight of 77 kg/mol. The sample of polyethylene with hexane branches crystallizes at high temperatures, and the polyethylene with butane branches only crystallizes at lower temperatures.

Figure 6.27 is the experimentally accessible temperature–concentration phase diagram for this binary blend of these two short-chain branched polyethylenes at atmospheric pressure. The similarity with the diagram in Fig. 6.26 is evident. Using specifically modified scattering and optical microscopy techniques, the liquid–liquid phase separation can be observed, as shown in Fig. 6.27. The equilibrium melting temperatures were deduced by extrapolation. In this figure, the solid crystal-melt phase boundary reflects the melting temperature of the crystals formed by polyethylene, with hexane branches decreasing with decreased concentration. This solid crystal-melt phase boundary (the melting temperature depression) intersects the binodal curve of the liquid–liquid phase separation. Within the phase separation region, the melting temperatures are invariant with varying concentrations (Wang *et al.*, 2002). This example implies that the crystallization started after completion of the liquid–liquid phase separation with two distinct concentrations.

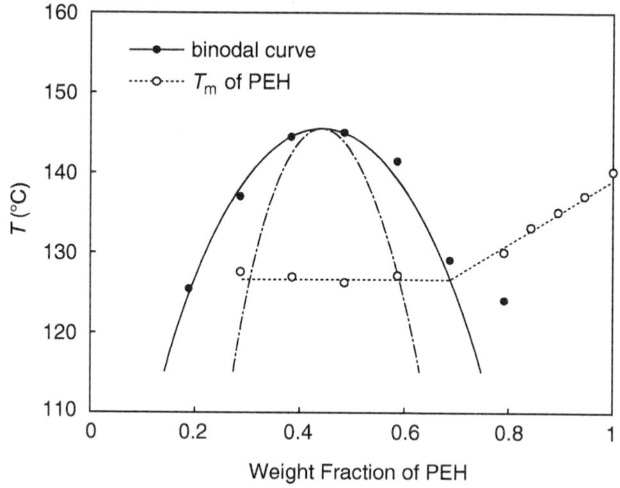

Figure 6.27 Temperature–concentration phase diagram of a binary blend of a polyethylene with hexane branches with a branch density of 9 per 1000 carbons and a polyethylene with butane branches with a branch density of 77 per 1000 carbons. The liquid–liquid phase separation is upper critical solution temperature bounded by the binodal curve. The crystalline solid-melt boundary line reflects the melting temperatures of the crystals formed by polyethylene with hexane branches and is seen to decrease outside of the binodal curve with decreasing concentration, but remains invariant within the binodal curve. The spinodal curve was calculated based on the Flory-Huggins formalism [replotted from Wang *et al.* (2002), with permission].

This study has brought up several interesting issues which need to be further investigated. For example, what would the crystallization behavior of the polyethylene with hexane branches be if the liquid–liquid phase separation was at a different stage of phase separation, such as in the initial or intermediate stage, as described in Section 3.2 of Chapter 4, where the concentration in the phase-separated domains is transient? Furthermore, if we could monitor the overall crystallization and growth rates throughout the liquid–liquid phase separation at different supercoolings, quantitative understanding of the interplay and competition between crystallization and phase separation could be achieved. It can be expected that the crystallization kinetics would also be affected by the phase-separated morphology. This interplay is particularly true when the phase separation and the phase-separated morphology are in different stages of development. When the crystallizable polyethylene with hexane branches is confined in very small spaces, such as in droplets, the crystallization rates slow down. Moreover, as long as the phase-separated domains do not reach an invariant concentration (in the late stage of phase separation), the crystal growth rate cannot be linear. On the other hand, the crystallization rates may also speed up after phase separation and morphological development are in the late stage.

Some of these interesting issues have been investigated using identical or similar polyethylene blends. The major difficulty for these studies lie in how to precisely identify the crystallization and liquid–liquid phase separation processes and then individually measure their kinetics. For example, in the same series of polyethylene

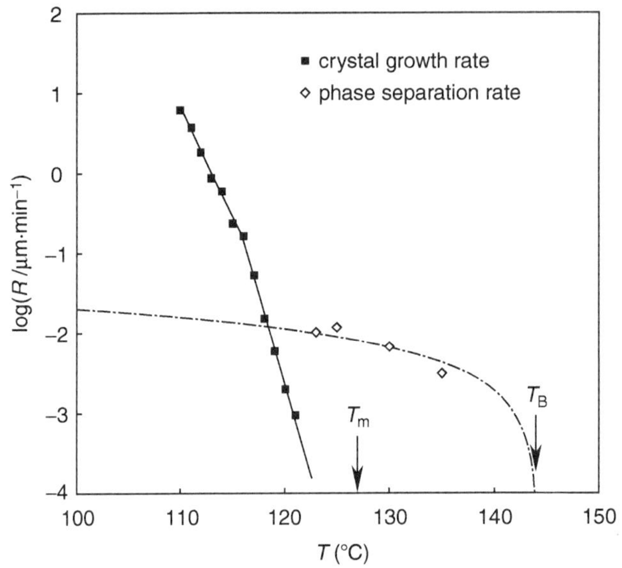

Figure 6.28 Crystal growth rates and liquid–liquid phase separation rates for a binary blend consisting of a polyethylene with hexane branches having a branch density of 9 per 1000 carbons (50%) and a polyethylene with butane branches having a branch density of 77 per 1000 carbons (50%). The solid squares represent the crystal growth rates and the open diamonds are the growth rates of the characteristic length of the liquid–liquid phase separation in the coarsening stage. T_B is the binodal temperature, and T_m is the melting temperature of the polyethylene crystals with hexane branches having a branch density of 9 per 1000 carbons. The two rate lines cross over at 118 °C [replotted from Shimizu *et al.* (2004), with permission].

blend samples, Shimizu *et al.* (2004) measured the crystal growth rates using optical microscopy by quenching the samples from the homogeneous melt to different crystallization temperatures. They also monitored the growth rates of the liquid–liquid phase separation in the coarsening stage by utilizing a small-angle light scattering technique. Figure 6.28 shows a plot which puts two sets of kinetic data together to illustrate the relationship between these two processes for a sample containing 50% polyethylene with hexane branches, and another, 50% polyethylene with butane branches. When the temperatures are far below the cross-over temperature of 118 °C, the crystallization is faster, and it dominates the final morphological development. Much above 118 °C, the phase coarsening process is in control. Near 118 °C, these two processes compete with each other (Shimizu *et al.*, 2004).

A specific experimental design can also be used to ensure that the liquid–liquid phase separation in polyethylene blends is carried out before the crystallization: a two-step isothermal process. The experiment calls first for isothermal annealing of the blend at a temperature either above or within the binodal curve boundary, yet above the melting temperature of the polyethylene crystals where crystallization is not able to take place. When the annealing temperature is within the binodal curve boundary, the liquid–liquid phase separation morphology can be carried out to the late stage, and the equilibrium concentrations can be reached within both of the phases. In the second step, isothermal crystallization can then occur by quenching the sample to a lower crystallization temperature where the crystallization takes place rapidly. The concentration of the polymer in both phases cannot change because crystallization occurs very quickly. The phase morphologies are thus preserved.

Utilizing this two-step isothermal experiment, the crystallization kinetics and observed crystal morphology are used to determine the binodal and spinodal curves. Let us examine one series of linear, crystallizable polyethylene and low-temperature crystallizable polyethylene with randomly distributed octane branches (a branching density of 120 per 1000 carbons) in blends above 100 °C. Figure 6.29 shows two real-time atomic force microscopy phase images for the blend, with a concentration of 40% crystallizable linear polyethylene at two different annealing temperatures. For the first image in Fig. 6.29, the first step lowers the sample to 180 °C where it was annealed for 24 h. The second step further quenches the sample to 120 °C where the crystallization occurs rapidly. For the second image in the same figure, the first quenching lowers the sample temperature to 170 °C, where the system is then annealed for 24 h. The second step, identical to the previous case, crystallizes the sample at 120 °C. The phase morphologies revealed by the spherulitic decoration observed in these two images are completely different. The first image shows spherical domains which exemplify the nucleation and growth mechanism. This example indicates that the phase separation took place between the binodal and spinodal curves. On the other hand, the second image is a bicontinuous phase morphology that indicates the spinodal decomposition mechanism, thus revealing that the phase separation was carried out within the spinodal curve region.

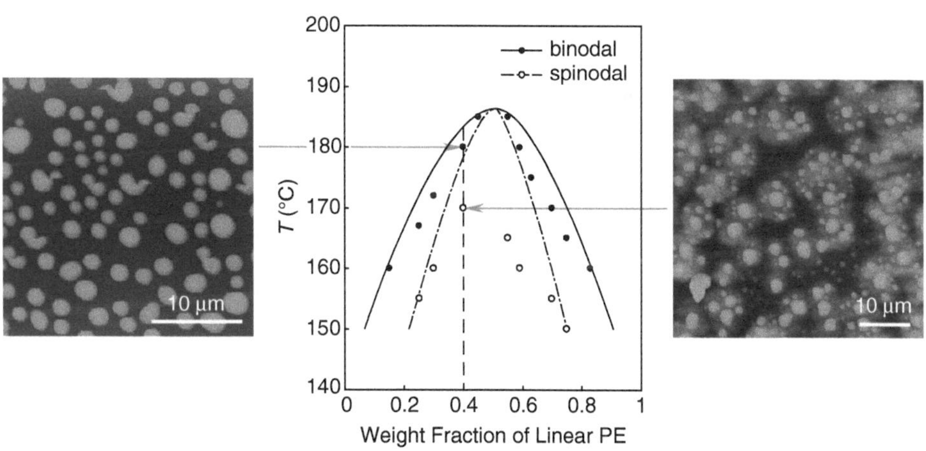

Figure 6.29 Temperature–concentration phase diagram of a binary blend consisting of a linear polyethylene and a polyethylene with octane branches at a branch density of 120 per 1000 carbons. The solid line is the binodal curve, and the dashed line is the spinodal curve. Two different phase morphologies were observed via tapping mode atomic force microscopy, utilizing the two-step isothermal experiment. In the first sample, the final morphology indicates the nucleation and growth mechanism; in the second sample, the morphology indicates the spinodal decomposition mechanism.

Now, the first issue is how to obtain the binodal curve, as shown in Fig. 6.29. We need to set up a series of different annealing temperatures for the first isothermal step, and each step needs to have a long coarsening time (~24 h). As soon as the binodal temperature is reached, the crystallization kinetics in the second isothermal step will speed up due to the liquid–liquid phase separation. Since the crystallizable polyethylene-rich phase is formed and coarsened during the initial annealing step, less "poisoning" is present to slow down the crystallization kinetics caused by the non-crystallizable, branched polyethylene (Section 1.5 of Chapter 4). Therefore, the annealing temperature where the crystallization kinetics shows a sudden increase is on the binodal curve for that specific concentration. Using blends with different concentrations, the binodal curve of the entire concentration range can be obtained.

The second issue then becomes: Can we experimentally delineate the spinodal curve? Note that the spinodal curve is usually calculated based on theory. As described above for the two atomic force microscopy images in Fig. 6.29, the spinodal curve can be identified as that stage when the liquid–liquid phase separated morphology changes from the droplet growth to the bicontinuous phase. If we carry out this type of experiment for different blend concentrations to detect the morphological change, reflecting the mechanism change from nucleation to spinodal decomposition, a complete phase diagram with both experimentally determined binodal and spinodal curves can be obtained, as shown in Fig. 6.29. This experiment indicates that when the crystallization is fast and the chain molecules cannot diffuse far during the second stage of the isothermal experiment (thus preventing further phase separation), the phase morphologies can be preserved by the crystallization, thus indicating the mechanism of liquid–liquid phase separation. The open circles in Fig. 6.29 represent the temperature level where the bicontinuous phase morphology being decorated by the crystal spherulitic morphology has been observed; while, the closed circles are where the droplet morphology is observed. Again, both the crystals and phase morphology are in "stable metastable" states. The crystal metastable state exists because the crystals are chain folded, and the metastable phase-separated morphology is due to the fact that two phases do not reach their ultimate two-layer equilibrium phase morphology.

Other interesting examples include binary blends of poly(vinylidene fluoride) with poly(ethyl acrylate) or poly(ε-caprolactone) with polystyrene-*block*-polyacrylonitrile. In the case of poly(vinylidene fluoride) and poly(ethyl acrylate) blends, the liquid–liquid phase separation occurs because of a lower critical solution temperature, and the curve intersects the crystalline solid-melt phase boundary. Figure 6.30 illustrates the temperature–concentration phase boundary diagram (Briber and Khoury, 1987; Endres *et al.*, 1985). In the case of poly(ε-caprolactone) and polystyrene-*block*-polyacrylonitrile binary blends, a closed miscibility loop with one lower and one virtual upper critical solution temperature has been reported. The crystallization process also intersects phase separation, as evidenced by light-scattering and transmission electron microscopy (Svoboda *et al.*, 1994).

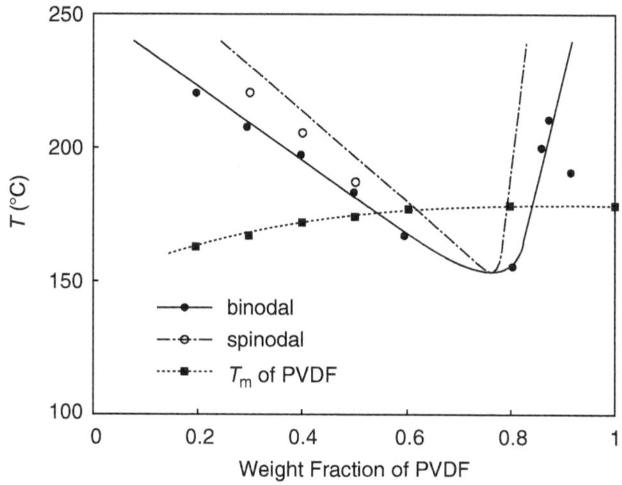

Figure 6.30 Temperature–concentration phase boundary diagram of a binary blend consisting of poly(vinylidene fluoride) and poly(ethyl acrylate) at atmospheric pressure. The liquid–liquid phase separation has a lower critical solution temperature and the crystalline solid-melt phase boundary intersects the binodal and spinodal curves [replotted from Endres *et al.* (1985), with permission].

3.4. Sequential liquid–liquid phase separation and crystallization in polymer blends

When liquid–liquid phase separation and crystallization in polymer–solvent systems takes place in sequence, it is the same as the first case described in Section 3.2 of this Chapter where the liquid–liquid phase separation with an upper critical solution temperature is buried beneath the crystalline solid–liquid phase boundary (Fig. 6.21C). The thermodynamic principles which determine polymer–solvent system behavior are also true for polymer binary blends. Furthermore, complicated sequences involving polymorphs and phase separation with both upper and lower critical solution temperatures may appear from time-to-time. We are focusing specifically on the consequences of the sequential liquid–liquid phase separation and crystallization regarding the final morphologies in polymer binary blends in which at least one component is crystallizable.

Figure 6.31 schematically shows a temperature–concentration phase diagram of a polymer binary blend in which one of the polymers is crystallizable. In this figure, we also illustrate the crystal and phase morphological consequences on the blend after the crystallization in different temperature and concentration regions. Since the crystalline solid-melt phase boundary is at the uppermost temperatures and the liquid–liquid phase separation region bounded by the binodal curve is located at lower temperatures, this liquid–liquid phase separation is entirely metastable. Certainly, within the crystal phase, the lamellar crystals themselves are metastable with respect to the extended chain equilibrium crystal. Therefore, the lamellar

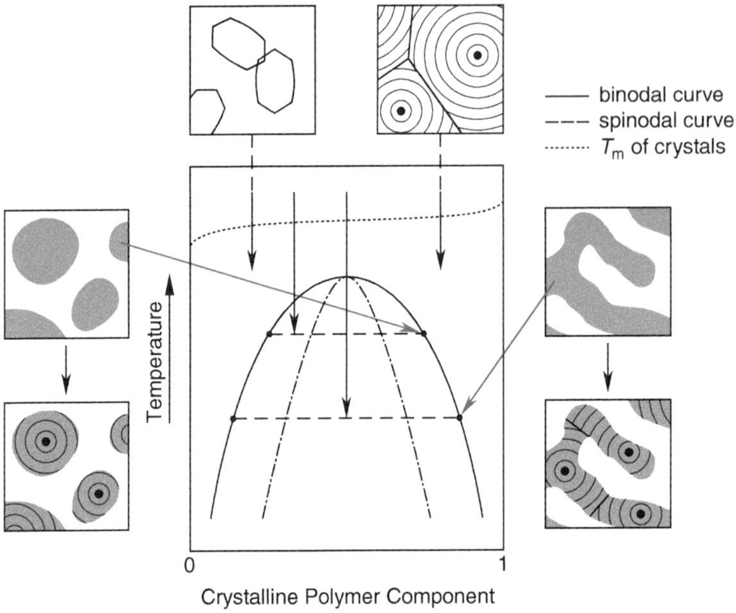

Figure 6.31 Schematic temperature–concentration phase diagram of a binary polymer blend with one crystallizable polymer. The uppermost line is the crystalline solid-melt phase boundary. The binodal curve is located at lower temperatures. Two different quenching processes are illustrated. One is at relatively low supercooling; thus, the temperature after quenching is located outside the binodal curve. The crystallization takes place from the homogeneous single phase melt. The crystal morphology is relatively uniform with segregation of the noncrystallizable phase on the micro-scale. The second is at a deeper quench, and the temperature is located within the binodal curve. If the liquid–liquid phase separation is accomplished first and is followed by crystallization, the crystal morphology can exhibit a variety of forms within the crystallizable polymer-rich phase.

crystals are again "stable metastable." Since the liquid–liquid phase separation beneath the solid crystal-melt boundary cannot, in almost all cases, reach the ultimate equilibrium phase separation, it should be a "metastable metastable" phase.

We have at least two ways to investigate the crystallization as described in Section 3.2 of this Chapter, based on different degrees of supercooling and different concentrations. When the blend is quenched into the temperature and concentration region outside of the binodal curve, the crystallizable component will be grown from the homogeneous single phase melt. The crystal morphology can then range from lamellar single crystals in one extreme to spherulitic textures in another, which reflect the macroscopic and/or microscopic segregation of the noncrystallizable component. When the quenching process brings the system into the phase separation region and the liquid–liquid phase separation can occur before the crystallization, we can actually observe the droplet phase morphology that is generated by this liquid–liquid phase separation. Note that this type of phase morphology can only be obtained when the blend is brought into the metastable region between the binodal and spinodal curves where a nucleation and growth mechanism dominates the phase separation. The sizes of these droplets are determined by the kinetics of the coarsening process and how long this process is allowed to take place before the crystallization starts. Ideally, the crystallization occurs in the crystallizable polymer-rich phase (assuming it is within the droplets), and these droplets are filled by the crystal morphology, as shown in Fig. 6.31. Outside of the droplets, very few crystals can be observed since the crystallizable polymer is in the minority domains.

On the other hand, when the blend is brought into the region bound by the spinodal curve, a bicontinuous phase morphology will replace the droplets due to the spinodal decomposition mechanism for the liquid–liquid phase separation. Further crystallization only takes place in the bicontinuous phase that is crystallizable polymer-rich. Practically, these morphological observations have more variety than those observed in polymer–solvent binary mixtures described in Section 3.2 of this Chapter, although the underlying thermodynamics are identical.

Experimental observations of polymer blends involving one or two crystalline components are extensive. However, most of the studies are focused on how crystallization behavior changes in the blends compared to the corresponding homopolymers and where the non-crystallizable components are located, such as between the crystalline lamellae or in front of the crystalline aggregates. Little attention has been paid to the phenomenon of sequential liquid–liquid phase separation and crystallization in polymer blends. As a result, only a few studies can be found in the literature. The first example is that of a binary blend of poly(vinylidene fluoride) and poly(1,4-butylene adipate). The temperature–concentration phase diagram at atmospheric pressure is shown in Fig. 6.32 (Fujita *et al.*, 1996; Penning and Manley, 1996a,b). It is interesting that in this binary blend, the liquid–liquid phase separation takes place at the uppermost temperature region, indicating a lower critical solution temperature behavior. This example indicates that at high temperatures liquid–liquid phase separation occurs, while at low temperatures both polymers are miscible. Below the lower critical solution temperature binodal curve, two crystalline solid-melt phase boundaries for both poly(vinylidene fluoride) and poly(1,4-butylene adipate) appear. At the lowest temperatures, single glass transition temperatures can be observed for each blend composition—additional evidence indicating that these two polymers form a mixed phase in the amorphous state.

In this blend system, the liquid–liquid phase separation is stable in the high temperature region. If this phase separation does not reach its final equilibrium at high temperatures above the binodal curve, it is again a "stable metastable" phase. So far, most attention has been paid to the study of poly(vinylidene fluoride) and poly(1,4-butylene adipate) crystallization from the homogeneous single phase melt, namely, below the binodal curve. Their separated lamellar crystal phases, miscible amorphous phase after crystallization (Liu *et al.*, 1997, 2000) and the composition effect on crystal morphologies (Isayeva *et al.*, 1998) have been reported.

One issue which has not been sufficiently addressed is how the crystallization would take place if it started in the phase-separated region. Can we retain the metastable phase morphologies of droplets formed by the nucleation-limited process

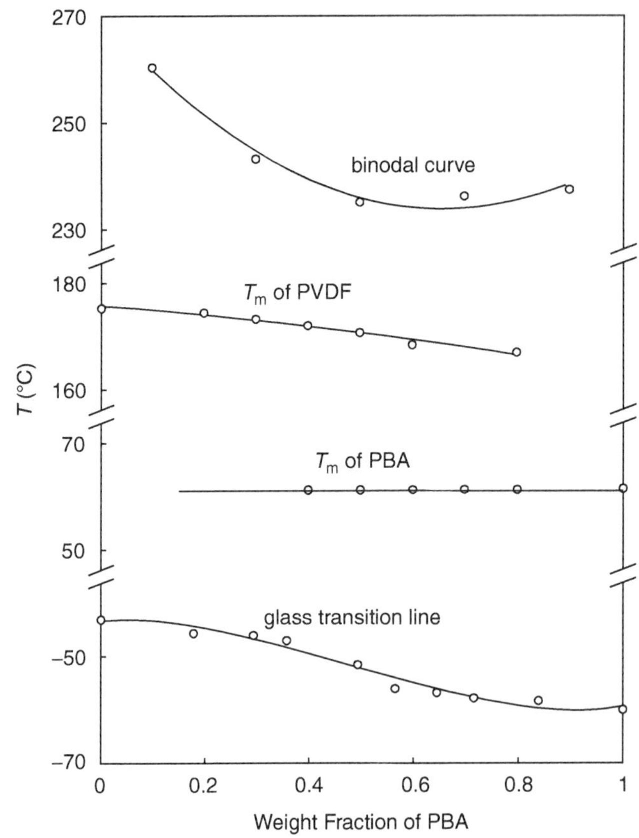

Figure 6.32 Temperature–concentration phase diagram of a binary blend consisting of poly (vinylidene fluoride) and poly(1,4-butylene adipate). The uppermost line is the binodal curve for the liquid–liquid phase separation with the lower critical solution temperature. Underneath the binodal curve is the crystalline solid-melt phase boundary of poly(vinylidene fluoride), followed by that of poly(1,4-butylene adipate). The lowest temperature line is for the glass transition of the miscible blend [replotted from Penning and Manley (1996a), with permission].

or the bicontinuous phase formed by the spinodal decomposition process after crystallization? As long as the crystallization rate is fast enough, retention of the phase morphology may be possible. A similar example is the binary blends of crystalline poly(ε-caprolactone) and polycarbonate, for which a liquid–liquid phase separation also possessing a lower critical solution temperature behavior was reported (Cheung and Stein, 1994; Cheung *et al.*, 1994).

On the other hand, a series of blends of isotactic polypropylene with hydrogenated polystyrene-*block*-polybutadiene were studied. In this series of polymer blends, isotactic polypropylene had a weight average molecular weight of 350 kg/mol and a polydispersity of 7, while the hydrogenated polystyrene-*block*-polybutadiene possessed a weight average molecular weight of 300 kg/mol with a styrene content of 10%. The blends exhibited upper critical solution temperature phase behavior, as determined by real-time, small angle light scattering experiments. Based on transmission electron microscopy observations, it was found that different crystal morphologies and sizes were identified above and below the binodal curve, indicating that the liquid–liquid phase separation had occurred (Otsuka *et al.*, 1998).

An interesting example is a series of binary blends of crystalline poly(vinylidene fluoride) with a number average molecular weight of 70 kg/mol, and poly(methyl methacrylate) with a weight average molecular weight of 110 kg/mol and a polydispersity of 2.2. These blends were also found to possess a lower critical solution temperature around 350 °C (Bernstein *et al.*, 1977). Below this temperature, it was determined that both components are miscible in the amorphous state. However, it was also discovered that, based on light scattering and morphological observations, these blends might also exhibit an upper critical solution temperature at low temperatures, as shown in Fig. 6.33. Therefore, poly(vinylidene fluoride) can crystallize in the miscible melt of the blends in a relatively narrow temperature region between the melting temperatures and the binodal curve or below the binodal curve of the upper critical solution temperature (Saito *et al.*, 1987). When the crystallization of poly(vinylidene fluoride) took place from the miscible melt, a large spherulitic morphology was observed; while, only a modulated morphology was found when the blend was quenched into the binodal region. This

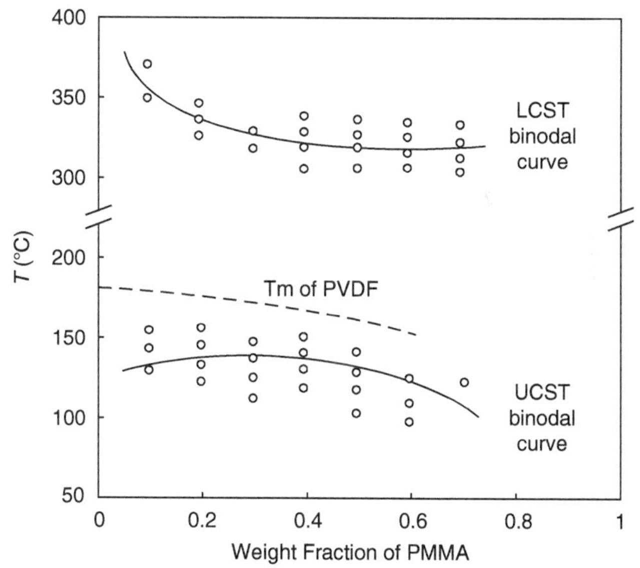

Figure 6.33 Temperature–concentration phase diagram of a binary blend consisting of poly(vinylidene fluoride) and poly(methyl methacrylate). The uppermost line is the binodal curve for the liquid–liquid phase separation of the lower critical solution temperature. Underneath the binodal curve is the crystalline solid-melt phase boundary of poly(vinylidene fluoride), followed by another binodal curve for the liquid–liquid phase separation of the upper critical solution temperature [replotted from Saito *et al.* (1987), with permission].

morphological transition occurred at the boundary of the binodal curve, as indicated in Fig. 6.33, providing evidence that an upper critical solution temperature liquid–liquid phase separation had occurred.

In fact, the crystallization of poly(vinylidene fluoride) may be designed to go through different pathways. It can be crystallized from the phase-separated melt by quenching the sample from above the lower critical solution temperature, around 350 °C, with different degrees of metastable phase separation or by crystallizing from the miscible melt above the melting temperature of the poly(vinylidene fluoride) crystals but below the lower critical solution temperature to lower temperatures of the miscible melt. To monitor the phase separation effects on poly(vinylidene fluoride) crystallization, one could also quench the sample from the phase-separated melt at high temperatures into the binodal region of the upper critical solution temperature at low temperatures. Complicated phase morphology, similar to Fig. 4.25C, may appear, and poly(vinylidene fluoride) crystallization in this type of phase morphology could give some interesting insights.

4. Liquid–Liquid Phase Separation Associated with Gelation and Crystallization

The possibility of whole hierarchies of metastabilities in polymers has been described and has already been featured in some of the foregoing aspects of liquid–liquid phase separation coupled with vitrification and crystallization. On the first level (frequently, in the classical sense), metastable states exist within each of these processes. On the second, and usually larger length scale, these processes also either intersect with each other or one overrides the other on the pathway to ultimate stability.

In what follows, we shall specifically discuss a feature in the liquid–liquid phase-separated morphologies that is directly associated with the polymer-rich phase. This association leads to the issue of connectivity in this phase. If the vitrified polymer-rich phase becomes connected throughout the macroscopic scale, a gel is formed. In fact, it is through the topic of physical gelation—the experimental observation which a solution of amorphous polystyrene can set as a gel upon cooling (Arnauts and Berghmans, 1987; Hikmet *et al.*, 1988)—that this entire subject area has come to recent prominence, although much is still under further investigation. Originally, the connectivity required for gelation was considered as being provided by chain molecules which need to be long enough to become incorporated into more than one vitrified polymer-rich domain (Keller, 1995). The long, solvated chains form "rubbery" bridges between these domains. Through experiments with nearly monodisperse amorphous polystyrene, the molecular weight requirement for establishing the connectivity to form a gel was found to follow the anticipated scaling relation (Arnauts *et al.*, 1993; Callister *et al.*, 1990). Such gels have a "wobbly" consistency in keeping with the usual concept of a gel.

In addition to such solvated chain connectivity, overall connectivity can also be established through the continuity of the glassy state. In this case, the "gel" is formed by connections on a larger length scale. They will be stiff, robust, and glass-like, as in the rather self-evident case where phase continuity arises through the high concentration of polymer (the case of matrix inversion). The gel will be fragile in the case where the phase continuity arises through spinodal decomposition in liquid–liquid phase separation, particularly at low polymer concentrations in which the morphology is a fine, continuous, glassy network. In fact, under such conditions gelation can also serve to identify the spinodal curve in the phase diagram (Arnauts and Berghmans, 1987, 1993; Keller, 1995). Specifically, the two kinds of gel, namely "wobbly" and "stiff," could be distinguished through the dramatic differences in mechanical properties which, when followed as a function of concentration at a particular temperature, can also serve to delineate the spinodal curve (Callister *et al.*, 1990).

As is the case with all the morphologies described previously, those leading to connectivity, and therefore gels, correspond to metastable states. In fact, all physical gels formed by chemically uniform homopolymers and arising through a phase transition, whether crystallization or liquid–liquid phase separation, are necessarily in a metastable state where some types of energy (either free energy barriers or intervening of another transition process) must act to arrest the phase transformation from going to completion. Therefore, these gels are at a stage of bicontinuous connectivity. In the present and conceptually simplest case, this process is vitrification.

Now we will consider more complicated hierarchical metastable states on different length scales. For instance, liquid–liquid phase separation can be coupled with gelation and crystallization in one or more components during cooling. In this case, phase separation, gelation, and crystallization compete with each other, and the ultimate morphology is often determined not only by the sequence of phase transitions, but also by the interactions between the molecules or segments during the phase transitions. It is known that the crystallization process has stronger interactions, while the liquid–liquid phase separation process has weaker interactions.

On the other hand, gelation virtually stops large-scale molecular diffusion of polymers. Figure 6.34 shows a representative example of isotactic polystyrene in *trans*-decalin. Here, the uppermost phase boundary corresponds to the formation of folded chain single crystals with the chains in the long-established, normal 3_1 helical structure (Natta *et al.*, 1960). Such stable crystals (ignoring for the present that the folded chain state is, in itself, metastable) precipitate upon cooling, giving rise to a turbid suspension. If crystals form in sufficient concentration, the system can become a molecularly connected gel (gel I in Fig. 6.34). However, when cooling is not sufficiently slow, the liquid reaches a new crystal phase boundary before the above-mentioned crystallization has had a chance to take place. At or somewhat below the new crystal–liquid phase boundary shown in Fig. 6.34, a new observation becomes evident, namely the whole system sets as a transparent gel very rapidly, even at low concentrations. For this reason, this subject has come to the forefront in connection with gelation studies. For our present purposes, however, gelation is not the main point of emphasis; it merely serves as a simple indicator of a phase change having taken place far below the usual crystalline solid–liquid phase boundary.

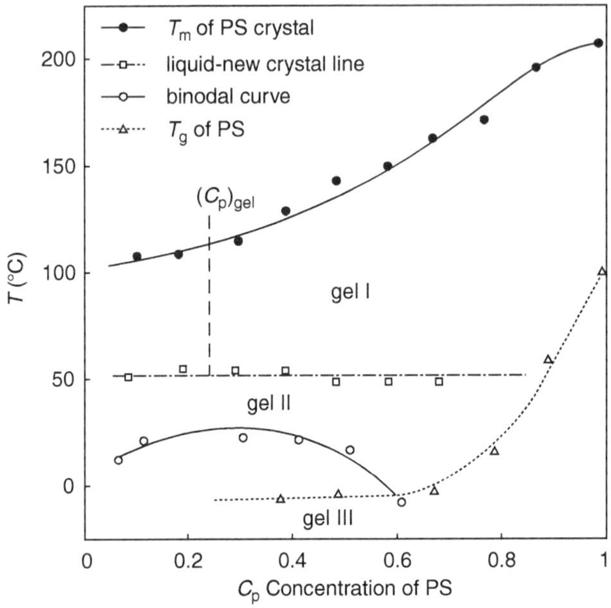

Figure 6.34 Temperature–concentration phase diagram for isotactic polystyrene in *trans*-decalin. The upper line represents the crystal–liquid boundary of a normal crystal with a 3_1 helical conformation. In the low temperature side, a liquid–liquid phase separation region can be observed, and it is bound by the binodal curve. A vitrification process also intersects the binodal curve providing a Berghmans' point. The new crystalline solid–liquid boundary in between the crystalline solid–liquid boundary and the binodal curve displays the intervention of another metastable phase. Each of these phase transformations can cause the gel formation. Gel I is formed between the crystalline solid-crystalline solid and new crystalline solid–liquid boundaries above a certain concentration $(C_p)_{gel}$. The liquid–liquid phase separation combined with the glass transition temperature forms gel III. Below the new crystal–liquid boundary, gel II is formed with its own characteristics [replotted from Keller (1995), with permission].

The detailed nature of the metastable state related to gels is far from fully resolved and is still the subject of some arguments and further investigations (Keller, 1995). It has the character of the crystal phase itself to create connectivity and, hence, is a source of gelation (gel II in Fig. 6.34). The crystalline nature is apparent in wide angle x-ray diffraction experiments, although in many cases, it is difficult to obtain such patterns. Atkins *et al.* obtained wide angle x-ray diffraction patterns of crystals in gels, which indicate a highly extended chain conformation with a possible 12_1 helix (1977, 1980). The resulting morphology corresponds to a fibrillar structure (Atkins *et al.*, 1984). We infer here that the isotactic polystyrene chains do not fold, but are, more or less, extended as a consequence of their association and packing upon crystallization. Therefore, the morphology favors the observed proclivity to form gels by the chain connectivity. The horizontal character of the new crystalline solid–liquid boundary in Fig. 6.34 suggests some kind of complex formation with the solvent or crystal solvates (Atkins *et al.*, 1984). This characteristic seems to be a common feature of this class of phenomena in similar systems. The stochiometry of the crystal solvates, however, is not yet established, and such complex formation and/or crystal solvates do not seem to be associated with a specific solvent. However, the basic features of this effect are displayed in a range of different solvents. The formation of the complex and/or crystal solvates is a newly emerging subject which is important in its own right; we will not give it further emphasis in the present subject of metastable states.

Moving to still lower temperatures in Fig. 6.34, we reach the liquid–liquid phase separation region bound by the binodal curve. This region is manifested through turbidity appearing in the initially transparent system. In contrast to the turbidity due to the 3_1 helix containing the crystals, which appears beneath the crystal–liquid phase boundary after a sufficiently long residence time at higher temperatures, the appearance of turbidity here is instantaneous and reversible with temperature, in compliance with the nature of liquid–liquid phase separation. As the new crystalline solid–liquid phase boundary cannot be traversed without some transformation occurring, the liquid–liquid phase separation itself is taking place within an already at least partially gelled system. Namely, chains or portions of the chains that have not yet taken part in the formation of the gel junctions (those corresponding to type gel II in Fig. 6.34) are involved in the liquid–liquid phase separation at the correspondingly lower temperature. This physical picture is likely to be supported by experimental observations.

For example, the system was being held for successively longer times in the temperature interval between the new crystal–liquid phase boundary and the liquid–liquid phase separation curve. The strength of the liquid–liquid phase separation on subsequent cooling was found to decrease progressively (Keller, 1995). Additionally, the process of liquid transformation to the new crystal consumes increasingly greater amounts of the material with increased residence time. This left less and less polymer for the liquid–liquid phase separation to occur on subsequent cooling.

Upon further cooling to below the liquid–liquid phase separation curve, the intersection with the glass transition temperature line is eventually reached, as revealed by invariance in the glass transition temperatures (detected by calorimetric measurements). Here, we almost return to the case of intervention by vitrification described in Section 2 of this Chapter. The difference from the previously discussed cases is that in the present situation, the entire vitrifying system is in a metastable state which is deeply buried beneath both the crystal–liquid and new crystal–liquid phase boundaries. Also, the intersection of the binodal curve with the glass transition temperature line should lead to a gel, but this gel will now be within a gel that already formed when traversing the new crystal–liquid phase boundary. Therefore, it is not expected to be apparent as gel formation itself, but merely as an observation of enhanced stiffness of an already stiff gel. So far, this phenomenon has received only a qualitative description; a more quantitative description is still awaited.

Regarding the subject of gelation, Fig. 6.34 contains information on three different sources of gel formation, either simultaneously or separately in the same system. In this figure, the turbid gel I corresponds to the formation of molecular connectivity between suspended single crystals (with chain folding), consisting of 3_1 helices formed from solutions of sufficiently high concentration, $(C_p)_{gel}$, beneath the crystal–liquid phase boundary. The transparent gel II arises through the intrinsic gel-forming ability of the extended chain type fibrous crystals, consisting possibly of 12_1 helices beneath the new crystal–liquid phase boundary. This is the most prominent of the three gelation processes. The third type of gel, gel III, is due to the intersection of the liquid–liquid phase separation by the vitrification in the presence of sufficient phase connectivity, which has been discussed separately in Section 2 of this Chapter.

The complexity and richness in phase behavior arising from the range of possible metastable states, even when confined to classical metastable states, is apparent from the preceding discussions. It appears that regarding systems such as those represented in Fig. 6.34, we are still in the beginning of mapping the process. This step is only the first toward reaching a comprehensive understanding of different metastable states and their relationships. There is an increasing number of indications that at sufficiently high degrees of supercooling, yet still at or above the new crystal–liquid phase boundary, the conformation of the random coil itself undergoes a coil-to-helix transformation before the onset of crystallization. This conformational change has been observed in the case of isotactic polystyrene in cyclohexanol, as described in Fig. 6.34. Others include the cases of isotactic poly (methyl methacrylate) in toluene (Berghmans *et al.*, 1994) and in 2-butanone (Buyse and Berghmans, 2000), syndiotactic polystyrene in *cis*-decalin (Deberdt and Berghmans, 1993; Roels *et al.*, 1997) and in bromoform (De Rudder *et al.*, 2002), and syndiotactic poly(methyl methacrylate) in toluene (Buyse *et al.*, 1998), based on careful spectroscopic identifications. These examples would mean that at sufficiently high supercooling, the helix, most likely stabilized by association with solvent, is stable with respect to the random coil but is metastable with respect to the ultimately stable crystal represented by the crystalline solid–liquid phase boundary. Accordingly, it would be these newly formed helices that would produce the gel (gel II) by association when passing through the new crystal–liquid phase boundary. Therefore, we would have a case in which the two ingredients of crystallization, developing a regular chain conformation and fitting these chains into a lattice, would occur consecutively, as opposed to simultaneously as normally envisaged. Full confirmation of this concept is still required. The remaining uncertainties at the moment lie in the problem of making sure that helix formation actually precedes the association (the gel state), which is still problematic because once the helix has formed, the association is very fast. Even so, the above possibility is likely and, if real, is of potential importance.

General principles shown in Fig. 6.34 are certainly not only illustrating the case of isotactic polystyrene in *trans*-decalin. An increasing number of examples have been reported, which fall into the discussion of Fig. 6.34. These examples include syndiotactic poly(methyl methacrylate) in toluene (Berghmans *et al.*, 1994), isotactic poly(methyl methacrylate) in 2-butanone (Buyse and Berghmans, 2000), and syndiotactic polystyrene in several solvents, such as *cis*-decalin, *o*-xylene, some chlorobenzenes, and bromoform (Berghmans and Deberdt, 1994; De Rudder *et al.*, 2002; Deberdt and Berghmans, 1993, 1994; Roels *et al.*, 1994, 1997). Another aspect of the study described here is to link these phase behaviors with biopolymers in solutions like carrageenans, gelatin, and others. In most cases, the solvent in these systems is water (Bongaerts *et al.*, 2000).

Finally, a comment should be specifically made about the competition among the phase transformation kinetics. As previously stated, crystallization on passing through the crystalline solid–liquid phase boundary is usually very slow in these systems, which is the general reason why the various buried metastable states can be attained. However, there is more to it than merely a sufficiently fast cooling rate to outpace the increasing crystallization rate with increasing supercooling. It has been found that in the case of isotactic polystyrene, the actual rate of the stable 3_1 helix crystal formation does not accelerate within the full temperature range between the crystal–liquid and new crystal–liquid phase boundaries, but rather, after passing through a maximum, the rate goes to a minimum that reaches negligible values before the new crystalline solid–liquid processes take off. Therefore, it seems that both processes hamper each other in the temperature range where their regions of stability start overlapping.

Such a possibility has been described in connection with region II of Fig. 3.8 in Section 3 of Chapter 3, and has been previously supported by the observations of a pronounced minimum in the crystallization rates of oligomers such as *n*-alkanes (Boda *et al.*, 1997; Hobbs *et al.*, 2001; Hosier *et al.*, 2000; Morgan *et al.*, 1998; Organ *et al.*, 1989, 1997; Putra and Ungar, 2003; Sutton *et al.*, 1996; Ungar and Keller, 1986, 1987; Ungar and Zeng, 2001; Ungar *et al.*, 2000, 2005) and low molecular weight poly(ethylene oxide) at the transition region between once-folded and extended chain crystals (Cheng and Chen, 1991), as described in Section 1.5 of Chapter 4. We are now seeing the same effect, a rate minimum in the overall transformation process, in the presently discussed polymer–solvent system, underpinning both the reality and generality of the underlying considerations.

Let us now briefly examine the last example which represents another type of binary mixture and is potentially interesting in biosystems but can fall into the phase behavior discussed in this section. It is a binary mixture of water-soluble and uncrystallizable poly(vinyl methyl ether) with deuterium oxide. The phase behavior of this system is interesting since the deuterium oxide solvent can crystallize, rather than the polymer. Based on small-angle neutron scattering experiments, this system possesses two lower critical solution temperatures at high temperatures, having bimodality of the liquid–liquid phase separation with two stable critical points (Nies *et al.*, 2005), as shown in Fig. 6.35. Moreover, at low temperatures, a narrow upper critical solution temperature can also be found in the poly(vinyl methyl ether) concentration region between 0.6 and 1.0 (Nies *et al.*, 2006), within which heterogeneous two-phase structures form due to the liquid–liquid phase separation. In this figure, the crystalline solid–liquid phase boundary of deuterium oxide is also included in addition to the glass transition temperature depression with increasing concentration of deuterium oxide. These data are derived from the work of Van Durme *et al.* (2005). This crystalline solid–liquid phase boundary is in between the lower and upper critical solution temperatures, and it should be close to equilibrium. At a sufficiently high polymer concentration of above ~0.6, the solvent ceases to crystallize (Meeussen *et al.*, 2000; Zhang *et al.*, 2003). This finding was originally attributed to the strong interaction between the polymer and solvent caused by hydrogen-bonded complexes (Maeda, 2001; Meeussen *et al.*, 2000).

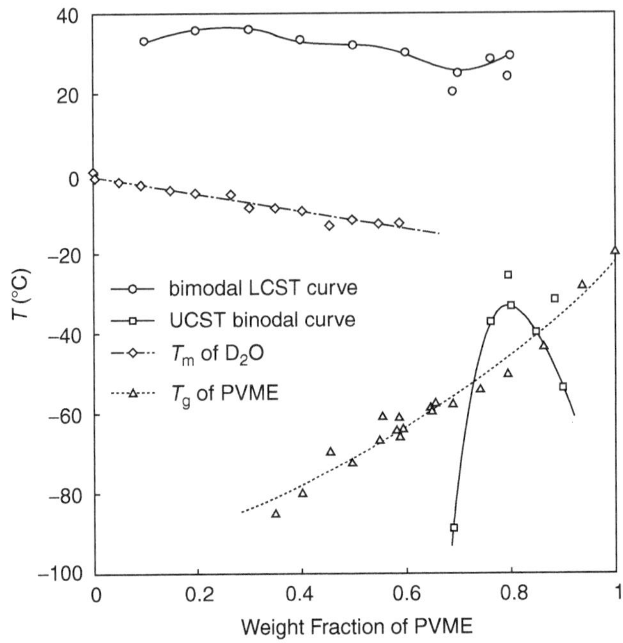

Figure 6.35 Temperature–concentration phase diagram for binary mixtures of poly(vinyl methyl ether) with deuterium oxide. Above 20 °C, there is a lower critical solution temperature binodal curve (spherical symbols experimentally observed data). Slightly below 0 °C, it is the crystalline solid–liquid boundary for deuterium oxide. This boundary ends at around a 0.6 concentration of poly(vinyl methyl ether) (triangle symbols). In a narrow concentration region between 0.6 and 1, a liquid–liquid phase separation with an upper critical solution temperature can be observed below −30 °C (square symbols). Finally, the glass transition temperature decreases with concentration of the polymer as shown by triangle symbols [replotted from Nies *et al.* (2006), with permission].

However, recent theoretical calculations do not support this complex formation. The cause of this crystallization is thus attributed to the newly discovered upper critical solution temperature phase separation (Nies *et al.*, 2006). Irrespective of the origin of this heterogeneous structure, two different "gels" may be expected. One of the "gels" is below the crystal–liquid phase boundary after the crystallization of deuterium oxide when the polymer concentration is lower than 0.6, while the other forms by liquid–liquid phase separation. These two "gels" must exhibit different mechanical responses. Most importantly, they are not in thermodynamic equilibrium, but are at various stages of metastability. The phase behavior described in Fig. 6.35 again provides different pathways to illustrate hierarchical metastable states, yet the investigation is just starting.

The glass transition temperature depression passes through the upper critical solution temperature region, as shown in Fig. 6.35. In principle, within this narrow concentration region, there should be two glass transition temperatures that are representative of the two phase-separated domains with different compositions. Both of these glass transition temperatures must be constant with respect to concentration

(Section 2 of this Chapter). However, since the concentration region for this upper critical solution temperature is narrow, the glass transition temperatures are close enough to overlap causing one relatively broad transition to be observed. The apparent decrease of this glass transition temperature may be due to the different composition partitions of these two phases from liquid–liquid phase separation.

REFERENCES AND FURTHER READING

Arnauts, J., and Berghmans, H. (1987). Amorphous thermoreversible gels of atactic polystyrene. *Polym. Commun.* **28**, 66–68.

Arnauts, J., Berghmans, H., and Koningsveld, R. (1993). Structure formation in solutions of atactic polystyrene in trans-decalin. *Die Makromol. Chem.* **194**, 77–85.

Atkins, E. D. T., Isaac, D. H., Keller, A., and Miyasaka, K. (1977). Analysis of anomalous X-ray diffraction effects of isotactic polystyrene gels and its implications for chain conformation and isomeric homogeneity. *J. Polym. Sci. Polym. Phys. Edn.* **15**, 211–226.

Atkins, E. D. T., Isaac, D. H., and Keller, A. (1980). Conformation of polystyrene with special emphasis to the near all-trans extended-chain model relevant in polystyrene gels. *J. Polym. Sci. Polym. Phys. Edn.* **18**, 71–82.

Atkins, E. D. T., Hill, M. J., Jarvis, D. A., Keller, A., Sarhene, E., and Shapiro, J. S. (1984). Structural studies on gels from isotactic polystyrene. *Colloid Polym. Sci.* **262**, 22–45.

Bassett, D. C., and Turner, B. (1972). New high-pressure phase in chain-extended crystallization of polythene. *Nat. Phys. Sci.* **240**, 146–148.

Bassett, D. C., and Turner, B. (1974a). On chain-extended and chainfolded crystallization of polyethylene. *Phil. Mag.* **29**, 285–307.

Bassett, D. C., and Turner, B. (1974b). On the phenomenology of chain-extended crystallization in polyethylene. *Phil. Mag.* **29**, 925–955.

Berghmans, H., and Deberdt, F. (1994). Phase behavior and structure formation in solutions of vinyl polymers. *Phil. Trans.: Phys. Sci. Eng.* **348**, 117–130.

Berghmans, M., Thijs, S., Cornette, M., Berghmans, H., De Schryver, F. C., Moldenaers, P., and Mewis, J. (1994). Thermoreversible gelation of solutions of syndiotactic poly(methyl methacrylate) in toluene: A two-step mechanism. *Macromolecules* **27**, 7669–7676.

Berghmans, S., Mewis, J., Berghmans, H., and Meijer, H. (1995). Phase behavior and structure formation in solutions of poly(2,6-dimethyl-1,4-phenylene ether). *Polymer* **36**, 3085–3091.

Bernstein, R. E., Cruz, C. A., Paul, D. R., and Barlow, J. W. (1977). LCST behaviors in polymer blends. *Macromolecules* **10**, 681–686.

Boda, E., Ungar, G., Brooke, G. M., Burnett, S., Mohammed, S., Proctor, D., and Whiting, M. C. (1997). Crystallization rate minima in a series of *n*-alkanes from $C_{194}H_{390}$ to $C_{294}H_{590}$. *Macromolecules* **30**, 4674–4678.

Bongaerts, K., Paoletti, S., Denef, B., Vanneste, K., Cuppo, F., and Reynaers, H. (2000). Light scattering investigation of ι-carrageenan aqueous solutions. Concentration dependence of association. *Macromolecules* **33**, 8709–8719.

Briber, R. M., and Khoury, F. (1987). The phase diagram and morphology of blends of poly (vinylidene fluoride) and poly(ethyl acrylate). *Polymer* **28**, 38–46.

Brückner, S., and Meille, S. V. (1989). Non-parallel chains in crystalline (γ-isotactic polypropylene. *Nature* **340**, 455–457.

Buyse, K., and Berghmans, H. (2000). Thermoreversible gelation of solutions of isotactic poly(methyl methacrylate) in 2-butanone. *Polymer* **41**, 1045–1053.

Buyse, K., Berghmans, H., Bosco, M., and Paoletti, S. (1998). Mechanistic aspects of the thermoreversible gelation of syndiotactic poly(methyl methacrylate) in toluene. *Macromolecules* **31**, 9224–9230.

Callister, S., Keller, A., and Hikmet, R. M. (1990). On thermoreversible gels: Their classification, relation to phase transitions and vitrification, their morphology and properties. *Die Makromol. Chem. Macromol. Symposia* **39**, 19–54.

Cartier, L., Okihara, T., and Lotz, B. (1998). The α'' "superstructure" of syndiotactic polystyrene: A frustrated structure. *Macromolecules* **31**, 3303–3310.
Cheng, S. Z. D., and Chen, J. H. (1991). Nonintegral and integral folding crystal growth in low-molecular mass poly(ethylene oxide) fractions. III. Linear crystal growth rates and crystal morphology. *J. Polym. Sci. Polym. Phys. Edn.* **29**, 311–327.
Cheng, S. Z. D., Zhu, L., Li, C. Y., Honigfort, P. S., and Keller, A. (1999). Size effect of metastable states on semicrystalline polymer structures and morphologies. *Thermochim. Acta* **332**, 105–113.
Cheung, Y. W., and Stein, R. S. (1994). Critical analysis of the phase behavior of poly(ε-caprolactone) (PCL)/polycarbonate (PC) blends. *Macromolecules* **27**, 2512–2519.
Cheung, Y. W., Stein, R. S., Lin, J. S., and Wignall, G. D. (1994). Small-angle scattering investigation of poly(ε-caprolactone)/polycarbonate blends. 2. Small-angle X-ray and light scattering study of semicrystalline/semicrystalline and semicrystalline/amorphous blend morphologies. *Macromolecules* **27**, 2520–2528.
Crist, B., and Hill, M. J. (1997). Recent development in phase separation of polyolefin melt blends. *J. Polym. Sci. Polym. Phys. Edn.* **35**, 2329–2353.
De Rosa, C. (1996). Crystal structure of the trigonal modification (α form) of syndiotactic polystyrene. *Macromolecules* **29**, 8460–8465.
De Rosa, C., Rapacciuolo, M., Guerra, G., Petraccone, V., and Corradini, P. (1992). On the crystal-structure of the orthorhombic form of syndiotactic polystyrene. *Polymer* **33**, 1423–1428.
De Rudder, J., Berghmans, H., and Arnauts, J. (1999). Phase behavior and structure formation in the system syndiotactic polystyrene/cyclohexanol. *Polymer* **40**, 5919–5928.
De Rudder, J., Berghmans, H., De Schryver, F. C., Bosco, M., and Paoletti, S. (2002). Gelation mechanism of syndiotactic polystyrene in bromoform. *Macromolecules* **35**, 9529–9535.
Deberdt, F., and Berghmans, H. (1993). Phase behavior of syndiotactic polystyrene-decalin. *Polymer* **34**, 2192–2201.
Deberdt, F., and Berghmans, H. (1994). Phase behavior of syndiotactic polystyrene-*o*-xylene. *Polymer* **35**, 1694–1704.
Dafay, R., Prigogine, I., Bellemans, A., and Everett, D. H. (1966). "Surface Tension and Adsorption." Wiley, New York.
Endres, B., Garbella, R. W., and Wendorff, J. H. (1985). Studies on phase separation and coarsening in blends of poly(vinylidene fluoride) and poly(ethyl acrylate). *Colloid Polym. Sci.* **263**, 361–371.
Evans, R. (1990). Fluids adsorbed in narrow pores: Phase equilibria and structure. *J. Phys.: Condens. Matter* **2**, 8989–9007.
Finter, J., and Wegner, G. (1981). The relation between phase transition and crystallization behavior of 1,4-trans-poly(butadiene). *Die Makromol. Chem.* **182**, 1859–1874.
Foresta, T., Piccarolo, S., and Goldbeck-Wood, G. (2001). Competition between α and γ phases in isotactic polypropylene: Effects of ethylene content and nucleating agents at different cooling rates. *Polymer* **42**, 1167–1176.
Frank, F. C., and Keller, A. (1988). Two-fluid phase separation: Modified by a glass transition. *Polym. Commun.* **29**, 186–189.
Fujita, K., Kyu, T., and Manley, R. St. J. (1996). Miscible blends of two crystalline polymers. 3. Liquid-liquid phase separation in blends of poly(vinylidene fluoride)/poly(butylene adipate). *Macromolecules* **29**, 91–96.
Garber, C. A., and Geil, P. H. (1966). Solution crystallization of poly-3,3-bis(chloromethyl)-oxacyclobutane. *J. Appl. Phys.* **37**, 4034–4040.
Geil, P. H., Anderson, F. R., Wunderlich, B., and Arakawa, T. (1964). Morphology of polyethylene crystallized from the melt under pressure. *J. Polym. Sci. Part A* **2**, 3707–3720.
Guerra, G., Vitagliano, V. M., De Rosa, C., Petraccone, V., and Corradini, P. (1990). Polymorphism in melt crystallized syndiotactic polystyrene samples. *Macromolecules* **23**, 1539–1544.
Guerra, G., De Rosa, C., Vitagliano, V. M., Petraccone, V., and Corradini, P. (1991). Effects of blending on the polymorphic behavior of melt-crystallized syndiotactic polystyrene. *J. Polym. Sci. Polym. Phy. Edn.* **29**, 265–271.
Gutzow, I., and Toschev, S. (1968). Non-steady state nucleation in the formation of isotropic and anisotropic phases. *Kristall und Technik* **3**, 485–497.

Hay, I. L., and Keller, A. (1965). Polymer deformation in terms of spherulites. *Kolloid-Zeitschrift* **204**, 43–74.

Hikmet, R. M., Callister, S., and Keller, A. (1988). Thermoreversible gelation of atactic polystyrene: Phase transformation and morphology. *Polymer* **29**, 1378–1388.

Hikosaka, M. (1987). Unified theory of nucleation of folded-chain crystals and extended-chain crystals of linear-chain polymers. *Polymer* **28**, 1257–1264.

Hikosaka, M. (1990). Unified theory of nucleation of folded-chain crystals (FCCs) and extended-chain crystals (ECCs) of linear-chain polymers. 2. Origin of FCC and ECC. *Polymer* **31**, 458–468.

Hikosaka, M., Rastogi, S., Keller, A., and Kawabata, H. (1992). Investigations on the crystallization of polyethylene under high pressure: Role of mobile phase, lamellar thickening growth, phase transformations, and morphology. *J. Macromol. Sci. Part B: Phys.* **31**, 87–131.

Hikosaka, M., Okada, H., Toda, A., Rastogi, S., and Keller, A. (1995). Dependence of the lamellar thickness of an extended-chain single crystal of polyethylene on the degree of supercooling and the pressure. *J. Chem. Soc. Faraday Trans.* **91**, 2573–2579.

Hikosaka, M., Amano, K., Rastogi, S., and Keller, A. (1997). Lamellar thickening growth of an extended chain single crystal of polyethylene. 1. Pointers to a new crystallization mechanism of polymers. *Macromolecules* **30**, 2067–2074.

Hikosaka, M., Amano, K., Rastogi, S., and Keller, A. (2000). Lamellar thickening growth of an extended chain single crystal of polyethylene (II): ΔT dependence of lamellar thickening growth rate and comparison with lamellar thickening. *J. Mater. Sci.* **35**, 5157–5168.

Ho, R.-M., Lin, C.-P., Tsai, H.-Y., and Woo, E.-M. (2000). Metastability study of syndiotactic polystyrene polymorphism. *Macromolecules* **33**, 6517–6526.

Ho, R.-M., Lin, C.-P., Hseih, P.-Y., Chung, T.-M., and Tsai, H.-Y. (2001). Isothermal crystallization-induced phase transition of syndiotactic polystyrene polymorphism. *Macromolecules* **34**, 6727–6736.

Hobbs, J. K., Hill, M. J., and Barham, P. J. (2001). Crystallization and isothermal thickening of single crystals of $C_{246}H_{494}$ in dilute solution. *Polymer* **42**, 2167–2176.

Hosier, I. L., Bassett, D. C., and Vaughan, A. S. (2000). Spherulitic growth and cellulation in dilute blends of monodisperse long n-alkanes. *Macromolecules* **33**, 8781–8790.

Isayeva, I., Kyu, T., and Manley, R. St. J. (1998). Phase transitions, structure evolution, and mechanical properties of blends of two crystalline polymers: Poly(vinylidene fluoride) and poly(butylene adipate). *Polymer* **39**, 4599–4608.

Iwayanagi, S., Sakurai, I., Sakurai, T., and Seto, T. (1968). X-ray structure analysis of trans-1, 4-polybutadiene. *J. Macromol. Sci. Part B: Phys.* **2**, 163–177.

Keller, A. (1995). Introductory lecture: Aspects of polymer gels. *Faraday Discuss.* **101**, 1–49.

Keller, A., and Cheng, S. Z. D. (1998). The role of metastability in polymer phase transitions. *Polymer* **39**, 4461–4487.

Keller, A., Hikosaka, M., Rastogi, S., Toda, A., Barham, P. J., and Goldback-Wood, G. (1994). An approach to the formation and growth of new phases with application to polymer crystallization: Effect of finite size, metastability, and Ostwald's rule of stages. *J. Mater. Sci.* **29**, 2579–2604.

Keller, A., Hikosaka, M., and Rastogi, S. (1996). The role of metastability in phase transformations: New pointers through polymer mesophases. *Phys. Scr.* **T66**, 243–247.

Khoury, F., and Barnes, J. D. (1972). Formation of curved polymer crystals. Poly(4-methyl-1-pentene). *J. Res. Natl. Bur. Stand. Sect. A: Phys. Chem.* **76**, 225–252.

Khoury, F., and Barnes, J. D. (1974a). Formation of curved polymer crystals. Poly(oxymethylene). *J. Res. Natl. Bur. Stand. Sect. A: Phys. Chem.* **78**, 95–128.

Khoury, F., and Barnes, J. D. (1974b). Formation of curved polymer crystals. Poly(chlorotrifluoroethylene). *J. Res. Natl. Bur. Stand. Sect. A: Phys. Chem.* **78**, 363–373.

Kojima, Y., Usuki, A., Kawasumi, M., Okada, A., Kurauchi, T., Kamigaito, O., and Kaji, K. (1994). Fine structure of nylon-6-clay hybrid. *J. Polym. Sci. Polym. Phys. Edn.* **32**, 625–630.

Kuo, C.-C., and Woodward, A. E. (1984). Morphology and properties of trans-1,4-polyisoprene crystallized from solution. *Macromolecules* **17**, 1034–1041.

Liu, L.-Z., Chu, B., Penning, J. P., and Manley, R. St. J. (1997). A synchrotron SAXS study of miscible blends of semicrystalline poly(vinylidene fluoride) and semicrystalline poly(1,4-butylene adipate). *Macromolecules* **30**, 4398–4404.

Liu, L.-Z., Chu, B., Penning, J. P., and Manley, R. St. J. (2000). A synchrotron SAXS study of miscible blends of semicrystalline poly(vinylidenefluoride) and semicrystalline poly(1,4-butylene adipate). II. Crystallization, morphology, and PBA inclusion in PVDF spherulites. *J. Polym. Sci. Polym. Phys. Edn.* **38**, 2296–2308.

Lotz, B, Wittmann, J. C., and Lovinger, A. J. (1996). Structure and morphology of poly(propylenes): A molecular analysis. *Polymer* **37**, 4979–4992.

Maeda, Y. (2001). IR spectroscopic study on the hydration and the phase transition of poly(vinyl methyl ether) in water. *Langmuir* **17**, 1737–1742.

Meeussen, F., Bauwens, Y., Moerkerke, R., Nies, E., and Berghmans, H. (2000). Molecular complex formation in the system poly(vinyl methyl ether)/water. *Polymer* **41**, 3737–3743.

Morgan, R. L., Barham, P. J., Hill, M. J., Keller, A, and Organ, S. J. (1998). The crystallization of the n-alkane $C_{294}H_{590}$ from solution: Inversion of crystallization rates, crystal thickening, and effects of supersaturation. *J. Macromol. Sci. Part B: Phys.* **37**, 319–338.

Nakajima, A., Fujiwara, H., and Hamada, F. (1966). Phase relationships and thermodynamic interactions in linear polyethylene-diluent systems. *J. Polym. Sci. Polym. Phys. Edn.* **4**, 507–518.

Natta, G., Corradini, P., and Bassi, I. W. (1960). Crystal structure of isotactic polystyrene. *Del Nuovo Cimento, Supplemento* **15**, 68–82.

Nies, E., Ramzi, A., Berghmans, H., Li, T., Heenan, R. K., and King, S. M. (2005). Composition fluctuations, phase behavior, and complex formation in poly(vinyl methyl ether)/D_2O investigated by small-angle neutron scattering. *Macromolecules* **38**, 915–924.

Nies, E., Li, T., Berghmans, H., Heenan, R. K., and King, S. M. (2006). Upper critical solution temperature phase behavior, composition fluctuations, and complex formation in poly(vinyl methyl ether)/D_2O solutions: Small-angle neutron-scattering experiments and Wertheim lattice thermodynamics perturbation theory predictions. *J. Phys. Chem. B* **110**, 5321–5329.

Nojima, S., Satoh, K., and Ashida, T. (1991). Morphology formation by combined effect of crystallization and phase separation in a binary blend of poly(ε-caprolactone) and polystyrene oligomer. *Macromolecules* **24**, 942–947.

Organ, S. J., Ungar, G., and Keller, A. (1989). Rate minimum in solution crystallization of long paraffins. *Macromolecules* **22**, 1995–2000.

Organ, S. J., Barham, P. J., Hill, M. J., Keller, A., and Morgan, R. L. (1997). A study of the crystallization of the n-alkane $C_{246}H_{494}$ from solution: Further manifestations of the inversion of crystallization rates with temperature. *J. Polym. Sci. Polym. Phys. Edn.* **35**, 1775–1791.

Ostwald, W. (1897). Studien über die Bildung und Umwandlung fester Körper. *Zeitschrift für Physikalische Chemie, Stöchiometrie und Verwandschaftslehre* **22**, 289–300.

Otsuka, N., Yang, Y., Saito, H., Inoue, T., and Takemura, Y. (1998). Phase behavior and morphology development in a blend of isotactic polypropylene and hydrogenated poly(styrene-co-butadiene). *Polymer* **39**, 1533–1538.

Paul, D. R., and Bucknall, C. B., eds. (2000). "Polymer Blends." John Wiley, New York.

Penning, J. P., and Manley, R. St. J. (1996a). Miscible blends of two crystalline polymers. 1. Phase behavior and miscibility in blends of poly(vinylidene fluoride) and poly(1,4-butylene adipate). *Macromolecules* **29**, 77–83.

Penning, J. P., and Manley, R. St. J. (1996b). Miscible blends of two crystalline polymers. 2. Crystallization kinetics and morphology in blends of poly(vinylidene fluoride) and poly(1,4-butylene adipate). *Macromolecules* **29**, 84–90.

Pradère, P., and Thomas, E. L. (1990). Antiphase boundaries and ordering defects in syndiotactic polystyrene crystals. *Macromolecules* **23**, 4954–4958.

Price, C., Harris, P. A., Holton, T. J., and Stubbersfield, R. B. (1975). Growth of lamellar crystals of poly(γ-benzyl-L-glutamate). *Polymer* **16**, 69–71.

Price, C., Holton, T. J., and Stubbersfield, R. B. (1979). Crystallization of poly(λ-benzyl-L-glutamate) from dilute solutions of hexafluoroisopropanol. *Polymer* **20**, 1059–1061.

Prime, R. B., and Wunderlich, B. (1969). Extended-chain crystals, III. Size distribution of polyethylene crystals grown under elevated pressure. *J. Polym. Sci. Polym. Phys. Edn.* **7**, 2061–2072.

Putra, E. G. R., and Ungar, G. (2003). In situ solution crystallization study of n-$C_{246}H_{494}$: Self-poisoning and morphology of polymethylene crystals. *Macromolecules* **36**, 5214–5225.

Rastogi, S., and Kurelec, L. (2000). Polymorphism in polymers, its implications for polymer crystallization. *J. Mater. Sci.* **35**, 5121–5138.

Rastogi, S., and Ungar, G. (1992). Hexagonal columnar phase in 1,4-trans-polybutadiene: Morphology, chain extension, and isothermal phase reversal. *Macromolecules* **25**, 1445–1452.

Rastogi, S., Hikosaka, M., Kawabata, H., and Keller, A. (1991). Role of mobile phases in the crystallization of polyethylene. 1. Metastability and lateral growth. *Macromolecules* **24**, 6384–6391.

Rees, D. V., and Bassett, D. C. (1968). Origin of extended-chain lamellae in polyethylene. *Nature* **219**, 368–370.

Rees, D. V., and Bassett, D. C. (1971). Crystallization of polyethylene at elevated pressures. *J. Polym. Sci. Polym. Phys. Edn.* **9**, 385–406.

Richards, R. B. (1946). The phase equilibria between a crystalline polymer and solvents. I. The effect of polymer chain length on the solubility and swelling of polythene. *Trans. Faraday Soc.* **42**, 10–20.

Roels, T., Deberdt, F., and Berghmans, H. (1994). Solvent quality and phase stability in syndiotactic polystyrene-solvent systems. *Macromolecules* **27**, 6216–6220.

Roels, T., Rostogi, S., De Rudder, J., and Berghmans, H. (1997). Temperature induced structural changes in syndiotactic polystyrene/cis-decalin systems. *Macromolecules* **30**, 7939–7944.

Saito, H., Fujita, Y., and Inoue, T. (1987). Upper critical solution temperature behavior in poly (vinylidene fluoride)/poly(methyl metharylate) blends. *Polym. J. (Jpn.)* **19**, 405–412.

Schaaf, P., Lotz, B., and Wittmann, J. C. (1987). Liquid-liquid phase separation and crystallization in binary polymer systems. *Polymer* **28**, 193–200.

Shimizu, K., Wang, H., Wang, Z. G., Matsuba, G., Kim, H., and Han, C. C. (2004). Crystallization and phase separation kinetics in blends of linear low-density polyethylene copolymers. *Polymer* **45**, 7061–7069.

Suehiro, K., and Takayanagi, M. (1970). Structural studies of the high temperature form of trans-1,4-polybutadiene crystal. *J. Macromol. Sci. Part B: Phys.* **4**, 39–46.

Sutton, S. J., Vaughan, A. S., and Bassett, D. C. (1996). On the morphology and crystallization kinetics of monodisperse polyethylene oligomers crystallized from the melt. *Polymer* **37**, 5735–5738.

Svoboda, P., Kressler, J., Chiba, T., Inoue, T., and Kammer, H.-W. (1994). Light-scattering and TEM analyses of virtual upper critical solution temperature behavior in PCL/SAN blends. *Macromolecules* **27**, 1154–1159.

Tanaka, H., and Nishi, T. (1985). New types of phase separation behavior during the crystallization process in polymer blends with phase diagram. *Phys. Rev. Lett.* **55**, 1102–1105.

Tanaka, H., and Nishi, T. (1989). Local phase separation at the growth front of a polymer spherulite during crystallization and nonlinear spherulitic growth in a polymer mixture with a phase diagram. *Phys. Rev. A* **39**, 783–794.

Tosaka, M., Tsuji, M., Kohjiya, S., Cartier, L., and Lotz, B. (1999). Crystallization of syndiotactic polystyrene in β-form. 4. Crystal structure of melt-grown modification. *Macromolecules* **32**, 4905–4911.

Turner-Jones, A., Aizlewood, J. M., and Beckett, D. R. (1964). Crystalline forms of isotactic polypropylene. *Die Makromol. Chem.* **75**, 134–158.

Ungar, G. (1986). From plastic crystal paraffins to liquid crystal polyethylene. *Macromolecules* **19**, 1317–1324.

Ungar, G., and Keller, A. (1986). Time-resolved synchrotron X-ray study of chain-folded crystallization of long paraffins. *Polymer* **27**, 1835–1844.

Ungar, G., and Keller, A. (1987). Inversion of the temperature dependence of crystallization rates due to onset of chain folding. *Polymer* **28**, 1899–1907.

Ungar, G., and Zeng, X.-B. (2001). Learning polymer crystallization with the aid of linear, branched and cyclic model compounds. *Chem. Rev.* **101**, 4157–4188.

Ungar, G., Mandal, P., Higgs, P. G., de Silva, D. S. M., Boda, E., and Chen, C. M. (2000). Dilution wave and negative-order crystallization kinetics of chain molecules. *Phys. Rev. Lett.* **85**, 4397–4400.

Ungar, G, Putra, E. G. R., de Silva, D. S. M., Shcherbina, M. A., and Waddon, A. J. (2005). The effect of self-poisoning on crystal morphology and growth rates. *Adv. Polym. Sci.* **180**, 45–87.

Utracki, L. A. (2002). "Polymer Blends Handbook." Kluwer Academic Publishers, Dordrecht.

Van Durme, K., Loozen, E., Nies, E., and Van Mele, B. (2005). Phase behavior of poly(vinyl methyl ether) in deuterium oxide. *Macromolecules* **38**, 10234–10243.

Vaughan, A. S., Ungar, G., Bassett, D. C., and Keller, A. (1985). On hexagonal phases of paraffins and polyethylene. *Polymer* **26**, 726–732.

Wang, H., Shimizu, K., Hobbie, E. K., Wang, Z.-G., Meredith, J. C., Karim, A., Amis, E. J., Hsiao, B. S., Hsieh, E. T., and Han, C. C. (2002). Phase diagram of a nearly isorefractive polyolefin blend. *Macromolecules* **35**, 1072–1078.

Wittmann, J. C., Hodge, A. M., and Lotz, B. (1983). Epitaxial crystallization of polymers onto benzoic acid: Polyethylene and paraffins, aliphatic polyesters, and polyamides. *J. Polym. Sci. Polym. Phys. Edn.* **21**, 2495–2509.

Wunderlich, B. (1980). "Macromolecular Physics. Volume III. Crystal Melting." Academic Press, New York.

Wunderlich, B., and Arakawa, T. (1964). Polyethylene crystallized from the melt under elevated pressure. *J. Polym. Sci. Part A* **2**, 3697–3706.

Wunderlich, B., and Davidson, T. (1969). Extended-chain crystals. I. General crystallization conditions and review of pressure crystallization of polyethylene. *J. Polym. Sci. Polym. Phys. Edn.* **7**, 2043–2050.

Wunderlich, B., and Melillo, L. (1968). Morphology and growth of extended chain crystals of polyethylene. *Die Makromo. Chem.* **118**, 250–264.

Xu, J.-R., and Woodward, A. E. (1986). Further morphological studies of trans-1,4-polyisoprene crystallized from solution. *Macromolecules* **19**, 1114–1118.

Yasuniwa, M., Nakafuku, C., and Takemura, T. (1973). Melting and crystallization process of polyethylene under high pressure. *Poly. J. (Jpn.)* **4**, 526–533.

Yasuniwa, M., Enoshita, R., and Takemura, T. (1976). X-ray studies of polyethylene under high pressure. *Jpn. J. Appl. Phys.* **15**, 1421–1428.

Zhang, J., Bergé, B., Meeussen, F., Nies, E., Berghmans, H., and Shen, D. Y. (2003). Influence of the interactions in aqueous mixtures of poly(vinyl methyl ether) on the crystallization behavior of water. *Macromolecules* **36**, 9145–9153.

CHAPTER SEVEN

Outlook: A Personal View

Before addressing the current and future role of polymer physics in the development of the fundamental science and future technologies, let us briefly review how polymer physics has developed. Polymer physics naturally began with the search for the consequences of long-chain connectivity through experimentation and theoretical understanding. The pioneering experimental research was carried out in polymer dilute solutions which are in *thermodynamic equilibrium*. From these experiments, the approximate size of polymers and their behavior under different environmental conditions could be evaluated. As polymers became more pervasive in daily life, it became imperative to move beyond understanding polymer solution properties in order to investigate polymers in their solid states and determine how to both extend the range of physical properties, such as tensile strength, impact strength, and elasticity, and exactly tailor these properties for a particular application. Accordingly, as plastics technology developed, it became evident that both polymer structures and dynamics were critical to determining polymer properties and to fabricating products from polymeric materials; therefore, it was essential to establish structure–dynamics–property relationships in order to realize the full promise of polymeric materials.

Phase transitions in polymers play a central role not only in the scientific understanding of condensed matter, but also in their technological development. The phenomena researched in this field are fundamentally the consequence of structures and dynamics. At a transition point, both structures and dynamics exhibit sudden changes. The relatively mature science of structural analysis, developed more than one hundred years ago, is used to study organized synthetic and natural polymeric materials. Through the determination of their ordered structures, morphologies, and phase partitions (like crystallinity), it is now possible to correlate these parameters with physical properties. It is presumptuous to assume, however, that the research activities in this area are "traditional" or "classical." Although the methodology and procedure for determining these structures is well established, there is much to be understood about how hierarchical structures affect material properties. As such, studying the transformation, thermodynamics, and kinetics of hierarchical structures, in particular, in low-dimensional and confined environments, is critical to the continued development of this field. On the other hand, dynamics is an area (originally associated with polymer processing) that has been established relatively recently. Dynamics research explores insights into segmental, molecular, and cluster mobility in different phase structures, dimensions, and sizes.

Although polymer physics developed relying on the tremendous fundamental knowledge established for small and simple molecules, this subject offers unique opportunities to control structures and dynamics on broader length and time scales at easily accessible conditions. Polymer physics thus becomes one of the core components of condensed matter physics and solid state physical chemistry because of its unique structures and dynamics. The role of this book is to utilize the concepts of *metastable states* and *metastability* to link polymer phase transition phenomena and show that metastability is a rich area for future research.

The field of polymer research fulfilled the difficult task of moving polymeric materials from fundamental scientific study to a central technological platform in our everyday lives. To accomplish this, chemists provided the methods to create new, well-defined polymers and macromolecules. Physicists and engineers have utilized their knowledge of structural analysis and dynamics characterization to probe the materials' physical changes and their correlation with material properties. However, it was discovered that the atomically precise tailoring of structures on the sub-nanometer length scale alone was not sufficient to achieve specifically tailored material properties. The future of polymer physics lies in determining the effects of structures and dynamics on multiple length and time scales in both thermodynamically stable and metastable states, understanding the free energy landscapes in phase transitions, learning to control the structural and dynamic changes in different states and their correlation to material properties, as well as optimizing the properties relevant to new opportunities.

To achieve this goal, we need to continue to learn how to tailor polymeric structures and dynamics on even greater length and time scales than have currently been achieved in the bulk and at surfaces or interfaces. The key unsolved areas in the structures and dynamics of polymers should be identified and explored through experimentation and theoretical development. It is also important to correlate how these areas relate to the material's properties. Unfortunately, there are several experimental limitations that must be addressed. All of the issues center around a combination of small dimensions, weak interactions, and short-lived, transient metastable states.

The success of these research areas relies on the following developments. First, the absolute identification of chemical structures, sequences, and functionalities of single and multiple chains needs to be realized. It is important to observe and understand how each of these structures passively and actively responds to different environments at different dimensions and how the structure affects those responses. The next step is to characterize and determine the role of atomic and molecular weak physical interactions, including both polar and nonpolar interactions. These interactions play a central role in self-assembling hierarchical structures. Finally, in order to determine the structure–dynamics–property relationships of polymer materials, we must improve real-time, *in situ* diffraction and scattering experiments to identify transient and permanent changes of structure, morphology, and dynamics for complex systems on multiple length and time scales. These experimental observations will help us to understand whether the systems are at equilibrium, in a metastable state, or at a transition. The primary challenges to resolve the identified issues involve the integration of experimental measurements with theory. We need to further develop our basic knowledge and experimental methodology to account for reduced material dimensionality.

To address these challenges, the following developments should be of high priority. We need precisely defined model systems designed and synthesized for structural and dynamic characterization at different length and time scales. We need to advance new and nontraditional techniques to characterize structures, morphologies, dynamics, and interfaces of polymers. Such innovations will occur by improving spatial resolution to extrapolate local events to global behavior and increasing temporal resolution to determine intermediate states in the material's response to stimuli. A critical technical breakthrough will be to combine improved spatial and temporal resolution to monitor both physical and chemical phenomena in low-dimensional spaces, not only in single-component, but also in multiple-component systems and hybrid materials. Such inventions will enable us to obtain local chemical and physical information and guide our research, not only in synthetic polymers, but also for bio-macromolecules. In addition, we need to design methods to better measure complete sets of physical properties, such as electron and light transport and biocompatibility.

One way to determine hierarchical structure–dynamics–property relationships is through multidisciplinary efforts involving synthetic chemists, physicists, and engineers to construct functionalized structures and dynamics with different dimensions in order to understand their formation mechanisms. In particular, the structural and dynamic behavior changes *at the transitions* need to be probed and explained. The role polymer physics plays in other research activities is critically dependent on the further development of fundamental understanding and frontier breakthroughs.

New developments in polymeric materials will require long-term efforts in both fundamental research and education. Polymer education has become ever more engaged with other science and engineering disciplines. How do we maintain the identity of polymer science and engineering, yet train our students to succeed in a multidisciplinary environment? One of the key issues is how to balance the presentation of basic knowledge and new frontier disciplines in education. We are aware that any new frontier discipline can not succeed without a fundamental understanding of broad scientific principles. As such, this book is an attempt to design a new approach to introduce phase transitions in polymers and to demonstrate that a central subset of condensed polymer physics involves *metastable states*. Hopefully, this book helps the reader realize the major role polymer physics plays in current scientific research and leads to new efforts in education and multidisciplinary collaboration.

INDEX

Q

R

S

FSC
www.fsc.org
MIX
Papier | Fördert
gute Waldnutzung
FSC® C083411

Kolibri 360 GmbH
Ferdinand-Jühlke-Straße 7
99095 Erfurt, Deutschland
produktsicherheit@kolibri360.de